PLANET EARTH

Readings from
**SCIENTIFIC
AMERICAN**

PLANET EARTH

With Introductions by
Frank Press
Massachusetts Institute of Technology

Raymond Siever
Harvard University

W. H. Freeman and Company
San Francisco

Most of the SCIENTIFIC AMERICAN articles in
PLANET EARTH are available as separate Offprints. For
a complete list of more than 900 articles now available
as Offprints, write to W. H. Freeman and Company,
660 Market Street, San Francisco, California 94104.

Library of Congress Cataloging in Publication Data

Press, Frank, comp.
    Planet earth; readings from Scientific American.

    1. Earth sciences—Addresses, essays, lectures.
2. Force and energy—Addresses, essays, lectures.
3. Pollution—Addresses, essays, lectures.   I. Siever,
Raymond, joint comp.   II. Scientific American.
III. Title.
QE35.P73   550′.8   74–14919
ISBN 0–7167–0507–9
ISBN 0–7167–0506–0 (pbk.)

Printed in the United States of America

9 8 7 6 5 4 3 2 1

PREFACE

Earth, along with the other planets in our solar system, was born some 5 billion years ago. Each planet is unique in its size, distance from the sun, and chemical composition, and, because of these differences, each followed a particular course of evolution from its beginnings to its present state. Earth is the one planet on which environmental conditions were hospitable, where the biological evolution of over 1 million species could occur.

In these articles, reprinted from the *Scientific American*, the authors explore the dynamic processes that govern our planet's internal workings, shape its surface, control its climate, and make available the material and energy resources crucial to the sustenance of life in all its forms, from the most simple to the most advanced.

A collection such as this one has a useful function. The reader can conveniently sample many topics, each self-contained and up to date. Of necessity there are gaps—even twenty-eight different specialists will not cover the entire story of our planet. However, taken together, the articles impart something of the flavor, diversity, and excitement of modern earth sciences. The reader willing to invest more time can find a more orderly development of the subject in such books as our own—*Earth* (San Francisco: W. H. Freeman and Company, 1974).

Editors of collections are always faced with the necessity of selecting a few from a long list of highly eligible articles. Our criterion was the article's pertinence to such contemporary problems of mankind as energy and material shortages and pollution. We also recognized that a proper development and understanding of these topics could only be achieved by providing some background information on the physical and chemical bases of our environment and the internal elements of our planet. Finally, no modern collection dealing with earth would be complete without a discussion of plate tectonics and space exploration, subjects that have profoundly affected the way earth scientists view their planet.

The index to this reader was prepared with the able assistance of Evelyn Stinchfield.

Frank Press
Raymond Siever

30 May 1974

CONTENTS

Note on cross-references: References to articles included in this book are noted by the title of the article and the page on which it begins; references to articles that are available as Offprints, but are not included here, are noted by the article's title and Offprint number; references to articles published by SCIENTIFIC AMERICAN, but which are not available as Offprints, are noted by the title of the article and the month and year of its publication.

PLANET EARTH

ENERGY

Although specialists knew it was coming for years, public awareness of the energy shortage began only in the 1970s when workers were laid off, thermostats lowered, and gasoline stations began to limit sales. Man's accelerating demand for energy temporarily exceeded his ability to supply it. Respite may be found in the future as a result of new exploration for oil and gas, coal liquefaction and gasification, and oil extraction from tar sands and oil shales—all spurred by prices reflecting a more realistic valuation of energy. However, fossil fuels will not offer a long-term solution. Their reserves are finite and they will most certainly be depleted in a few centuries.

In this section we have included three articles to illustrate the energy problem. The first evaluates the several sources of energy available to man in the short and long term. The second traces the flow of energy from diverse sources through an industrial economy like that of the United States. Industrial, household, commercial, and transportation uses are considered and the efficiency of the whole operation is evaluated. The third deals with geothermal power—one of the new energy sources that will be required to augment and eventually replace fossil fuels.

The Energy Resources of the Earth

INTRODUCTION

In this article, "The Energy Resources of the Earth," M. K. Hubbert takes stock of the likely sources of energy available to man in the near, intermediate, and long terms. Fossil fuels are being depleted at an accelerating rate and will serve us for a few more centuries.

Over the long term, tidal energy, geothermal heat, solar energy, nuclear fission, and nuclear fusion are the obvious sources for meeting the world's energy needs. The last three offer almost inexhaustible supplies. Nuclear fission is already here and the first breeder reactors, which will conserve uranium, are being built. Although present-day reactors have problems of faulty design, possible accidents, and environmental damage, solutions can surely be found fairly soon.

Although solar energy is currently being used to generate heat and electricity on a small scale (cookers in India, water heaters in Israel, satellite power supplies), an entirely new technology is needed to realize the full potential of this source. Some experts believe that we will make hot water and heat buildings with sunlight in significant amounts in a matter of decades, but large-scale power generation is between fifty and a hundred years away. A novel proposal recently advanced is to convert solar energy to fuel by growing immense crops of sugar cane that can then be converted into gas for combustion.

Nuclear fusion, which involves the combining of light nuclei at extremely high temperatures, has only been achieved in significant amounts in the H-bomb. Controlled fusion with the release of useful energy may, like solar power, be between fifty and a hundred years away. It would supply a clean, almost limitless source of energy for all our needs—electricity to run our factories and heat our homes, fuel for our cars, planes, and ships.

The big question is whether we can make the transfer from fossil fuels to new energy sources without disrupting economies by energy shortages. Perhaps we can—with the help of conservation, population control, international cooperation, and greater investment in research and development.

The Energy Resources
of the Earth

by M. King Hubbert
September 1971

*They are solar energy (current and stored), the tides,
the earth's heat, fission fuels and possibly fusion fuels.
From the standpoint of human history the epoch of the
fossil fuels will be quite brief*

Energy flows constantly into and out of the earth's surface environment. As a result the material constituents of the earth's surface are in a state of continuous or intermittent circulation. The source of the energy is preponderantly solar radiation, supplemented by small amounts of heat from the earth's interior and of tidal energy from the gravitational system of the earth, the moon and the sun. The materials of the earth's surface consist of the 92 naturally occurring chemical elements, all but a few of which behave in accordance with the principles of the conservation of matter and of nontransmutability as formulated in classical chemistry. A few of the elements or their isotopes, with abundances of only a few parts per million, are an exception to these principles in being radioactive. The exception is crucial in that it is the key to an additional large source of energy.

A small part of the matter at the earth's surface is embodied in living organisms: plants and animals. The leaves of the plants capture a small fraction of the incident solar radiation and store it chemically by the mechanism of photosynthesis. This store becomes the energy supply essential for the existence of the plant and animal kingdoms. Biologically stored energy is released by oxidation at a rate approximately equal to the rate of storage. Over millions of

years, however, a minute fraction of the vegetable and animal matter is buried under conditions of incomplete oxidation and decay, thereby giving rise to the fossil fuels that provide most of the energy for industrialized societies.

It is difficult for people living now, who have become accustomed to the steady exponential growth in the consumption of energy from the fossil fuels, to realize how transitory the fossil-fuel epoch will eventually prove to be when it is viewed over a longer span of human history. The situation can better be seen in the perspective of some 10,000 years, half before the present and half afterward. On such a scale the complete cycle of the exploitation of the world's fossil fuels will be seen to encompass perhaps 1,300 years, with the principal segment of the cycle (defined as the period during which all but the first 10 percent and the last 10 percent of the fuels are extracted and burned) covering only about 300 years.

What, then, will provide industrial energy in the future on a scale at least as large as the present one? The answer lies in man's growing ability to exploit other sources of energy, chiefly nuclear at present but perhaps eventually the much larger source of solar energy. With this ability the energy resources now at hand are sufficient to sustain an industrial operation of the present magnitude for another millennium or longer. Moreover, with such resources of energy the limits to the growth of industrial activity are no longer set by a scarcity of energy but rather by the space and material limitations of a finite earth together with the principles of ecology. According to these principles both biological and industrial activities tend to increase exponentially with time, but the resources of the entire earth are not sufficient to sustain such an increase of

any single component for more than a few tens of successive doublings.

Let us consider in greater detail the flow of energy through the earth's surface environment [*see illustration on next two pages*]. The inward flow of energy has three main sources: (1) the intercepted solar radiation; (2) thermal energy, which is conveyed to the surface of the earth from the warmer interior by the conduction of heat and by convection in hot springs and volcanoes, and (3) tidal energy, derived from the combined kinetic and potential energy of the earth-moon-sun system. It is possible in various ways to estimate approximately how large the input is from each source.

In the case of solar radiation the influx is expressed in terms of the solar constant, which is defined as the mean rate of flow of solar energy across a unit of area that is perpendicular to the radiation and outside the earth's atmosphere at the mean distance of the earth from the sun. Measurements made on the earth and in spacecraft give a mean value for the solar constant of 1.395 kilowatts per square meter, with a variation of about 2 percent. The total solar radiation intercepted by the earth's diametric plane of 1.275×10^{14} square meters is therefore 1.73×10^{17} watts.

The influx of heat by conduction from the earth's interior has been determined from measurements of the geothermal gradient (the increase of temperature with depth) and the thermal conductivity of the rocks involved. From thousands of such measurements, both on land and on the ocean beds, the average rate of flow of heat from the interior of the earth has been found to be about .063 watt per square meter. For the earth's surface area of 510×10^{12} square meters the total heat flow amounts to

RESOURCE EXPLORATION is beginning to be aided by airborne side-looking radar pictures such as the one on the opposite page made by the Aero Service Corporation and the Goodyear Aerospace Corporation. The technique has advantage of "seeing" through cloud cover and vegetation. This picture, which was made in southern Venezuela, extends 70 miles from left to right.

some 32×10^{12} watts. The rate of heat convection by hot springs and volcanoes is estimated to be only about 1 percent of the rate of conduction, or about $.3 \times 10^{12}$ watts.

The energy from tidal sources has been estimated at 3×10^{12} watts. When all three sources of energy are expressed in the common unit of 10^{12} watts, the total power influx into the earth's surface environment is found to be $173,035 \times 10^{12}$ watts. Solar radiation accounts for 99.98 percent of it. Another way of stating the sun's contribution to the energy budget of the earth is to note that at $173,000 \times 10^{12}$ watts it amounts to 5,000 times the energy input from all other sources combined.

About 30 percent of the incident solar energy ($52,000 \times 10^{12}$ watts) is directly reflected and scattered back into space as short-wavelength radiation. Another 47 percent ($81,000 \times 10^{12}$ watts) is absorbed by the atmosphere, the land surface and the oceans and converted directly into heat at the ambient surface temperature. Another 23 percent ($40,000 \times 10^{12}$ watts) is consumed in the evaporation, convection, precipitation and surface runoff of water in the hydrologic cycle. A small fraction, about 370×10^{12} watts, drives the atmospheric and oceanic convections and circulations and the ocean waves and is eventually dissipated into heat by friction. Finally, an even smaller fraction—about

40×10^{12} watts—is captured by the chlorophyll of plant leaves, where it becomes the essential energy supply of the photosynthetic process and eventually of the plant and animal kingdoms.

Photosynthesis fixes carbon in the leaf and stores solar energy in the form of carbohydrate. It also liberates oxygen and, with the decay or consumption of the leaf, dissipates energy. At any given time, averaged over a year or more, the balance between these processes is almost perfect. A minute fraction of the organic matter produced, however, is deposited in peat bogs or other oxygen-deficient environments under conditions that prevent complete decay and loss of energy.

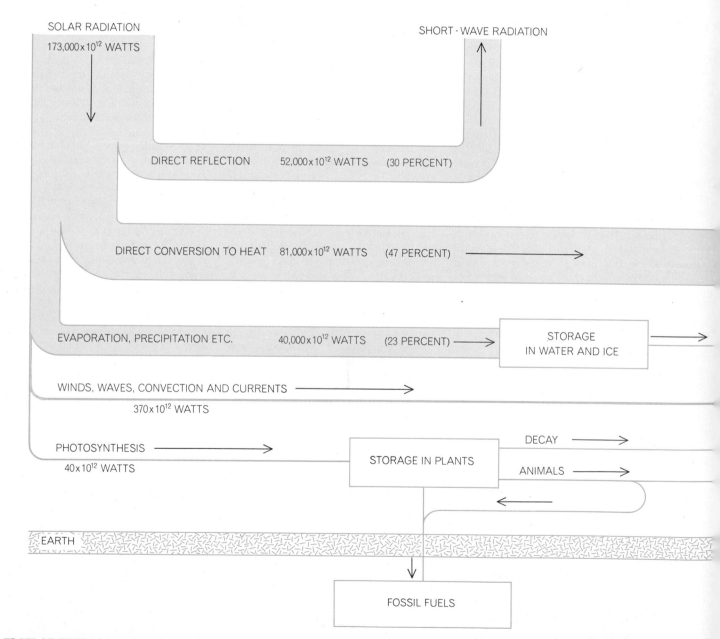

SOLAR RADIATION
173,000×10¹² WATTS

SHORT - WAVE RADIATION

DIRECT REFLECTION   52,000×10¹² WATTS   (30 PERCENT)

DIRECT CONVERSION TO HEAT   81,000×10¹² WATTS   (47 PERCENT)

EVAPORATION, PRECIPITATION ETC.   40,000×10¹² WATTS   (23 PERCENT)   STORAGE IN WATER AND ICE

WINDS, WAVES, CONVECTION AND CURRENTS
370×10¹² WATTS

PHOTOSYNTHESIS   STORAGE IN PLANTS   DECAY   ANIMALS
40×10¹² WATTS

EARTH

FOSSIL FUELS

FLOW OF ENERGY to and from the earth is depicted by means of bands and lines that suggest by their width the contribution of each item to the earth's energy budget. The principal inputs are solar radiation, tidal energy and the energy from nuclear, thermal and gravitational sources. More than 99 percent of the input is solar radiation. The apportionment of incoming solar radiation is

Little of the organic material produced before the Cambrian period, which began about 600 million years ago, has been preserved. During the past 600 million years, however, some of the organic materials that did not immediately decay have been buried under a great thickness of sedimentary sands, muds and limes. These are the fossil fuels: coal, oil shale, petroleum and natural gas, which are rich in energy stored up chemically from the sunshine of the past 600 million years. The process is still continuing, but probably at about the same rate as in the past; the accumulation during the next million years will probably be a six-hundredth of the amount built up thus far.

Industrialization has of course withdrawn the deposits in this energy bank with increasing rapidity. In the case of coal, for example, the world's consumption during the past 110 years has been about 19 times greater than it was during the preceding seven centuries. The increasing magnitude of the rate of withdrawal can also be seen in the fact that the amount of coal produced and consumed since 1940 is approximately equal to the total consumption up to that time. The cumulative production from 1860 through 1970 was about 133 billion metric tons. The amount produced before 1860 was about seven million metric tons.

Petroleum and related products were not extracted in significant amounts before 1880. Since then production has increased at a nearly constant exponential rate. During the 80-year period from 1890 through 1970 the average rate of increase has been 6.94 percent per year, with a doubling period of 10 years. The cumulative production until the end of 1969 amounted to 227 billion (227×10^9) barrels, or 9.5 trillion U.S. gallons. Once again the period that encompasses most of the production is notably brief. The 102 years from 1857 to 1959 were required to produce the first half of the cumulative production; only the 10-year period from 1959 to 1969 was required for the second half.

Examining the relative energy contributions of coal and crude oil by comparing the heats of combustion of the respective fuels (in units of 10^{12} kilowatt-hours), one finds that until after 1900 the contribution from oil was barely significant compared with the contribution from coal. Since 1900 the contribution from oil has risen much faster than that from coal. By 1968 oil represented about 60 percent of the total. If the energy from natural gas and natural-gas liquids had been included, the contribution from petroleum would have been about 70 percent. In the U.S. alone 73 percent of the total energy produced from fossil fuels in 1968 was from petroleum and 27 percent from coal.

Broadly speaking, it can be said that the world's consumption of energy for industrial purposes is now doubling approximately once per decade. When confronted with a rate of growth of such magnitude, one can hardly fail to wonder how long it can be kept up. In the case of the fossil fuels a reasonably definite answer can be obtained. Their human exploitation consists of their being withdrawn from an essentially fixed initial supply. During their use as sources of energy they are destroyed. The complete cycle of exploitation of a fossil fuel must therefore have the following characteristics. Beginning at zero, the rate of production tends initially to increase exponentially. Then, as difficulties of discovery and extraction increase, the production rate slows in its growth, passes one maximum or more and, as the resource is progressively depleted, declines eventually to zero.

If known past and prospective future rates of production are combined with a reasonable estimate of the amount of a fuel initially present, one can calculate the probable length of time that the fuel can be exploited. In the case of coal reasonably good estimates of the

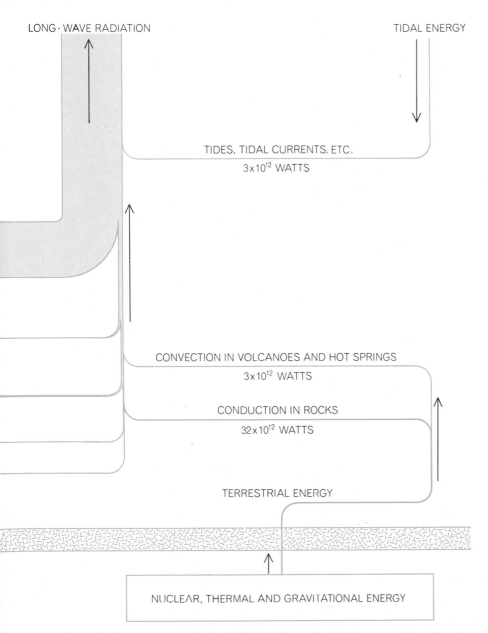

LONG-WAVE RADIATION

TIDAL ENERGY

TIDES, TIDAL CURRENTS, ETC.
3×10^{12} WATTS

CONVECTION IN VOLCANOES AND HOT SPRINGS
3×10^{12} WATTS

CONDUCTION IN ROCKS
32×10^{12} WATTS

TERRESTRIAL ENERGY

NUCLEAR, THERMAL AND GRAVITATIONAL ENERGY

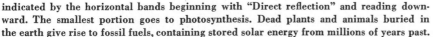

indicated by the horizontal bands beginning with "Direct reflection" and reading downward. The smallest portion goes to photosynthesis. Dead plants and animals buried in the earth give rise to fossil fuels, containing stored solar energy from millions of years past.

amount present in given regions can be made on the basis of geological mapping and a few widely spaced drill holes, inasmuch as coal is found in stratified beds or seams that are continuous over extensive areas. Such studies have been made in all the coal-bearing areas of the world.

The most recent compilation of the present information on the world's initial coal resources was made by Paul Averitt of the U.S. Geological Survey. His figures [*see illustration below*] represent minable coal, which is defined as 50 percent of the coal actually present. Included is coal in beds as thin as 14 inches (36 centimeters) and extending to depths of 4,000 feet (1.2 kilometers) or, in a few cases, 6,000 feet (1.8 kilometers).

Taking Averitt's estimate of an initial supply of 7.6 trillion metric tons and assuming that the present production rate of three billion metric tons per year does not double more than three times, one can expect that the peak in the rate of production will be reached sometime between 2100 and 2150. Disregarding the long time required to produce the first 10 percent and the last 10 percent, the length of time required to produce the middle 80 percent will be roughly

the 300-year period from 2000 to 2300.

Estimating the amount of oil and gas that will ultimately be discovered and produced in a given area is considerably more hazardous than estimating for coal. The reason is that these fluids occur in restricted volumes of space and limited areas in sedimentary basins at all depths from a few hundred meters to more than eight kilometers. Nonetheless, the estimates for a given region improve as exploration and production proceed. In addition it is possible to make rough estimates for relatively undeveloped areas on the basis of geological comparisons between them and well-developed regions.

The most highly developed oil-producing region in the world is the coterminous area of the U.S.: the 48 states exclusive of Alaska and Hawaii. This area has until now led the world in petroleum development, and the U.S. is still the leading producer. For this region a large mass of data has been accumulated and a number of different methods of analysis have been developed that give fairly consistent estimates of the degree of advancement of petroleum exploration and of the amounts of oil and gas that may eventually be produced.

One such method is based on the principle that only a finite number of oil or gas fields existed initially in a given region. As exploration proceeds the shallowest and most evident fields are usually discovered first and the deeper and more obscure ones later. With each discovery the number of undiscovered fields decreases by one. The undiscovered fields are also likely to be deeper, more widely spaced and better concealed. Hence the amount of exploratory activity required to discover a fixed quantity of oil or gas steadily increases or, conversely, the average amount of oil or gas discovered for a fixed amount of exploratory activity steadily decreases.

Most new fields are discovered by what the industry calls "new-field wildcat wells," meaning wells drilled in new territory that is not in the immediate vicinity of known fields. In the U.S. statistics have been kept annually since 1945 on the number of new-field wildcat wells required to make one significant discovery of oil or gas ("significant" being defined as one million barrels of oil or an equivalent amount of gas). The discoveries for a given year are evaluated only after six years of subsequent development. In 1945 it required 26

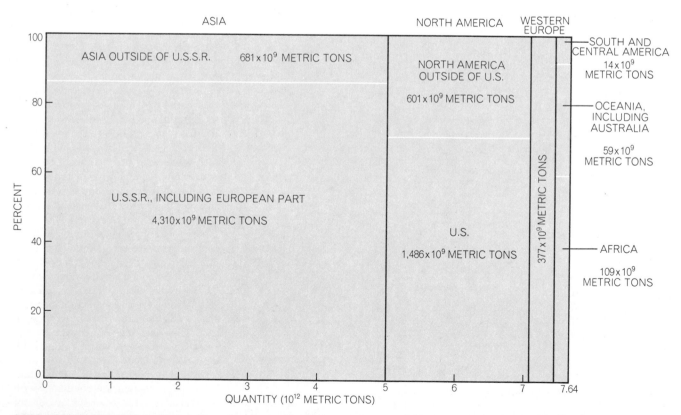

COAL RESOURCES of the world are indicated on the basis of data compiled by Paul Averitt of the U.S. Geological Survey. The figures represent the total initial resources of minable coal, which is defined as 50 percent of the coal actually present. The horizontal scale gives the total supply. Each vertical block shows the apportionment of the supply in a continent. From the first block, for example, one can ascertain that Asia has some 5×10^{12} metric tons of minable coal, of which about 86 percent is in the U.S.S.R.

new-field wildcat wells to make a significant discovery; by 1963 the number had increased to 65.

Another way of illuminating the problem is to consider the amount of oil discovered per foot of exploratory drilling. From 1860 to 1920, when oil was fairly easy to find, the ratio was 194 barrels per foot. From 1920 to 1928 the ratio declined to 167 barrels per foot. Between 1928 and 1938, partly because of the discovery of the large East Texas oil field and partly because of new exploratory techniques, the ratio rose to its maximum of 276 barrels per foot. Since then it has fallen sharply to a nearly constant rate of about 35 barrels per foot. Yet the period of this decline coincided with the time of the most intensive research and development in petroleum exploration and production in the history of the industry.

The cumulative discoveries in the 48 states up to 1965 amounted to 136 billion barrels. From this record of drilling and discovery it can be estimated that the ultimate total discoveries in the coterminous U.S. and the adjacent continental shelves will be about 165 billion barrels. The discoveries up to 1965 therefore represent about 82 percent of the prospective ultimate total. Making

due allowance for the range of uncertainty in estimates of future discovery, it still appears that at least 75 percent of the ultimate amount of oil to be produced in this area will be obtained from fields that had already been discovered by 1965.

For natural gas in the 48 states the present rate of discovery, averaged over a decade, is about 6,500 cubic feet per barrel of oil. Assuming the same ratio for the estimated ultimate amount of 165 billion barrels of crude oil, the ultimate amount of natural gas would be about 1,075 trillion cubic feet. Combining the estimates for oil and gas with the trends of production makes it possible to estimate how long these energy resources will last. In the case of oil the period of peak production appears to be the present. The time span required to produce the middle 80 percent of the ultimate cumulative production is approximately the 65-year period from 1934 to 1999—less than the span of a human lifetime. For natural gas the peak of production will probably be reached between 1975 and 1980.

The discoveries of petroleum in Alaska modify the picture somewhat. In particular the field at Prudhoe Bay appears likely by present estimates to contain

about 10 billion barrels, making it twice as large as the East Texas field, which was the largest in the U.S. previously. Only a rough estimate can be made of the eventual discoveries of petroleum in Alaska. Such a speculative estimate would be from 30 to 50 billion barrels. One must bear in mind, however, that 30 billion barrels is less than a 10-year supply for the U.S. at the present rate of consumption. Hence it appears likely that the principal effect of the oil from Alaska will be to retard the rate of decline of total U.S. production rather than to postpone the date of its peak.

Estimates of ultimate world production of oil range from 1,350 billion barrels to 2,100 billion barrels. For the higher figure the peak in the rate of world production would be reached about the year 2000. The period of consumption of the middle 80 percent will probably be some 58 to 64 years, depending on whether the lower or the higher estimate is used [see bottom illustration on page 13].

A substantial but still finite amount of oil can be extracted from tar sands and oil shales, where production has barely begun. The largest tar-sand deposits are in northern Alberta; they have total recoverable reserves of about 300 billion

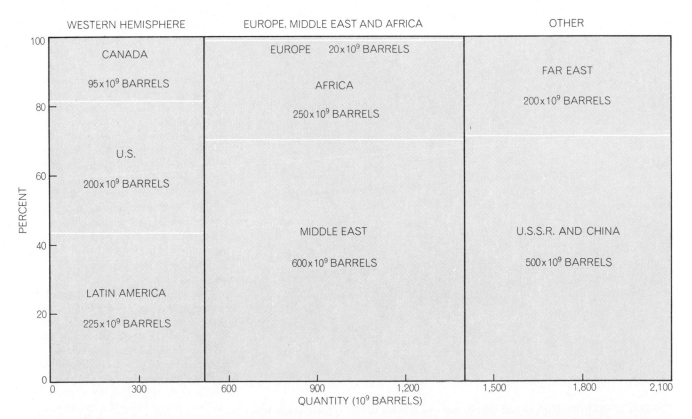

PETROLEUM RESOURCES of the world are depicted in an arrangement that can be read in the same way as the diagram of coal supplies on the opposite page. The figures for petroleum are derived from estimates made in 1967 by W. P. Ryman of the Standard Oil Company of New Jersey. They represent ultimate crude-oil production, including oil from offshore areas, and consist of oil already produced, proved and probable reserves, and future discoveries. Estimates as low as 1,350 × 10⁹ barrels have also been made.

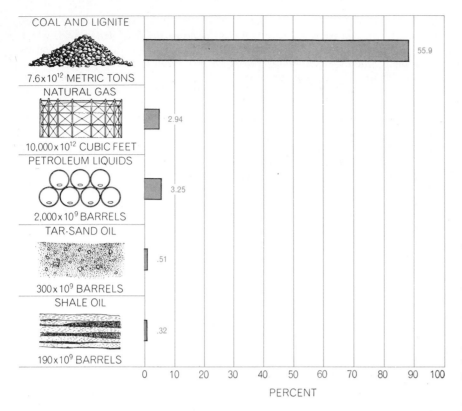

ENERGY CONTENT of the world's initial supply of recoverable fossil fuels is given in units of 10^{15} thermal kilowatt-hours (*color*). Coal and lignite, for example, contain 55.9 × 10^{15} thermal kilowatt-hours of energy and represent 88.8 percent of the recoverable energy.

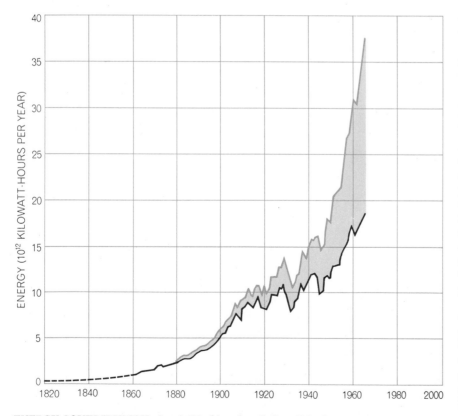

ENERGY CONTRIBUTION of coal (*black*) and coal plus oil (*color*) is portrayed in terms of their heat of combustion. Before 1900 the energy contribution from oil was barely significant. Since then the contribution from oil (*shaded area*) has risen much more rapidly than that from coal. By 1968 oil represented about 60 percent of the total. If the energy from natural gas were included, petroleum would account for about 70 percent of the total.

barrels. A world summary of oil shales by Donald C. Duncan and Vernon E. Swanson of the U.S. Geological Survey indicated a total of about 3,100 billion barrels in shales containing from 10 to 100 gallons per ton, of which 190 billion barrels were considered to be recoverable under 1965 conditions.

Since the fossil fuels will inevitably be exhausted, probably within a few centuries, the question arises of what other sources of energy can be tapped to supply the power requirements of a moderately industrialized world after the fossil fuels are gone. Five forms of energy appear to be possibilities: solar energy used directly, solar energy used indirectly, tidal energy, geothermal energy and nuclear energy.

Until now the direct use of solar power has been on a small scale for such purposes as heating water and generating electricity for spacecraft by means of photovoltaic cells. Much more substantial installations will be needed if solar power is to replace the fossil fuels on an industrial scale. The need would be for solar power plants in units of, say, 1,000 megawatts. Moreover, because solar radiation is intermittent at a fixed location on the earth, provision must also be made for large-scale storage of energy in order to smooth out the daily variation.

The most favorable sites for developing solar power are desert areas not more than 35 degrees north or south of the Equator. Such areas are to be found in the southwestern U.S., the region extending from the Sahara across the Arabian Peninsula to the Persian Gulf, the Atacama Desert in northern Chile and central Australia. These areas receive some 3,000 to 4,000 hours of sunshine per year, and the amount of solar energy incident on a horizontal surface ranges from 300 to 650 calories per square centimeter per day. (Three hundred calories, the winter minimum, amounts when averaged over 24 hours to a mean power density of 145 watts per square meter.)

Three schemes for collecting and converting this energy in a 1,000-megawatt plant can be considered. The first involves the use of flat plates of photovoltaic cells having an efficiency of about 10 percent. A second possibility is a recent proposal by Aden B. Meinel and Marjorie P. Meinel of the University of Arizona for utilizing the hothouse effect by means of selective coatings on pipes carrying a molten mixture of sodium and potassium raised by solar energy to a temperature of 540 degrees Celsius. By

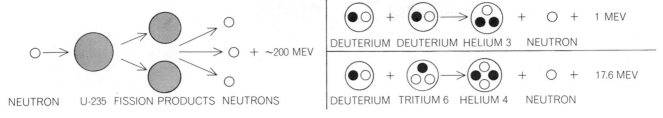

$$\text{DEUTERIUM} + \text{DEUTERIUM} \rightarrow \text{HELIUM 3} + \text{NEUTRON} + 1\ \text{MEV}$$

$$\text{DEUTERIUM} + \text{TRITIUM 6} \rightarrow \text{HELIUM 4} + \text{NEUTRON} + 17.6\ \text{MEV}$$

NEUTRON U-235 FISSION PRODUCTS NEUTRONS $+ \sim 200\ \text{MEV}$

FISSION AND FUSION REACTIONS hold the promise of serving as sources of energy when fossil fuels are depleted. Present nuclear-power plants burn uranium 235 as a fuel. Breeder reactors now under development will be able to use surplus neutrons from the fission of uranium 235 (*left*) to create other nuclear fuels: plutonium 239 and uranium 233. Two promising fusion reactions, deuterium-deuterium and deuterium-tritium, are at right. The energy released by the various reactions is shown in million electron volts.

means of a heat exchanger this heat is stored at a constant temperature in an insulated chamber filled with a mixture of sodium and potassium chlorides that has enough heat capacity for at least one day's collection. Heat extracted from this chamber operates a conventional steam-electric power plant. The computed efficiency for this proposal is said to be about 30 percent.

A third system has been proposed by Alvin F. Hildebrandt and Gregory M. Haas of the University of Houston. It entails reflecting the radiation reaching a square-mile area into a solar furnace and boiler at the top of a 1,500-foot tower. Heat from the boiler at a temperature of 2,000 degrees Kelvin would be converted into electric power by a magnetohydrodynamic conversion. An energy-storage system based on the hydrolysis of water is also proposed. An overall efficiency of about 20 percent is estimated.

Over the range of efficiencies from 10 to 30 percent the amount of thermal power that would have to be collected for a 1,000-megawatt plant would range from 10,000 to 3,300 thermal megawatts. Accordingly the collecting areas for the three schemes would be 70, 35 and 23 square kilometers respectively. With the least of the three efficiencies the area required for an electric-power capacity of 350,000 megawatts—the approximate capacity of the U.S. in 1970—would be 24,500 square kilometers, which is somewhat less than a tenth of the area of Arizona.

The physical knowledge and technological resources needed to use solar energy on such a scale are now available. The technological difficulties of doing so, however, should not be minimized.

Using solar power indirectly means relying on the wind, which appears impractical on a large scale, or on the streamflow part of the hydrologic cycle. At first glance the use of streamflow appears promising, because the world's total water-power capacity in suitable sites is about three trillion watts, which

approximates the present use of energy in industry. Only 8.5 percent of the water power is developed at present, however, and the three regions with the greatest potential—Africa, South America and Southeast Asia—are the least developed industrially. Economic problems therefore stand in the way of extensive development of additional water power.

Tidal power is obtained from the filling and emptying of a bay or an estuary that can be closed by a dam. The enclosed basin is allowed to fill and empty only during brief periods at high and low tides in order to develop as much power as possible. A number of promising sites exist; their potential capacities range from two megawatts to 20,000 megawatts each. The total potential tidal power, however, amounts to about 64 billion watts, which is only 2 percent of the world's potential water power. Only one full-scale tidal-electric plant has been built; it is on the Rance estuary on the Channel Island coast of France. Its capacity at start-up in 1966 was 240 megawatts; an ultimate capacity of 320 megawatts is planned.

Geothermal power is obtained by extracting heat that is temporarily stored in the earth by such sources as volcanoes and the hot water filling the sands of deep sedimentary basins. Only volcanic sources are significantly exploited at present. A geothermal-power operation has been under way in the Larderello area of Italy since 1904 and now has a capacity of 370 megawatts. The two other main areas of geothermal-power production are The Geysers in northern California and Wairakei in New Zealand. Production at The Geysers began in 1960 with a 12.5-megawatt unit. By 1969 the capacity had reached 82 megawatts, and plans are to reach a total installed capacity of 400 megawatts by 1973. The Wairakei plant began operation in 1958 and now has a capacity of 290 megawatts, which is believed to be about the maximum for the site.

Donald E. White of the U.S. Geological Survey has estimated that the stored thermal energy in the world's major geothermal areas amounts to about 4×10^{20} joules. With a 25 percent conversion factor the production of electrical energy would be about 10^{20} joules, or three million megawatt-years. If this energy, which is depletable, were withdrawn over a period of 50 years, the average annual power production would be 60,000 megawatts, which is comparable to the potential tidal power.

Nuclear power must be considered under the two headings of fission and fusion. Fission involves the splitting of nuclei of heavy elements such as uranium. Fusion involves the combining of light nuclei such as deuterium. Uranium 235, which is a rare isotope (each 100,000 atoms of natural uranium include six atoms of uranium 234, 711 atoms of uranium 235 and 99,283 atoms of uranium 238), is the only atomic species capable of fissioning under relatively mild environmental conditions. If nuclear energy depended entirely on uranium 235, the nuclear-fuel epoch would be brief. By breeding, however, wherein by absorbing neutrons in a nuclear reactor uranium 238 is transformed into fissionable plutonium 239 or thorium 232 becomes fissionable uranium 233, it is possible to create more nuclear fuel than is consumed. With breeding the entire supply of natural uranium and thorium would thus become available as fuel for fission reactors.

Most of the reactors now operating or planned in the rapidly growing nuclear-power industry in the U.S. depend essentially on uranium 235. The U.S. Atomic Energy Commission has estimated that the uranium requirement to meet the projected growth rate from 1970 to 1980 is 206,000 short tons of uranium oxide (U_3O_8). A report recently issued by the European Nuclear Energy Agency and the International Atomic Energy Agency projects requirements of 430,000 short tons of uranium

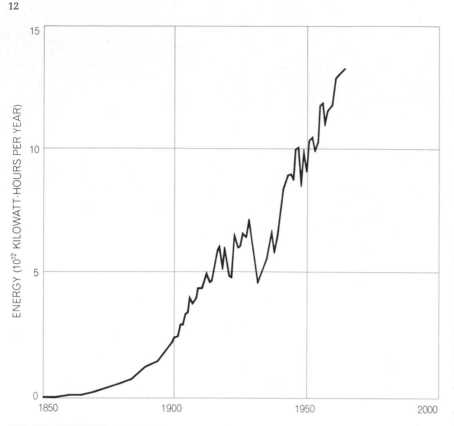

U.S. PRODUCTION OF ENERGY from coal, from petroleum and related sources, from water power and from nuclear reactors is charted for 120 years. The petroleum increment includes natural gas and associated liquids. The dip at center reflects impact of Depression.

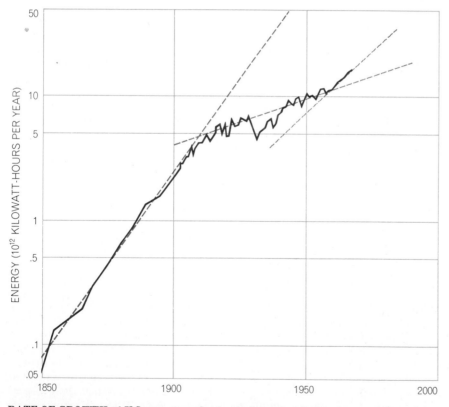

RATE OF GROWTH of U.S. energy production is shown by plotting on a semilogarithmic scale the data represented in the illustration at the top of the page. Broken lines show that the rise had three distinct periods. In the first the growth rate was 6.91 percent per year and the doubling period was 10 years; in the second the rate was 1.77 percent and the doubling period was 39 years; in the third the rate was 4.25 percent with doubling in 16.3 years.

oxide for the non-Communist nations during the same period.

Against these requirements the AEC estimates that the quantity of uranium oxide producible at $8 per pound from present reserves in the U.S. is 243,000 tons, and the world reserves at $10 per pound or less are estimated in the other report at 840,000 tons. The same report estimates that to meet future requirements additional reserves of more than a million short tons will have to be discovered and developed by 1985.

Although new discoveries of uranium will doubtless continue to be made (a large one was recently reported in northeastern Australia), all present evidence indicates that without a transition to breeder reactors an acute shortage of low-cost ores is likely to develop before the end of the century. Hence an intensive effort to develop large-scale breeder reactors for power production is in progress. If it succeeds, the situation with regard to fuel supply will be drastically altered.

This prospect results from the fact that with the breeder reactor the amount of energy obtainable from one gram of uranium 238 amounts to 8.1×10^{10} joules of heat. That is equal to the heat of combustion of 2.7 metric tons of coal or 13.7 barrels (1.9 metric tons) of crude oil. Disregarding the rather limited supplies of high-grade uranium ore that are available, let us consider the much more abundant low-grade ores. One example will indicate the possibilities.

The Chattanooga black shale (of Devonian age) crops out along the western edge of the Appalachian Mountains in eastern Tennessee and underlies at minable depths most of Tennessee, Kentucky, Ohio, Indiana and Illinois. In its outcrop area in eastern Tennessee this shale contains a layer about five meters thick that has a uranium content of about 60 grams per metric ton. That amount of uranium is equivalent to about 162 metric tons of bituminous coal or 822 barrels of crude oil. With the density of the rock some 2.5 metric tons per cubic meter, a vertical column of rock five meters long and one square meter in cross section would contain 12.5 tons of rock and 750 grams of uranium. The energy content of the shale per square meter of surface area would therefore be equivalent to about 2,000 tons of coal or 10,000 barrels of oil. Allowing for a 50 percent loss in mining and extracting the uranium, we are still left with the equivalent of 1,000 tons of coal or 5,000 barrels of oil per square meter.

Taking Averitt's estimate of 1.5 trillion metric tons for the initial minable coal in the U.S. and a round figure of 250 billion barrels for the petroleum liquids, we find that the nuclear energy in an area of about 1,500 square kilometers of Chattanooga shale would equal the energy in the initial minable coal; 50 square kilometers would hold the energy equivalent of the petroleum liquids. Adding natural gas and oil shales, an area of roughly 2,000 square kilometers of Chattanooga shale would be equivalent to the initial supply of all the fossil fuels in the U.S. The area is about 2 percent of the area of Tennessee

and a very small fraction of the total area underlain by the shale. Many other low-grade deposits of comparable magnitude exist. Hence by means of the breeder reactor the energy potentially available from the fissioning of uranium and thorium is at least a few orders of magnitude greater than that from all the fossil fuels combined.

David J. Rose of the AEC, reviewing recently the prospects for controlled fusion, found the deuterium-tritium reaction to be the most promising. Deuterium is abundant (one atom to each

6,700 atoms of hydrogen), and the energy cost of separating it would be almost negligible compared with the amount of energy released by fusion. Tritium, on the other hand, exists only in tiny amounts in nature. Larger amounts must be made from lithium 6 and lithium 7 by nuclear bombardment. The limiting isotope is lithium 6, which has an abundance of only 7.4 percent of natural lithium.

Considering the amount of hydrogen in the oceans, deuterium can be regarded as superabundant. It can also be extracted easily. Lithium is much less

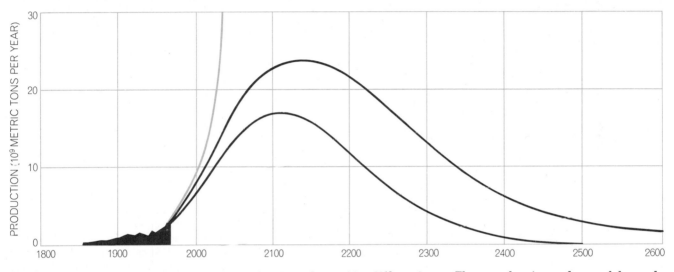

CYCLE OF WORLD COAL PRODUCTION is plotted on the basis of estimated supplies and rates of production. The top curve reflects Averitt's estimate of 7.6×10^{12} metric tons as the initial supply of minable coal; the bottom curve reflects an estimate of 4.3×10^{12} metric tons. The curve that rises to the top of the graph shows the trend if production continued to rise at the present rate of 3.56 percent per year. The amount of coal mined and burned in the century beginning in 1870 is shown by the black area at left.

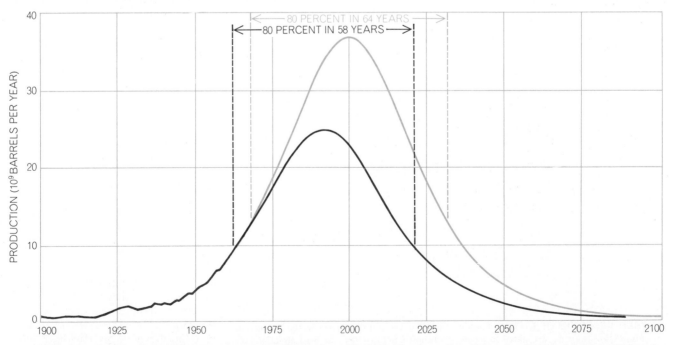

CYCLE OF WORLD OIL PRODUCTION is plotted on the basis of two estimates of the amount of oil that will ultimately be produced. The colored curve reflects Ryman's estimate of $2,100 \times 10^9$ barrels and the black curve represents an estimate of $1,350 \times 10^9$ barrels.

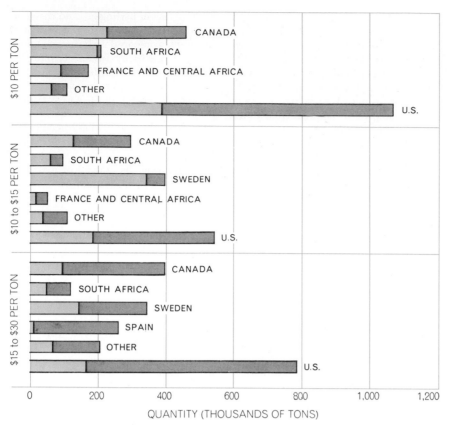

WORLD RESERVES OF URANIUM, which would be the source of nuclear power derived from atomic fission, are given in tons of uranium oxide (U_3O_8). The colored part of each bar represents reasonably assured supplies and the gray part estimated additional supplies.

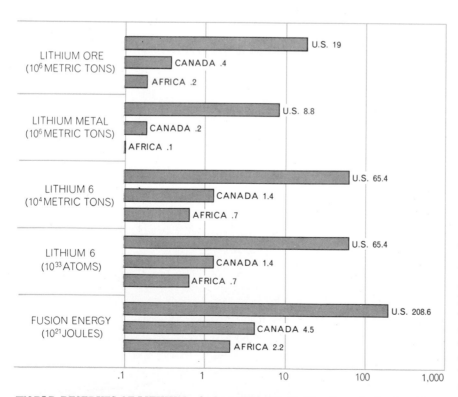

WORLD RESERVES OF LITHIUM, which would be the limiting factor in the deuterium-tritium fusion reaction, are stated in terms of lithium 6 because it is the least abundant isotope. Even with this limitation the energy obtainable from fusion through the deuterium-tritium reaction would almost equal the energy content of the world's fossil-fuel supply.

abundant. It is produced from the geologically rare igneous rocks known as pegmatites and from the salts of saline lakes. The measured, indicated and inferred lithium resources in the U.S., Canada and Africa total 9.1 million tons of elemental lithium, of which the content of lithium 6 would be 7.42 atom percent, or 67,500 metric tons. From this amount of lithium 6 the fusion energy obtainable at 3.19×10^{-12} joule per atom would be 215×10^{21} joules, which is approximately equal to the energy content of the world's fossil fuels.

As long as fusion power is dependent on the deuterium-tritium reaction, which at present appears to be somewhat the easier because it proceeds at a lower temperature, the energy obtainable from this source appears to be of about the same order of magnitude as that from fossil fuels. If fusion can be accomplished with the deuterium-deuterium reaction, the picture will be markedly changed. By this reaction the energy released per deuterium atom consumed is 7.94×10^{-13} joule. One cubic meter of water contains about 10^{25} atoms of deuterium having a mass of 34.4 grams and a potential fusion energy of 7.94×10^{12} joules. This is equivalent to the heat of combustion of 300 metric tons of coal or 1,500 barrels of crude oil. Since a cubic kilometer contains 10^9 cubic meters, the fuel equivalents of one cubic kilometer of seawater are 300 billion tons of coal or 1,500 billion barrels of crude oil. The total volume of the oceans is about 1.5 billion cubic kilometers. If enough deuterium were withdrawn to reduce the initial concentration by 1 percent, the energy released by fusion would amount to about 500,000 times the energy of the world's initial supply of fossil fuels!

Unlimited resources of energy, however, do not imply an unlimited number of power plants. It is as true of power plants or automobiles as it is of biological populations that the earth cannot sustain any physical growth for more than a few tens of successive doublings. Because of this impossibility the exponential rates of industrial and population growth that have prevailed during the past century and a half must soon cease. Although the forthcoming period of stability poses no insuperable physical or biological difficulties, it can hardly fail to force a major revision of those aspects of our current social and economic thinking that stem from the assumption that the growth rates that have characterized this temporary period can somehow be made permanent.

2

The Flow of Energy
in an Industrial Society

INTRODUCTION

One measure of a society's affluence in today's world is the per capita consumption of energy. Earl Cook, in his article, "The Flow of Energy in an Industrial Society," calculates that the United States, the world's richest nation, used $.06 \times 10^{18}$ BTU in 1970, about 1/3 of the world's energy consumption. If the rest of the world were to use energy at the same rate as the United States, the amount consumed would increase fifteenfold. The total world supply of fossil fuels, not counting depletion to date, is about 250×10^{18} BTU. If one allows for the aspirations for a higher standard of living in the underdeveloped world and for growth in the economies of the technologically advanced nations, the total demand might amount to 10^{18} BTU per year. At this rate most of the oil, gas, and coal supplies will be depleted within two or three centuries and sooner if population growth continues. Shortages and increased costs will face us in the near and intermediate terms, for the discovery of new deposits of oil and gas may not keep pace with demand. Methods of mining oil shales and coal without despoiling the environment, and the technology of liquefaction and gasification of these materials on a large scale have yet to be developed. The use of reserves of fossil fuels as political and economic weapons adds to the list of factors that portend uncertainty and dislocation in a world hungry for energy.

Over the long term, allowing time for the development of the necessary technology, the world's bulk energy needs can be met in several ways. Solar energy, nuclear fission using breeder reactors to conserve uranium, and nuclear fusion hold out the promise of almost inexhaustible supplies of energy. At this stage, perhaps in a few centuries, thermal pollution may set limits to man's use of energy. Waste heat, even if put to use, ultimately finds its way to the atmosphere and ocean. If man's energy consumption does not stabilize, thermal pollution could trigger climatic changes—for instance, in rainfall patterns—which could make the ice caps melt and the seas rise as a result.

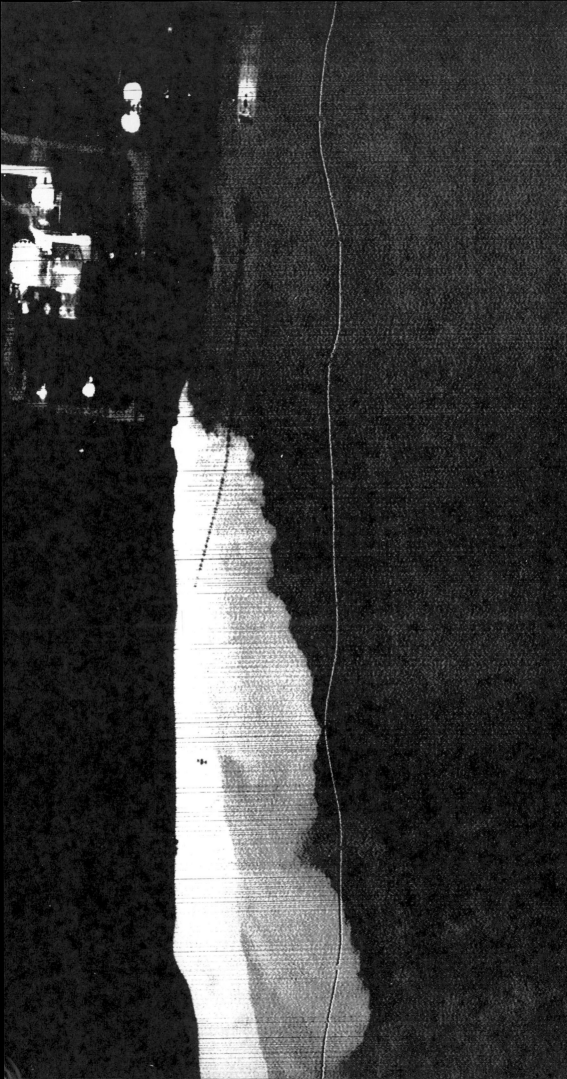

The Flow of Energy in an Industrial Society

by Earl Cook
September 1971

The U.S., with 6 percent of the world's population, uses 35 percent of the world's energy. In the long run the limiting factor in high levels of energy consumption will be the disposal of the waste heat

This article will describe the flow of energy through an industrial society: the U.S. Industrial societies are based on the use of power: the rate at which useful work is done. Power depends on energy, which is the ability to do work. A power-rich society consumes—more accurately, degrades—energy in large amounts. The success of an industrial society, the growth of its economy, the quality of the life of its people and its impact on other societies and on the total environment are determined in large part by the quantities and the kinds of energy resources it exploits and by the efficiency of its systems for converting potential energy into work and heat.

Whether by hunting, by farming or by burning fuel, man introduces himself into the natural energy cycle, converting energy from less desired forms to more desired ones: from grass to beef, from wood to heat, from coal to electricity. What characterizes the industrial societies is their enormous consumption of energy and the fact that this consumption is primarily at the expense of "capital" rather than of "income," that is, at the expense of solar energy stored in coal, oil and natural gas rather than of solar radiation, water, wind and muscle power. The advanced industrial societies, the U.S. in particular, are further characterized by their increasing dependence on electricity, a trend that has direct effects on gross energy consumption and indirect effects on environmental quality.

The familiar exponential curve of increasing energy consumption can be considered in terms of various stages of human development [*see illustration on next page*]. As long as man's energy consumption depended on the food he could eat, the rate of consumption was some 2,000 kilocalories per day; the domestication of fire may have raised it to 4,000 kilocalories. In a primitive agricultural society with some domestic animals the rate rose to perhaps 12,000 kilocalories; more advanced farming societies may have doubled that consumption. At the height of the low-technology industrial revolution, say between 1850 and 1870, per capita daily consumption reached 70,000 kilocalories in England, Germany and the U.S. The succeeding high-technology revolution was brought about by the central electric-power station and the automobile, which enable the average person to apply power in his home and on the road. Beginning shortly before 1900, per capita energy consumption in the U.S. rose at an increasing rate to the 1970 figure: about 230,000 kilocalories per day, or about 65×10^{15} British thermal units (B.t.u.) per year for the country as a whole. Today the industrial regions, with 30 percent of the world's people, consume 80 percent of the world's energy. The U.S., with 6 percent of the people, consumes 35 percent of the energy.

In the early stages of its development in western Europe industrial society based its power technology on income sources of energy, but the explosive growth of the past century and a half has been fed by the fossil fuels, which are not renewable on any time scale meaningful to man. Modern industrial society is totally dependent on high rates of consumption of natural gas, petroleum and coal. These nonrenewable fossil-fuel resources currently provide 96 percent of the gross energy input into the U.S. economy [*see top illustration on page 19*]. Nuclear power, which in 1970 accounted for only .3 percent of the total energy input, is also (with present reactor technology) based on a capital source of energy: uranium 235. The energy of falling water, converted to hydropower, is the only income source of energy that now makes any significant contribution to the U.S. economy, and its proportional role seems to be declining from a peak reached in 1950.

Since 1945 coal's share of the U.S. energy input has declined sharply, while both natural gas and petroleum have increased their share. The shift is reflected in import figures. Net imports of petroleum and petroleum products doubled between 1960 and 1970 and now constitute almost 30 percent of gross consumption. In 1960 there were no imports of natural gas; last year natural-gas imports (by pipeline from Canada and as liquefied gas carried in cryogenic tankers) accounted for almost 4 percent of gross consumption and were increasing.

The reasons for the shift to oil and gas are not hard to find. The conversion of railroads to diesel engines represented a large substitution of petroleum for coal. The rapid growth, beginning during World War II, of the national

HEAT DISCHARGE from a power plant on the Connecticut River at Middletown, Conn., is shown in this infrared scanning radiograph. The power plant is at upper left, its structures outlined by their heat radiation. The luminous cloud running along the left bank of the river is warm water discharged from the cooling system of the plant. The vertical oblong object at top left center is an oil tanker. The luminous spot astern is the infrared glow of its engine room. The dark streak between the tanker and the warm-water region is a breakwater. The irregular line running down the middle of the picture is an artifact of the infrared scanning system. The picture was made by HRB-Singer, Inc., for U.S. Geological Survey.

network of high-pressure gas-transmission lines greatly extended the availability of natural gas. The explosion of the U.S. automobile population, which grew twice as fast as the human population in the decade 1960–1970, and the expansion of the nation's fleet of jet aircraft account for much of the increase in petroleum consumption. In recent years the demand for cleaner air has led to the substitution of natural gas or low-sulfur residual fuel oil for high-sulfur coal in many central power plants.

An examination of energy inputs by sector of the U.S. economy rather than by source reveals that much of the recent increase has been going into household, commercial and transportation applications rather than industrial ones [*see bottom illustration on opposite page*]. What is most striking is the growth of the electricity sector. In 1970 almost 10 percent of the country's useful work was done by electricity. That is not the whole story. When the flow of energy from resources to end uses is charted for 1970 [*see illustration on pages 20 and 21*], it is seen that producing that much electricity accounted for 26 percent of the gross consumption of energy, because of inefficiencies in generation and transmission. If electricity's portion of end-use consumption rises to about 25 percent by the year 2000, as is expected, then its generation will account for between 43 and 53 percent of the country's gross energy consumption. At that point an amount of energy equal to about half of the useful work done in the U.S. will be in the form of waste heat from power stations!

All energy conversions are more or less inefficient, of course, as the flow diagram makes clear. In the case of electricity there are losses at the power plant, in transmission and at the point of application of power; in the case of fuels consumed in end uses the loss comes at the point of use. The 1970 U.S. gross consumption of 64.6×10^{15} B.t.u. of energy (or 16.3×10^{15} kilocalories, or 19×10^{12} kilowatt-hours) ends up as 32.8×10^{15} B.t.u. of useful work and 31.8×10^{15} B.t.u. of waste heat, amounting to an overall efficiency of about 51 percent.

The flow diagram shows the pathways of the energy that drives machines, provides heat for manufacturing processes and heats, cools and lights the country. It does not represent the total energy budget because it includes neither food nor vegetable fiber, both of which bring solar energy into the economy through photosynthesis. Nor does it include environmental space heating by solar radiation, which makes life on the earth possible and would be by far the largest component of a total energy budget for any area and any society.

The minute fraction of the solar flux that is trapped and stored in plants provides each American with some 10,000 kilocalories per day of gross food production and about the same amount in the form of nonfood vegetable fiber. The fiber currently contributes little to the energy supply. The food, however, fu-

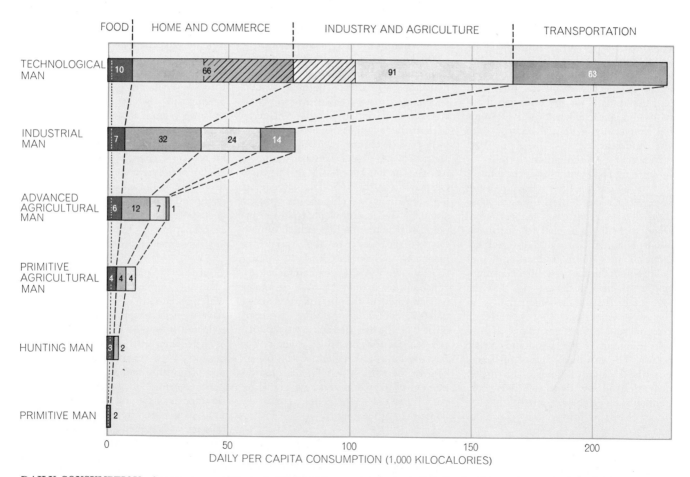

DAILY CONSUMPTION of energy per capita was calculated by the author for six stages in human development (and with an accuracy that decreases with antiquity). Primitive man (East Africa about 1,000,000 years ago) without the use of fire had only the energy of the food he ate. Hunting man (Europe about 100,000 years ago) had more food and also burned wood for heat and cooking. Primitive agricultural man (Fertile Crescent in 5000 B.C.) was growing crops and had gained animal energy. Advanced agricultural man (northwestern Europe in A.D. 1400) had some coal for heating, some water power and wind power and animal transport. Industrial man (in England in 1875) had the steam engine. In 1970 technological man (in the U.S.) consumed 230,000 kilocalories per day, much of it in form of electricity (*hatched area*). Food is divided into plant foods (*far left*) and animal foods (or foods fed to animals).

els man. Gross food-plant consumption might therefore be considered another component of gross energy consumption; it would add about 3×10^{15} B.t.u. to the input side of the energy-flow scheme. Of the 10,000 kilocalories per capita per day of gross production, handling and processing waste 15 percent. Of the remaining 8,500 kilocalories, some 6,300 go to feed animals that produce about 900 kilocalories of meat and 2,200 go into the human diet as plant materials, for a final food supply of about 3,100 kilocalories per person. Thus from field to table the efficiency of the food-energy system is 31 percent, close to the efficiency of a central power station. The similarity is not fortuitous; in both systems there is a large and unavoidable loss in the conversion of energy from a less desired form to a more desired one.

Let us consider recent changes in U.S. energy flow in more detail by seeing how the rates of increase in various sectors compare. Not only has energy consumption for electric-power generation been growing faster than the other sectors but also its growth rate has been increasing: from 7 percent per year in 1961–1965 to 8.6 percent per year in 1965–1969 to 9.25 percent last year [see top illustration on page 22]. The energy consumed in industry and commerce and in homes has increased at a fairly steady rate for a decade, but the energy demand of transportation has risen more sharply since 1966. All in all, energy consumption has been increasing lately at a rate of 5 percent per year, or four times faster than the increase in the U.S. population. Meanwhile the growth of the gross national product has tended to fall off, paralleling the rise in energy sectors other than fast-growing transportation and electricity. The result is a change in the ratio of total energy consumption to G.N.P. [see bottom illustration on page 22]. The ratio had been in a long general decline since 1920 (with brief reversals) but since 1967 it has risen more steeply each year. In 1970 the U.S. consumed more energy for each dollar of goods and services than at any time since 1951.

Electricity accounts for much of this decrease in economic efficiency, for several reasons. For one thing, we are substituting electricity, with a thermal efficiency of perhaps 32 percent, for many direct fuel uses with efficiencies ranging from 60 to 90 percent. Moreover, the fastest-growing segment of end-use consumption has been electric air conditioning. From 1967 to 1970 consumption for

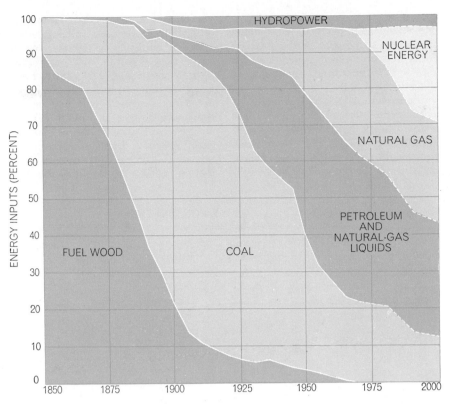

FOSSIL FUELS now account for nearly all the energy input into the U.S. economy. Coal's contribution has decreased since World War II; that of natural gas has increased most in that period. Nuclear energy should contribute a substantial percent within the next 20 years.

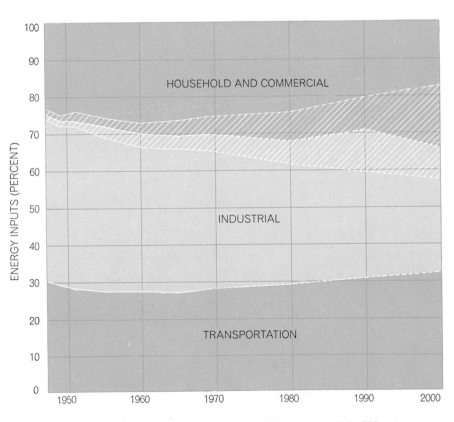

USEFUL WORK is distributed among the various end-use sectors of the U.S. economy as shown. The trend has been for industry's share to decrease, with household and commercial uses (including air conditioning) and transportation growing. Electricity accounts for an ever larger share of the work (hatched area). U.S. Bureau of Mines figures in this chart include nonenergy uses of fossil fuels, which constitute about 7 percent of total energy inputs.

air conditioning grew at the remarkable rate of 20 percent per year; it accounted for almost 16 percent of the total increase in electric-power generation from 1969 to 1970, with little or no multiplier effect on the G.N.P.

Let us take a look at this matter of efficiency in still another way: in terms of useful work done as a percentage of gross energy input. The "useful-work equivalent," or overall technical efficiency, is seen to be the product of the conversion efficiency (if there is an intermediate conversion step) and the application efficiency of the machine or device that does the work [see bottom illustration on page 23]. Clearly there is a wide range of technical efficiencies in energy systems, depending on the conversion devices. It is often said that electrical resistance heating is 100 percent efficient, and indeed it is in terms, say, of converting electrical energy to thermal energy at the domestic hot-water heater. In terms of the energy content of the natural gas or coal that fired the boiler that made the steam that drove the turbine that turned the generator that produced the electricity that heated the wires that warmed the water, however, it is not so efficient.

The technical efficiency of the total U.S. energy system, from potential energy at points of initial conversion to work at points of application, is about 50 percent. The economic efficiency of

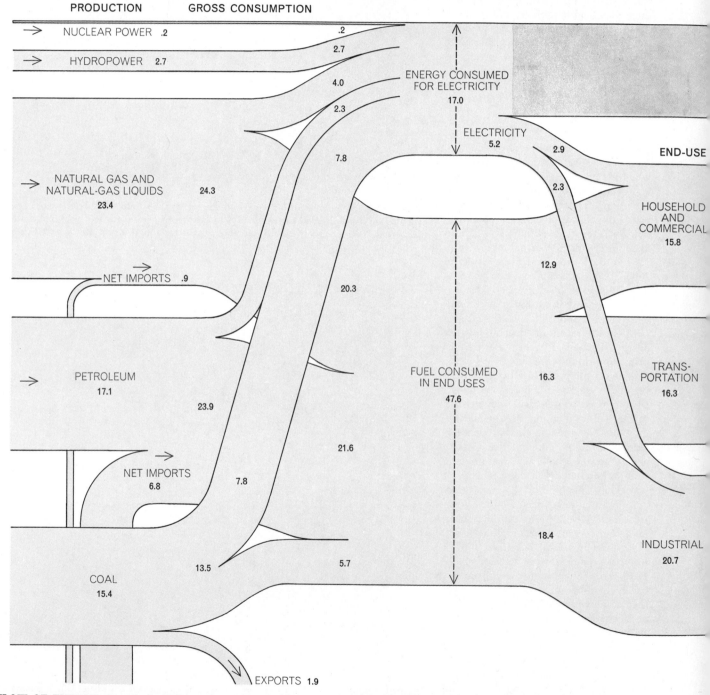

FLOW OF ENERGY through the U.S. system in 1970 is traced from production of energy commodities (*left*) to the ultimate conversion of energy into work for various industrial end products and waste heat (*right*). Total consumption of energy in 1970 was 64.6 × 10^15 British thermal units. (Adding nonenergy uses of fossil fuels, primarily for petrochemicals, would raise the total to 68.8 × 10^15 B.t.u.) The overall efficiency of the system was about 51 percent. Some of the fossil-fuel energy is consumed directly and

the system is considerably less. That is because work is expended in extracting, refining and transporting fuels, in the construction and operation of conversion facilities, power equipment and electricity-distribution networks, and in handling waste products and protecting the environment.

An industrial society requires not only a large supply of energy but also a high use of energy per capita, and the society's economy and standard of living are shaped by interrelations among resources, population, the efficiency of conversion processes and the particular applications of power. The effect of these interrelations is illustrated by a comparison of per capita energy consumption and per capita output for a number of countries [*see illustration on page 24*]. As one might expect, there is a strong general correlation between the two measures, but it is far from being a one-to-one correlation. Some countries (the U.S.S.R. and the Republic of South Africa, for example) have a high energy consumption with respect to G.N.P.; other countries (such as Sweden and New Zealand) have a high output with relatively less energy consumption. Such differences reflect contrasting combinations of energy-intensive heavy industry and light consumer-oriented and service industries (characteristic of different stages of economic development) as well as differences in the efficiency of energy use. For example, countries that still rely on coal for a large part of their energy requirement have higher energy inputs per unit of production than those that use mainly petroleum and natural gas.

A look at trends from the U.S. past is also instructive. Between 1800 and 1880 total energy consumption in the U.S. lagged behind the population increase, which means that per capita energy consumption actually declined somewhat. On the other hand, the American standard of living increased during this period because the energy supply in 1880 (largely in the form of coal) was being used much more efficiently than the energy supply in 1800 (largely in the form of wood). From 1900 to 1920 there was a tremendous surge in the use of energy by Americans but not a parallel increase in the standard of living. The ratio of energy consumption to G.N.P. increased 50 percent during these two decades because electric power, inherently less efficient, began being substituted for the direct use of fuels; because the automobile, at best 25 percent efficient, proliferated (from 8,000 in 1900 to 8,132,000 in 1920), and because mining and manufacturing, which are energy-intensive, grew at very high rates during this period.

Then there began a long period during which increases in the efficiency of energy conversion and utilization fulfilled about two-thirds of the total increase in demand, so that the ratio of energy consumption to G.N.P. fell to about 60 percent of its 1920 peak although per capita energy consumption continued to increase. During this period (1920–1965) the efficiency of electric-power generation and transmission almost trebled, mining and manufacturing grew at much lower rates and the services sector of the economy, which is not energy-intensive, increased in importance.

"Power corrupts" was written of man's control over other men but it applies also to his control of energy re-

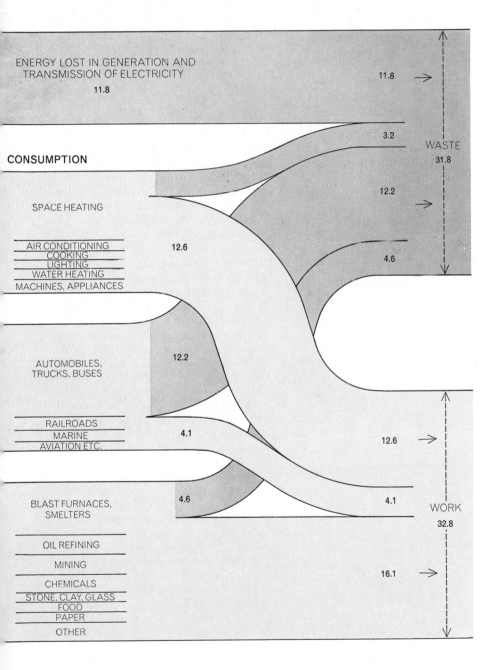

some is converted to generate electricity. The efficiency of electrical generation and transmission is taken to be about 31 percent, based on the ratio of utility electricity purchased in 1970 to the gross energy input for generation in that year. Efficiency of direct fuel use in transportation is taken as 25 percent, of fuel use in other applications as 75 percent.

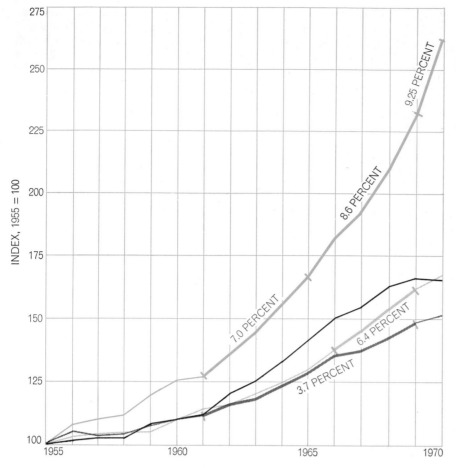

INCREASE IN CONSUMPTION of energy for electricity generation (*dark color*), transportation (*light color*) and other applications (*gray*) and of the gross national product (*black*) are compared. Annual growth rates for certain periods are shown beside heavy segments of curves. Consumption of electricity has a high growth rate and is increasing.

RATIO OF ENERGY CONSUMPTION to gross national product has varied over the years. It tends to be low when the G.N.P. is large and energy is being used efficiently, as was the case during World War II. The ratio has been rising steadily since 1965. Reasons include the increase in the use of air conditioning and the lack of advance in generating efficiency.

sources. The more power an industrial society disposes of, the more it wants. The more power we use, the more we shape our cities and mold our economic and social institutions to be dependent on the application of power and the consumption of energy. We could not now make any major move toward a lower per capita energy consumption without severe economic dislocation, and certainly the struggle of people in less developed regions toward somewhat similar energy-consumption levels cannot be thwarted without prolonging mass human suffering. Yet there is going to have to be some leveling off in the energy demands of industrial societies. Countries such as the U.S. have already come up against constraints dictated by the availability of resources and by damage to the environment. Another article in this issue considers the question of resource availability [see "The Energy Resources of the Earth," by M. King Hubbert, beginning on page 5]. Here I shall simply point out some of the decisions the U.S. faces in coping with diminishing supplies, and specifically with our increasing reliance on foreign sources of petroleum and petroleum products. In the short run the advantages of reasonable self-sufficiency must be weighed against the economic and environmental costs of developing oil reserves in Alaska and off the coast of California and the Gulf states. Later on such self-sufficiency may be attainable only through the production of oil from oil shale and from coal. In the long run the danger of dependence on dwindling fossil fuels—whatever they may be —must be balanced against the research and development costs of a major effort to shape a new energy system that is neither dependent on limited resources nor hard on the environment.

The environmental constraint may be more insistent than the constraint of resource availability. The present flow of energy through U.S. society leaves waste rock and acid water at coal mines; spilled oil from offshore wells and tankers; waste gases and particles from power plants, furnaces and automobiles; radioactive wastes of various kinds from nuclear-fuel processing plants and reactors. All along the line waste heat is developed, particularly at the power plants.

Yet for at least the next 50 years we shall be making use of dirty fuels: coal and petroleum. We can improve coal-combustion technology, we can build power plants at the mine mouth (so that the air of Appalachia is polluted instead of the air of New York City), we can make clean oil and gas from coal and oil

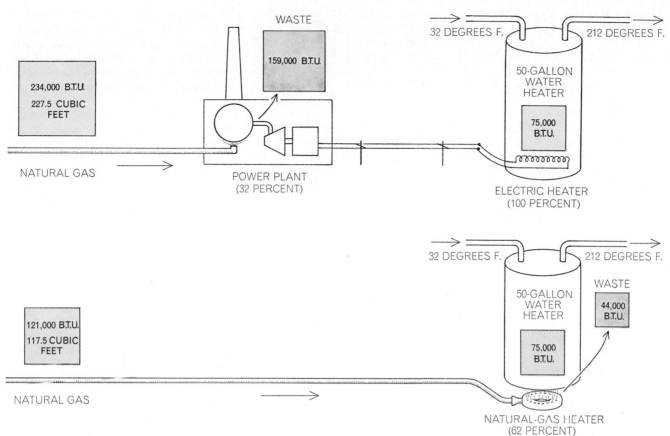

WASTE

159,000 B.T.U.

234,000 B.T.U.
227.5 CUBIC FEET

32 DEGREES F.   212 DEGREES F.

50-GALLON WATER HEATER

75,000 B.T.U.

NATURAL GAS

POWER PLANT
(32 PERCENT)

ELECTRIC HEATER
(100 PERCENT)

32 DEGREES F.   212 DEGREES F.

WASTE

44,000 B.T.U.

50-GALLON WATER HEATER

75,000 B.T.U.

121,000 B.T.U.
117.5 CUBIC FEET

NATURAL GAS

NATURAL-GAS HEATER
(62 PERCENT)

EFFICIENCIES OF HEATING WATER with natural gas indirectly by generating electricity for use in resistance heating (*top*) and directly (*bottom*) are contrasted. In each case the end result is enough heat to warm 50 gallons of water from 32 degrees Fahrenheit to 212 degrees. Electrical method requires substantially more gas even though efficiency at electric heater is nearly 100 percent.

from shale, and sow grass on the mountains of waste. As nuclear power plants proliferate we can put them underground, or far from the cities they serve if we are willing to pay the cost in transmission losses. With adequate foresight, caution and research we may even be able to handle the radioactive-waste problem without "undue" risk.

There are, however, definite limits to such improvements. The automobile engine and its present fuel simply cannot be cleaned up sufficiently to make it an acceptable urban citizen. It seems clear that the internal-combustion engine will be banned from the central city by the year 2000; it should probably be banned right now. Because our cities are shaped for automobiles, not for mass transit, we shall have to develop battery-powered

or flywheel-powered cars and taxis for inner-city transport. The 1970 census for the first time showed more metropolitan citizens living in suburbs than in the central city; it also showed a record high in automobiles per capita, with the greatest concentration in the suburbs. It seems reasonable to visualize the suburban two-car garage of the future with one car a recharger for "downtown" and

	PRIMARY ENERGY INPUT (UNITS)	SECONDARY ENERGY OUTPUT (UNITS)	APPLICATION EFFICIENCY (PERCENT)	TECHNICAL EFFICIENCY (PERCENT)
AUTOMOBILE				
INTERNAL-COMBUSTION ENGINE	100		25	25
FLYWHEEL DRIVE CHARGED BY ELECTRICITY	100	32	100	32
SPACE HEATING				
BY DIRECT FUEL USE	100		75	75
BY ELECTRICAL RESISTANCE	100	32	100	32
SMELTING OF STEEL				
WITH COKE	100	94	94	70
WITH ELECTRICITY	100	32	32	32

TECHNICAL EFFICIENCY is the product of conversion efficiency at an intermediate step (if there is one) and application efficiency at the device that does the work. Losses due to friction and heat are ignored in the flywheel-drive automobile data. Coke retains only about 66 percent of the energy of coal, but the energy recovered from the by-products raises the energy conservation to 94 percent.

the other, still gasoline-powered, for suburban and cross-country driving.

Of course, some of the improvement in urban air quality bought by excluding the internal-combustion engine must be paid for by increased pollution from the power plant that supplies the electricity for the nightly recharging of the downtown vehicles. It need not, however, be paid for by an increased draft on the primary energy source; this is one substitution in which electricity need not decrease the technical efficiency of the system. The introduction of heat pumps for space heating and cooling would be

another. In fact, the overall efficiency should be somewhat improved and the environmental impact, given adequate attention to the siting, design and operation of the substituting power plant, should be greatly alleviated.

If technology can extend resource availability and keep environmental deterioration within acceptable limits in most respects, the specific environmental problem of waste heat may become the overriding one of the energy system by the turn of the century.

The cooling water required by power

plants already constitutes 10 percent of the total U.S. streamflow. The figure will increase sharply as more nuclear plants start up, since present designs of nuclear plants require 50 percent more cooling water than fossil-fueled plants of equal size do. The water is heated 15 degrees Fahrenheit or more as it flows through the plant. For ecological reasons such an increase in water released to a river, lake or ocean bay is unacceptable, at least for large quantities of effluent, and most large plants are now being built with cooling ponds or towers from which much of the heat of the water is dissi-

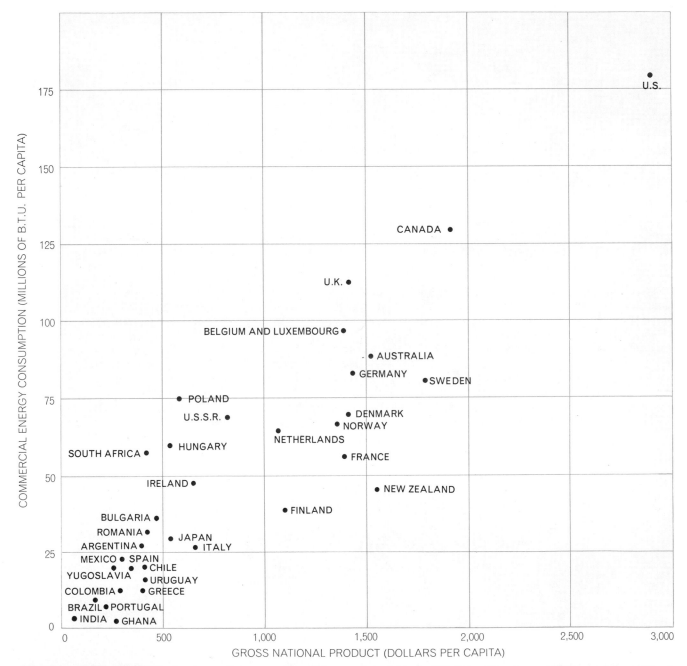

ROUGH CORRELATION between per capita consumption of energy and gross national product is seen when the two are plotted together; in general, high per capita energy consumption is a prerequisite for high output of goods and services. If the position plotted for the U.S. is considered to establish an arbitrary "line," some countries fall above or below that line. This appears to be related to a country's economic level, its emphasis on heavy industry or on services and its efficiency in converting energy into work.

pated to the atmosphere before the water is discharged or recycled through the plant. Although the atmosphere is a more capacious sink for waste heat than any body of water, even this disposal mechanism obviously has its environmental limits.

Many suggestions have been made for putting the waste heat from power plants to work: for irrigation or aquaculture, to provide ice-free shipping lanes or for space heating. (The waste heat from power generation today would be more than enough to heat every home in the U.S.!) Unfortunately the quantities of water involved, the relatively low temperature of the coolant water and the distances between power plants and areas of potential use are serious deterrents to the utilization of waste heat. Plants can be designed, however, for both power production and space heating. Such a plant has been in operation in Berlin for a number of years and has proved to be more efficient than a combination of separate systems for power production and space heating. The Berlin plant is not simply a conserver of waste heat but an exercise in fuel economy; its power capacity was reduced in order to raise the temperature of the heated water above that of normal cooling water.

With present and foreseeable technology there is not much hope of decreasing the amount of heat rejected to streams or the atmosphere (or both) from central steam-generating power plants. Two systems of producing power without steam generation offer some long-range hope of alleviating the waste-heat problem. One is the fuel cell; the other is the fusion reactor combined with a system for converting the energy released directly into electricity [see "The Con-

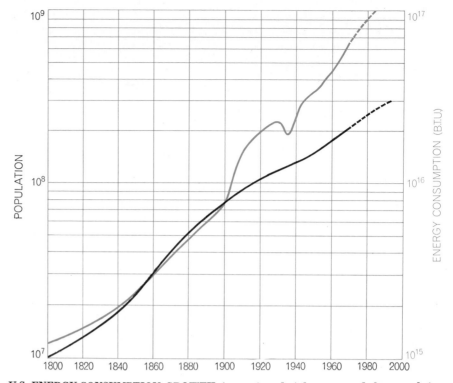

U.S. ENERGY-CONSUMPTION GROWTH (*curve in color*) **has outpaced the growth in population** (*black*) **since 1900, except during the energy cutback of the depression years.**

version of Energy," by Claude M. Summers; SCIENTIFIC AMERICAN Offprint 668]. In the fuel cell the energy contained in hydrocarbons or hydrogen is released by a controlled oxidation process that produces electricity directly with an efficiency of about 60 percent. A practical fusion reactor with a direct-conversion system is not likely to appear in this century.

Major changes in power technology will be required to reduce pollution and manage wastes, to improve the efficiency of the system and to remove the resource-availability constraint. Making

the changes will call for hard political decisions. Energy needs will have to be weighed against environmental and social costs; a decision to set a pollution standard or to ban the internal-combustion engine or to finance nuclear-power development can have major economic and political effects. Democratic societies are not noted for their ability to take the long view in making decisions. Yet indefinite growth in energy consumption, as in human population, is simply not possible.

3

Geothermal Power

INTRODUCTION

The energy crisis has accelerated the search for different sources of energy. Strictly speaking, none of the approaches being investigated is conceptually new, but new technology is required to extract energy on a large scale from solar radiation, the wind, the tides, ocean temperature differences, or the earth's heat. In his contribution, "Geothermal Power," Joseph Barnea discusses the hope that geothermal heat offers inexpensive, relatively pollution-free energy, a prospect that has triggered interest in exploitation by a number of companies, large and small.

Geothermal energy originates in the flow of heat from the earth's interior to the surface. The heat is produced mostly by the decay of radioactive elements, although some may come from an ancient, nearly molten earth. For these reasons temperature increases with depth in the interior of the earth. However, to be exploited efficiently, this energy must be tapped in those places where particularly high temperatures are found close to the surface. Such places are usually found in regions of recent volcanic activity, such as the western United States. Naturally occurring hot water or steam will be produced when rainwater percolates down to hot rock. If water does not percolate naturally, it may be injected into the hot zone to form the hot water or steam needed to drive electric generators. According to the United States Geological Survey, approximately 1.3 million acres of land, mostly in the western states, are potentially suitable for geothermal development, which may provide an energy source of between fifteen and thirty thousand megawatts.

Geothermal energy offers the advantages of long-lived supplies, competitive costs of extraction, no extensive pollution of the atmosphere by the products of combustion, and minimal defacement of the landscape. Suitable sites may be hard to find and tend to occur far from centers of population. Geothermal hot water and steam often contain corrosive salts and noxious gases, such as hydrogen sulfide, that pose technological difficulties.

Geologists and geophysicists are currently prospecting for potential sites, pilot programs to explore novel methods of heat extraction and conversion to electricity are underway, and environmental impact is being analyzed—all signs of growing interest. There is no question that the thermal energy is in the ground—more than man can use—but can it be extracted and converted cheaply, safely, and in sufficient quantity to make a dent on our energy shortage? That is the question that current research and development must answer.

Geothermal Power

by Joseph Barnea
January 1972

The pressure on energy resources has generated new interest in the earth's heat. The emphasis is on exploring for new geothermal areas and developing new ways to extract work from steam and hot water

An old source of power for man's work has begun to attract new interest. Natural underground reservoirs of steam and hot water are now being tapped on a significant scale, and it will come as a surprise to many people to learn that the harnessing of this geothermal energy has already reached an aggregate capacity of a million kilowatts in plants around the world. At the present rate of development it is likely that by the end of this decade the production of electric power from steam fields will be quadrupled.

The heat of many geothermal reservoirs comes from a large body of molten rock that has been pushed up into the earth's crust from great depths by geologic forces. This dome of magma heats the rocks in the crust near the surface, which in turn heats the water in fissured or porous rocks to a temperature of perhaps 500 degrees Fahrenheit. Being at depths of as much as six miles, the water is under high pressure and is therefore liquid. Where the hot water can escape through a fissure it begins to boil, and part of it flashes off as steam. The geothermal energy can be tapped by a well driven into the fissure or down to the porous layer.

Interest in this source of energy has quickened in the past few years. Recent explorations have revealed that the resource is larger and more extensive than had been supposed. A generation ago the hot springs and steam fields that had long been known in a few localities around the world were believed to be merely local freaks of nature. There is evidence now that reservoirs of steam and hot water are actually widespread in the earth's crust. Signs of their presence have been detected on most of the continents and on a number of islands. It seems possible that such fields will also

be found under the seas. Some of the explored fields are known to hold large quantities of energy. A single steam field in northern California, the Geysers field, is estimated to have a potential capacity of three million kilowatts, and surveys that have been made in the Imperial Valley of southern California have indicated a potential of 20 million kilowatts in that area.

The incentive for undertaking a major effort to tap geothermal fields has been heightened by projects showing that in addition to electric power they can yield other useful products. The geothermal steam or hot water can be applied to desalting seawater, to heating houses, greenhouses and swimming pools and to providing nonelectrical energy for refrigeration and air conditioning. Moreover, the hot water itself is a source of extractable minerals and can serve to provide potable water. These additional dividends increase the economic attractiveness of investment for the exploitation of this great earth resource, which up to now has served man mainly as a resort attraction (in the form of health spas) and as a somewhat esoteric and certainly minor source of power supporting small generating plants at a few sites around the globe.

Hot springs, where water from heated strata flows naturally to the surface, have of course been known and used since ancient times. The Romans developed these watering places for medical and recreational purposes all around the Mediterranean and to the outskirts of their empire as far as Bath in the British Isles; there were also medical spas in ancient Japan and elsewhere in the Far East. Hot springs still flourish as health resorts today in Japan, in France and other centers in continental Europe, in

Africa and in many other places outside the Anglo-Saxon world.

The realization that the steam in the crust might be tapped for power came at the beginning of this century. In 1904 the first electricity plant so powered was built and plugged into a steam field in northern Italy now known as the Larderello field. Over the following decades there was a slow and tentative growth of interest in geothermal energy. More plants were built on the Larderello field, and other small-scale projects for the use of natural steam or hot water for power, industrial purposes and heating were developed in Japan, Hungary, the U.S.S.R., Iceland, New Zealand and elsewhere. In the U.S. the first geothermal power plant, of 12,500-kilowatt capacity, was commissioned in 1960 on the Geysers field, which is by far the largest field yet discovered in the world.

To those investigators who early recognized the potentialities of geothermal energy the development of this resource has seemed agonizingly slow. There are several reasons why things have not gone more rapidly. Judging from the surface indications (the comparative rarity of hot springs or steam holes) the geothermal energy that might be available appeared to be highly localized and minor in amount. The explorations and the discovery of fields have been limited to those that show surface signs, because little information has been available on geological indications that might signal the presence of hidden fields. In the past such fields could be found only by speculative drilling, and it seemed that the expense of drilling would be justified only for fields located at a shallow depth. The paucity of research and information on the geothermal resources in the crust, the lack of guides for exploration and the shortage of trained specialists and

STEAM WELLS tap geothermal energy for the production of power at a plant operated by the Pacific Gas and Electric Company at The Geysers, about 90 miles north of San Francisco. Since 1960 the company has brought plant capacity at the site to 192,000 kilowatts.

LARDERELLO GEOTHERMAL FIELD in Italy has been used for generating electric power since 1904. It now has a capacity of 380,-000 kilowatts. The chimney-like structures at left are hyperbolic cooling towers that are associated with the power plant at the site.

technicians in this field have in the past combined to retard progress.

Nevertheless, the enterprise is now moving forward at an accelerating pace. It has been given impetus in the U.S. by the Geothermal Steam Act, adopted by Congress in December, 1970, which establishes the development of U.S. geothermal resources as a national goal. What is now needed is a worldwide expansion of efforts in research, exploration and the training of experts for this work.

The sources of usable geothermal energy in the earth fall into three classes: dry steam fields, wet steam fields and fields of lesser heat content consisting of water at temperatures below the boiling point (at atmospheric pressure). Each type of geothermal energy has its special uses and also capabilities for a variety of applications.

The dry steam fields are filled mainly with steam itself, under pressure and at relatively high temperatures. This steam is usable directly for the production of electric power. It can be piped right to the turbine and therefore simplifies the requirements for plant equipment; the investment in plant may be as low as $100 per kilowatt. In order to minimize piping costs the plant must be located close to the steam wells; moreover, since the steam emerges from the field at low pressure and large amounts of steam must be handled, the effective size of the turbines is limited. This means that the plant cannot be very large. The upper limit at present is about 55 megawatts. Power generators of this magnitude, each fed by 10 to 15 steam wells, are now being installed at the Geysers field in California.

The steam from a dry field can be put to uses other than power production. The water condensed from the steam after it has given up its energy can provide a supply of fresh water. In locations near the ocean or a saltwater lake the steam could be employed as a heating medium in distillation plants for producing potable water by subatmospheric boiling, the steam in this case being provided without any cost for fuel.

So far the existence of five important dry steam fields has been established: the Larderello field in Italy, the Geysers field in California, the Valle Caldera field in New Mexico and two fields in Japan. In the absence of systematic exploration it is not yet possible to estimate how many other such fields may lie hidden in the earth's crust.

On the basis of discoveries made to date, it seems that wet steam fields may

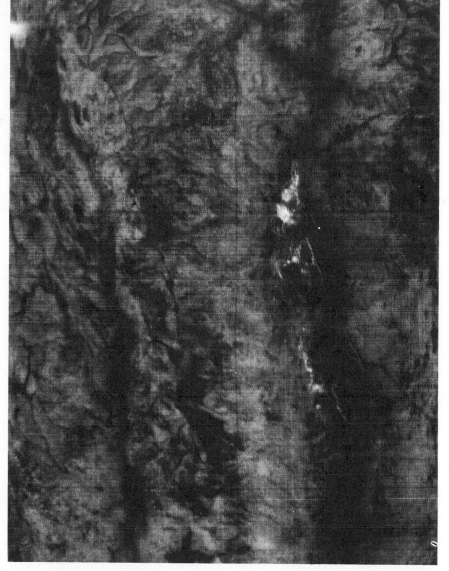

INFRARED VIEW of the steam field at The Geysers was obtained by the U.S. Geological Survey. The aerial photograph was made before dawn in order to minimize the effect of the sun on the temperature of the ground. Light areas at right center are geothermal areas.

be 20 times more abundant than dry steam fields. The wet field is filled with hot water (above its boiling point at atmospheric pressure) that does not become steam until the pressure is released by drilling into the field. The superheated water in the wet field, typically at temperatures ranging from 180 to 370 degrees Celsius (about 350 to 700 degrees Fahrenheit), flashes into a mixture of steam and water as it comes to the surface. About 10 to 20 percent of the discharge, by weight, is steam; the rest is hot water. The steam can be used for power production; the hot water has a multitude of potential uses.

The pioneering stages of the harnessing of geothermal energy have been marked by concentration on a single application of the yield from the wells. At

the Wairakei wet steam field in New Zealand, for example, the steam fraction is fed to a power-generating plant and the hot water is discarded into a river. In this respect geothermal energy has paralleled the history of the discovery of petroleum, which at first was used only for kerosene lamps. Now geothermal development is entering a more sophisticated stage through the analysis of its components and their combination into multipurpose projects.

There are already installations in which the steam of a wet steam field is devoted to the production of power and some of the hot geothermal water is distilled, without the addition of any more heat, to make fresh water. (Distillation is possible because the pressure in the flash-distillation plant is kept below at-

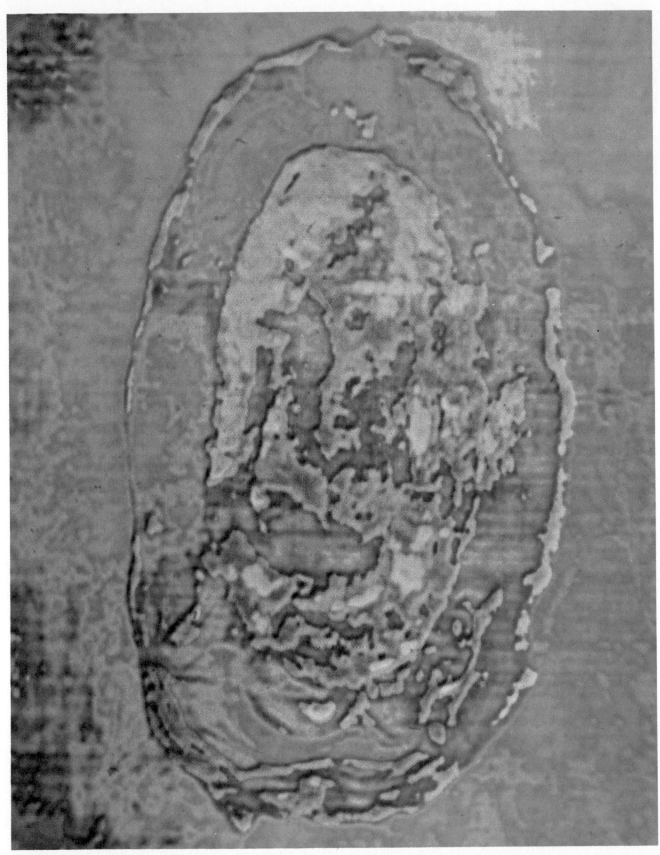

GEOTHERMAL SOURCE associated with a volcanic crater in the Rift Valley in Ethiopia was explored by means of infrared photography in a project carried out by the United Nations. The explorations were made by airplane using black-and-white infrared film, which shows hotter areas as white and cooler areas as progressively darker shades of gray. Promising photographs, such as the one shown here, were processed through a densitometer, which converts the density of the photograph in terms of ground temperature and applies a predetermined color scheme to indicate the differences. In this photograph the hottest areas are orange and the coolest ones are blue. The technique provided the first measurement of the extent of this geothermal source and range of temperatures in area.

mospheric pressure by exhaust pumps.) A further step is planned at a wet steam field recently discovered at El Tatio in Chile. There the Chilean government in cooperation with the United Nations is investigating the development of a facility that will generate three products [*see illustration on page 34*]. The steam will first be used to produce electricity. Hot water produced from the steam will go through a desalination plant, producing fresh water, and the effluent from the hot-water feed will be concentrated in a mineral-rich brine from which valuable minerals will be extracted in evaporation ponds.

Following the accidental discovery of mineral-rich geothermal brines in southern California and in the Red Sea the UN began a systematic search for mineral brines as part of its program of geothermal investigations. Two discoveries of potential economic importance have been made so far, one in Ethiopia and one in Chile.

At Kawerau in New Zealand a paper and pulp company is using the hot water from a wet steam field for heating in industrial processes. In Iceland the hot water from such fields has long been applied to industrial uses and household and district heating. In Japan the applications include uses in experimental fish-farming projects, cleaning, cooking, soil-heating and bathing. Househeating with hot-well water is being developed on a large scale in several countries, notably Japan, the U.S.S.R. and Hungary (where the cost of such heating is reported to be only a fourth of that with fuel-burning systems). In the U.S. househeating from hot wells is being applied on a small scale in Boise, Idaho, and Klamath Falls, Ore.

The use of geothermal water in air conditioning is based on a process that employs water as the refrigerant and a solution of lithium bromide as a low-temperature absorbent fluid. As in other refrigerating systems the refrigerant is vaporized, thereby extracting heat from the surroundings. Then, however, the refrigerant is taken up by the absorbent. External heat (in this case supplied by geothermal water) drives the refrigerant off the absorbent as a gas; the gas is condensed to liquid, which returns to the evaporator to begin the cycle again. Two Russian investigators of applications of geothermal energy, A. N. Tikhonov and I. M. Dvorov, recently reported that a machine of this kind, used in a system providing refrigeration in summer and heat in winter, is being mass-produced

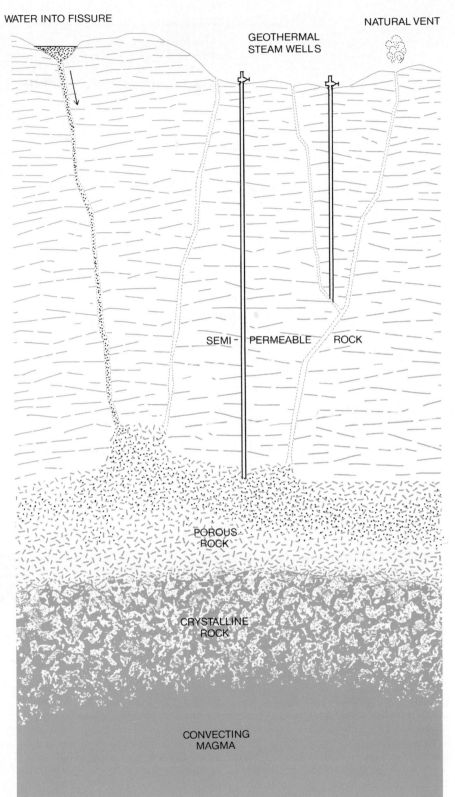

GEOLOGICAL SETTING of a geothermal energy source is portrayed. The heat comes from magma, or molten rock, that has been pushed up into the earth's crust. By convection of the magma the heat moves through crystalline rock to a layer of porous rock containing water that has percolated down from the ground, sometimes to great depths. Over the porous rock is relatively impermeable rock that serves as a cap to contain the heat. Being deep in the ground, the water is under high pressure and is therefore liquid, although its temperature may be some 500 degrees Fahrenheit. It expands and rises in a natural vent; as the pressure drops, water begins to boil and produce steam. A well can tap the vent or the porous layer.

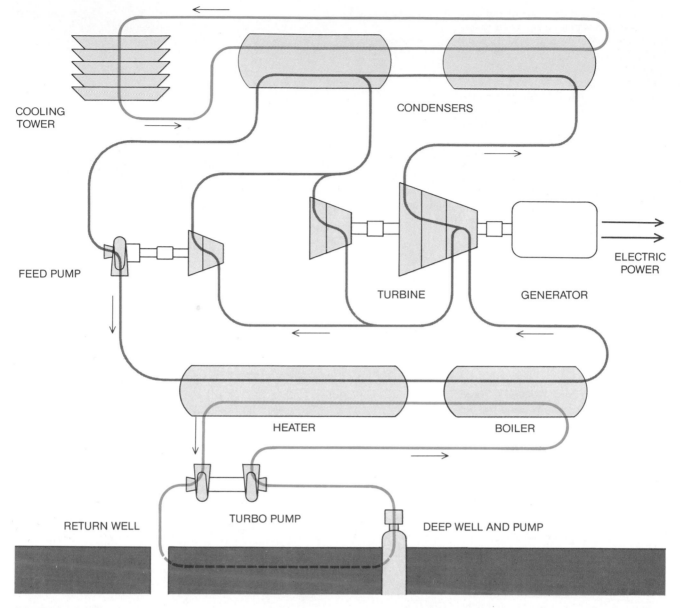

COOLING TOWER

CONDENSERS

FEED PUMP

TURBINE

GENERATOR

ELECTRIC POWER

HEATER

BOILER

RETURN WELL

TURBO PUMP

DEEP WELL AND PUMP

POWER PLANT using geothermal hot water instead of steam has been designed with a low-boiling-point heat absorbent such as Freon or isobutane as the driving fluid. Geothermal water is pumped through a heat-exchange system (*bottom*), where the absorbent takes up the heat. The absorbent evaporates and drives the system of turbines connected to the generator. Absorbent next goes to the condensers, where it is condensed into liquid again by water from the cooling tower and returned to heat exchanger for a new cycle.

in the U.S.S.R. Such a system has also been installed in a hotel in New Zealand, which reports that the energy cost, using geothermal water, is only a tenth of that for a system using electrically operated compressors.

The third type of geothermal field, called a low-temperature field, has only recently begun to receive attention. Fields in this class generally consist of large bodies of water in the range of 50 to 82 degrees C. (about 120 to 180 degrees F.). They are found in sedimentary deposits, notably in Hungary, where the field was discovered accidentally while drilling for petroleum was in progress. The hot water from this type of field

is most efficiently used for heating: in houses, greenhouses, mines in cold climates and industrial plants. The use of such water from low-temperature fields in the U.S.S.R. is reported to have represented a saving of about 15 million tons of fuel in 1970.

The new Geothermal Steam Act of the U.S. stresses the multipurpose approach in the development of geothermal energy resources. To this end it will be necessary to plan on a comprehensive scale, treating the problem with an approach like that for the development of an entire river basin. This means that we shall need planners who are acquainted with all the technologies and economic

considerations involved, from exploration to the numerous possible applications.

Much study has already been given to the costs of exploitation of geothermal energy for various purposes. Since a number of special factors are involved in this new technology, standards for estimating costs have not yet been developed; however, the UN, in response to a proposal made at the Symposium on the Development and Utilization of Geothermal Resources, which was held in Pisa in 1970, is expected to appoint a committee of experts to formulate uniform costing procedures, so that costs in various situations and various countries can be compared.

Some of the costs are already well known from experience. Drilling a steam well costs from $50 to $150 per meter, depending on conditions, so that the drilling cost of a field 1,000 meters deep will be between $50,000 and $150,000 for one well. There are also ready answers on the costs of piping, valves and the various items of equipment for a power plant. The cost of operation for delivery of the heat from a steam field to the plant is likewise well established; with proper management this cost is only about one to three cents per million British thermal units.

What, then, are the special costs? The most important ones are related to the question of the life expectancy of the available heat supply in a field or a given well. This obviously is difficult to estimate. There are reasons to believe, however, that with proper management a geothermal field will last for many years, particularly if it is recharged by ground water or by artificial injection of gas or geothermal effluent water. At the present stage of development I believe the lifetime of a typical field can prudently be assumed to be about 30 years for purposes of estimating the amortization of the investment used in developing it. To the initial investment we must add a special cost having to do with maintenance: the wells have to be cleaned regularly and sometimes even redrilled because of the precipitation of chemicals from the steam or hot water.

The experience thus far gained furnishes us with approximate cost figures for the various applications of geothermal energy. In a single-purpose installation producing only electric power at base load the cost of the power produced is between three and six mills per kilowatt-hour, including full amortization of all the investments over a reasonable period. In desalination plants the cost would probably be in the range of 20 to 50 cents per 1,000 gallons of freshwater yield—far below the costs of other desalination systems. For househeating, air conditioning and similar purposes the use of geothermal energy makes possible savings of up to 90 percent or more, as we have already noted. A hotel in the city of Rotorua in New Zealand reports that the operating cost of a heating and air-conditioning system based on the use of geothermal energy in a lithium bromide absorption installation is only 12 cents per million kilocalories, as against $2.40 per million kilocalories for an oil-burning system involving approximately the same investment in equipment.

These estimates are calculated from the experience of single-purpose facilities. With the development of multipurpose plants the dividends made possible by extraction of all the benefits in the crude outflow from the geothermal field (like the extraction of the various products from crude petroleum) should reduce the cost of the individual applications.

Not the least of the attractions of geothermal energy is that it can be used at little or no cost to the environment. Unlike fossil or fissionable fuels, it does not pollute the biosphere with combustion products or radiation; unlike hydropower systems, it does not flood fertile lands or generate stresses that may lead to earthquakes. It does present two hazards. The steam and hot water from many fields contain small amounts of boron and other chemicals, which can be harmful when discharged into streams. Trials at the Geysers field and at a geothermal field in El Salvador have shown, however, that the contaminated effluent can be injected back into the field without reducing production from the wells. There is reason to believe this problem will not be difficult to control. The other hazard is that the land may subside where large amounts of water are withdrawn from geothermal reservoirs. Some subsidence has occurred at the Wairakei field, which has been depleted of 70 million tons of water per year and as a result has changed in part from a wet to a dry steam field. This problem too can be controlled, by limiting withdrawals from the field to a safe rate and by recharging it with water, as is now done to prevent subsidence in petroleum fields.

The UN is taking an active interest in geothermal energy. In cooperation with the government of Italy it conducted the symposium on geothermal energy in Pisa, where it was demonstrated that geothermal applications are marked by considerable international interest and collaboration. Although the scale of this research is severely limited in funding, a variety of imaginative ideas are being explored.

One project is concerned with the possibility of producing electricity in low-temperature fields. The heat from the geothermal water is used in a heat exchanger to boil a secondary fluid with a

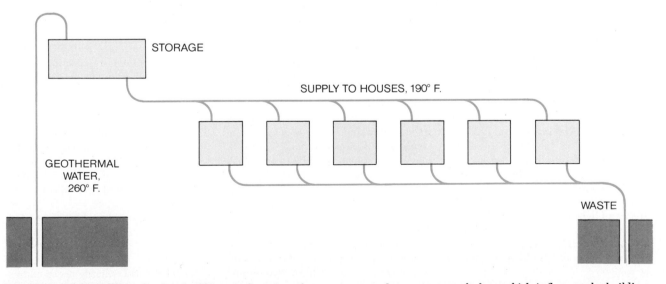

HEATING OF HOUSES and other buildings is done in a few places by a scheme such as the one shown here. Geothermal water is pumped to a storage tank, from which it flows to the buildings. Such systems are in use or being developed in several countries.

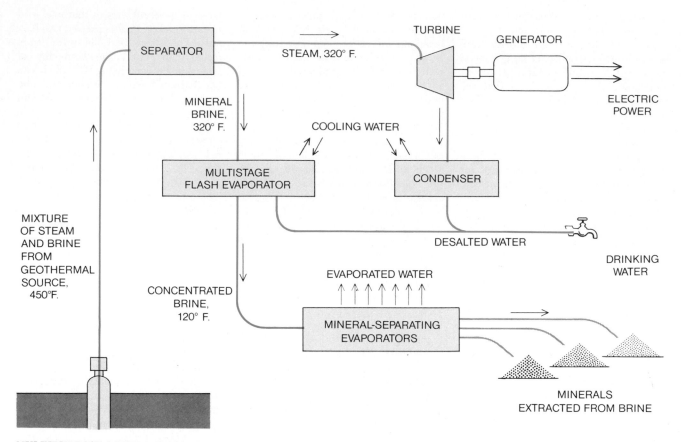

SEPARATOR

STEAM, 320° F.

TURBINE

GENERATOR

ELECTRIC POWER

MINERAL BRINE, 320° F.

COOLING WATER

MIXTURE OF STEAM AND BRINE FROM GEOTHERMAL SOURCE, 450°F.

MULTISTAGE FLASH EVAPORATOR

CONDENSER

DESALTED WATER

DRINKING WATER

CONCENTRATED BRINE, 120° F.

EVAPORATED WATER

MINERAL-SEPARATING EVAPORATORS

MINERALS EXTRACTED FROM BRINE

MULTIPURPOSE DEVELOPMENT based on geothermal energy is being designed by the UN and the government of Chile for a geothermal field recently discovered in Chile. In this case the geothermal source produces a mixture of steam and mineral-rich brine. The steam and brine are separated, and the steam drives a turbine to produce electric power while the brine is put through an evaporator that concentrates it, thereby producing desalted water. The concentrated brine goes to a separator that extracts the minerals.

low boiling point, which then drives the power turbine. Such a plant, installed on a field providing water at 81 degrees C., is in operation at Kamchatka in the U.S.S.R.; it uses Freon as the secondary fluid. Similar small plants have recently been built in Japan.

Among all the research needs the paramount one is the development of techniques of exploration to search the earth for geothermal reservoirs, hidden as well as visible. This will call for extensive geological, geochemical and geophysical studies and testing. (The UN recently resorted to infrared surveying in a large-scale search from the air for possible geothermal sites in Ethiopia and Kenya.) From the standpoint of geology, interest naturally focuses on areas underlain by rocks of high porosity, since these are likely to hold large quantities of water. From the standpoint of utility and benefit, one hopes to find geothermal reservoirs in arid areas where underground water itself, as well as energy and minerals, would be a boon to the region.

From surface indications alone it appears there are belts of geothermal reservoirs along the western side of the

Americas from Alaska all the way down to Chile, in the Middle East (Turkey) and East Africa throughout the African Rift Valley and in the Far East along the "Circle of Fire" of volcanic activity that surrounds the Pacific Ocean. In Turkey two-thirds of the country is believed to have geothermal potential, and there are good prospects for this resource in almost all the countries around the Mediterranean. The many spas of hot waters throughout Europe suggest that geothermal reservoirs should be widespread on that continent. Recent discoveries by drilling in Europe and elsewhere also indicate that a potential exists in many regions that had not previously been considered for exploration. The U.S. may have similar possibilities: drillers came on geothermal reservoirs in Louisiana and Texas recently during deep drilling for petroleum.

Inexpensive power and heating would be very helpful to many developing countries. Some of them, notably in Central America, are rich in geothermal resources—indeed, Central America has much more of this potential energy than it could use itself. Large-scale explora-

tion and development of its abundant geothermal fields would be very worthwhile, however, particularly for the region's economy, if the power potential were fully developed and marketed in the U.S. by way of long-distance transmission lines.

As new information becomes available the magnitude of geothermal energy resources is beginning to be appreciated. On the basis of a reconnaissance, which included airborne infrared scanning over a large area, carried out by the government of Ethiopia and the UN it has been estimated that a part of the Afar region in Ethiopia may have an exploitable geothermal potential sufficient to meet the present need for electric power for the whole of Africa. There are in addition other areas in Ethiopia that are believed to have a geothermal potential of similar magnitude.

At this stage it is impossible to estimate the magnitude of the exploitable resources of geothermal energy that lie hidden under our feet in the earth's crust. The world's energy needs and exciting recent discoveries, however, certainly warrant a great effort of exploration for this ready-made store of energy.

II

ENVIRONMENTAL CYCLES

Two parallel developments have characterized scientific thought about the surface of the earth in the past ten or fifteen years. One has been a growing understanding of the ways in which the atmosphere, oceans, and surface of the crust interact chemically, physically, and biologically, to recycle chemical elements constantly from one huge natural reservoir to another. The other, slightly newer to the general public, has been a deepening concern for the quality of the surface environment as a support for life. The new awareness was generated in part by quantitative analysis of surface geological, chemical, and biological cycles that show the finiteness of even so large a reservoir as the earth's entire surface.

The articles we have included in this section show some of the complexities of the earth's surface cycles and, at the same time, show the power of scientific analysis and abstraction to construct relatively simple models that not only explain how the system works but also may be used to predict the future. The first article analyzes the largest reservoir, the ocean, its dynamics, and its links with the other great reservoir, the atmosphere. The article on control of the water cycle takes the most important chemical compound in our existence, water, and shows how a quantitative understanding of the flow of water to atmosphere, ocean, and land surface might generate beneficial engineering modifications of the cycle. The three articles on the cycles of oxygen, carbon, and nitrogen introduce another aspect, cycling of elements through biological systems, and how the biological world affects and is affected by the surface environment. The last two articles are examples of concern for the degradation of the environment. One is on the ways in which atmospheric pollutants are distributed around the earth, linking all of the world's peoples inextricably by their wastes. The other discusses a toxic metal, mercury, follows the details of its cycle, and considers methods of dealing with such harmful elements.

4

The Atmosphere and the Ocean

INTRODUCTION

Since the first bottle with a note inside asking a return to the sender was thrown from a ship, eventually to be transported by ocean currents to distant shores, physical oceanographers have tried to establish the pattern of currents. At the turn of this century they began the attempt to explain them in terms of the wind as the main driver of ocean currents and the rotation of the earth as it affects those currents—the Coriolis effect. In its largest aspect the phenomenon is seen as the product of the interaction of two fluids on a rotating planet: the atmosphere's circulating as a response to the difference in heat between equator and poles, the ocean's coupled to the atmosphere by winds that create waves and currents.

In the article "The Atmosphere and the Ocean," R. W. Stewart gives a simple analysis of the Coriolis effect and the Ekman spiral, and describes how these tools of physical analysis allow us to understand smaller scale motions of water, such as upwelling and sinking in the thin top layer of the sea, as well as the larger scale ocean currents. In the complete oceanic circulation pattern, the Gulf Stream and the Kuroshio Current play the major roles, transporting huge volumes of warm equatorial water to high latitudes and thus exerting a strong influence on the climate of land masses adjacent to them. At the same time, surface currents influence a deeper circulation of the oceans that slowly stirs to great depths waters made denser by some combination of temperature and salinity while the lighter waters rise.

Observation of ocean currents reached new levels of scientific sophistication and international cooperation with the inauguration of the International Decade of Ocean Exploration, sponsored by the nations of the world. One program in particular, the Mid-Ocean Dynamic Experiment (the MODE project), has, by coordinating the operations of many ships in different positions, linked to one another by computerized teletype, yielded exciting discoveries of the behavior of huge turbulent eddies in the ocean. In such ways we are dramatically increasing our understanding of how the two great fluids of our surface earth environment operate and affect our lives in so many ways.

The Atmosphere and
the Ocean

by R. W. Stewart
September 1969

*The two are inextricably linked. The ocean's
circulation is driven by wind and by density
differences that largely depend on the air. The
atmospheric heat engine, in turn, is largely driven
by the sea*

The atmosphere drives the great ocean circulations and strongly affects the properties of seawater; to a large extent the atmosphere in turn owes its nature to and derives its energy from the ocean. Indeed, there are few phenomena of physical oceanography that are not somehow dominated by the atmosphere, and there are few atmospheric phenomena for which the ocean is unimportant. It is therefore hard to know where to start a discussion of the interactions of the atmosphere and the ocean, since in a way everything depends on everything else. One must break into this circle somewhere, and arbitrarily I shall begin by considering some of the effects of wind on ocean water.

When wind blows over water, it exerts a force on the surface in the direction of the wind. The mechanism by which it does so is rather complex and is far from being completely understood, but that it does it is beyond dispute. The ocean's response to this force is immensely complicated by a number of factors. The fact that the earth is rotating is of overriding importance. The presence of continental barriers across the natural directions of flow of the ocean complicates matters further. Finally there is the fact that water is a fluid, not a solid.

To simplify the picture somewhat, let us start by looking at what would happen to a slab of material resting on the surface of the earth. Let us further assume that the slab can move without friction. Consider the result of a sharp, brief impulse that sets the slab moving, say, due north [*see top illustration on page 40*]. Looked at by an observer on a rotating earth, any moving object is subject to a "Coriolis acceleration" directed exactly at a right angle to its motion. The magnitude of the acceleration increases with both the speed of the object's mo-

tion and the vertical component of the earth's rotation, and in the Northern Hemisphere it is directed to the right of the motion. An acceleration at right angles to the velocity is just what is required to cause motion in a circle, and in the illustration the center of the circle is due east of the original position of the slab. A circular motion of this kind is called an inertial oscillation, and something of this nature may sometimes happen in the ocean, since inertial oscillations are frequently found when careful observations are made with current meters.

An inertial oscillation requires exactly half a pendulum day for a full circle. (A pendulum day is the time required for a complete revolution of a Foucault pendulum. Like the Coriolis effect, it depends on the vertical component of the earth's rotation and therefore varies with latitude, being just under 24 hours at the poles and increasing to several days close to the Equator. To be precise, it is one sidereal—or star time—day divided by the sine of the latitude.) If there were a small amount of friction, the slab would gradually spiral to the center of the circle. Pushing it toward the north thus causes it to end up displaced to the east [*see bottom illustration on page 40*]. More generally, in the Northern Hemisphere a particle is moved to the right of the direction in which it is impelled, and in the Southern Hemisphere it is moved to the left.

Let us turn to what happens to our frictionless slab if, instead of giving it a short impulse, we give it a steady thrust. Again assume that the force is toward the north [*see upper illustration on page 41*]. Under the influence of this force the slab accelerates toward the north, but as soon as it starts to move it comes under the influence of the Coriolis effect

and its motion is deflected (in the Northern Hemisphere) to the right—to the east. As long as the slab has at least some component of velocity toward the north the force will continue to add energy to it and its speed will continue to increase. After a quarter of a pendulum day, however, it will be moving due east. In this position the applied force (which is to the north) is pushing at a right angle to the velocity (east), opposing the influence of the Coriolis effect, which is now trying to turn the slab toward the south.

If there has been no loss of energy because of friction, the slab is moving fast enough so that the Coriolis effect dominates, and it turns toward the south. Now there is a component of velocity opposing the applied force, which acts as a brake and takes energy from the motion. At the end of half a pendulum day the process has gone far enough to bring the slab to a full stop, at which point it is directly east of its starting point. If the force continues, it will again accelerate toward the north and the entire process is repeated, so that the slab performs a series of these looping (cycloidal) motions, each loop taking half a pendulum day to execute. Overall, then, a steady force on a frictionless body resting on a rotating earth causes it to move at right angles to the direction of the force. What is happening is that the force is balanced—on the average—by the Coriolis effect.

Now let us look at the situation when there is a certain amount of friction between the slab and the underlying surface [*see lower illustration on page 41*]. Any frictional drag reduces the speed attained by the slab, reducing the Coriolis effect until it is no longer entirely able to overcome the driving force. As a result if the force is toward the north, the slab will move in a more or less north-

Wind force: 4   Wind speed: 5½   Wave period: 5   Wave height: 1

Wind force: 5   Wind speed: 11½   Wave period: 6   Wave height: 2

Wind force: 6   Wind speed: 13   Wave period: 7   Wave height: 3

Wind force: 8   Wind speed: 18   Wave period: 6   Wave height: 5

Wind force: 9   Wind speed: 21   Wave period: 9   Wave height: 8

Wind force: 10   Wind speed: 27   Wave period: 9   Wave height: 7

easterly direction—more northerly if the friction is large, more easterly if it is small.

A body of water acts much like a set of such slabs, one on top of the other [see illustration on page 42]. Each slab is able to move largely independently of the others except for the frictional forces among them. If the top slab is pushed by the wind, it will, in the Northern Hemisphere, move in a direction somewhat to the right of the wind. It will exert a frictional force on the second slab down, which will then be set in motion in a direction still farther to the right. At each successive stage the force is somewhat reduced, so that not only does the direction change but also the speed is a bit less. A succession of such effects produces velocities for which the direction spirals as the depth increases. It is known as the Ekman spiral, after the pioneering Swedish oceanographer V. Walfrid Ekman, who first discussed it soon after the beginning of the century. At a certain depth both the current and the frictional forces associated with it become negligibly small. The entire layer above that depth, in which friction is important, is termed the Ekman layer. Since there is negligible friction between the Ekman layer and the water lying under it, the Ekman layer as a whole behaves like the frictionless slab discussed above: its average velocity must be at a right angle to the wind.

The frictional mechanism, which involves turbulence, has proved to be extraordinarily difficult to study either theoretically or through observations, and surprisingly little is known about it. The surface flow does appear to be somewhat to the right of the wind. Primitive theoretical calculations predict that its direction should be 45 degrees from the

EFFECT OF WIND on the surface of the sea is shown in a series of photographs made by the Meteorological Service of Canada. Much of the wind's momentum goes into generating waves rather than directly into making currents. The change in the surface as the wind increases is primarily a change in scale, except for the effect of surface tension: the waves break up more, making more whitecaps. For each photograph the wind force is given according to the Beaufort scale; the wind speed is given in meters per second, the wave period in seconds and the wave height in meters. (In the final photograph the waves are only about half as large as they might become if the force-10 wind, which had blown for less than nine hours, were to continue to blow.)

wind, but this theory is certainly inapplicable in detail. More complicated theoretical models have been attempted, but since almost nothing is known of the nature of turbulence in the presence of a free surface these models rest on weak ground. An educated guess, supported by rather flimsy observational evidence, suggests that the angle is much smaller, perhaps nearer to 10 degrees. All that seems fairly certain is that the average flow in the Ekman layer must be at a right angle to the wind and that there must be some kind of spiral in the current directions. We also believe the bottom of the Ekman layer lies 100 meters or so deep, within a factor of two or three. Of the details of the spiral, and of the turbulent mechanisms that determine its nature, we know very little indeed.

This Ekman-layer flow has some important fairly direct effects in several parts of the world. For example, along the coasts of California and Peru the presence of coastal mountains tends to deflect the low-level winds so that they blow parallel to the coast. Typically, in each case, they blow toward the Equator, and so the average Ekman flow—to the right off California and to the left off Peru—is offshore. As the surface water is swept away deeper water wells up to replace it. The upwelling water is significantly colder than the sun-warmed surface waters, somewhat to the discomfort of swimmers (and, since it is also well fertilized compared with the surface water, to the advantage of fishermen and birds).

The total amount of flow in the directly driven Ekman layer rarely exceeds a couple of tons per second across each meter of surface. That represents a substantial flow of water, but it is much less than the flow in major ocean currents. These are driven in a different way—also by the wind, but indirectly. To see how this works let us take a look at the North Atlantic [see bottom illustration on page 43]. The winds over this ocean, although they vary a good deal from time to time, have a most persistent characteristic: near 45 degrees north latitude or thereabouts the westerlies blow strongly from west to east, and at about 15 degrees the northeast trades blow, with a marked east-to-west component. The induced Ekman flow is to the right in each case, so that in both cases the water is pushed toward the region known as the Sargasso Sea, with its center at 30 degrees north. This "gathering together of the waters" leads not so much to a piling up (the surface level is only about a meter higher at

the center than at the edges) as a pushing down.

(If it were not for the continental boundaries, the piling up would be much more important. Because water tends to seek a level, the piled-up water would push north above 30 degrees and south below; the pushing force, like any other force in the Northern Hemisphere, would cause a flow to its right, so that in the northern part of the ocean a strong eastward flow would develop and in the southern part a strong westward one. On the earth as it now exists, however, these east-west flows are blocked by the continents; only in the Southern Ocean, around the Antarctic Continent, is such a flow somewhat free. In the absence of the continents the oceans, like the atmosphere, would be dominated by east-west motion. As it is, only a residue of such motion is possible, and it is the pushing down rather than the piling up of water that is important.)

The downward thrust of the surface waters presses down on the layers of water underneath [see illustration on page 44]. For practical purposes water is incompressible, so that pushing it down from the top forces it out at the sides. It must be remembered that this body of underlying water is rotating with the earth. As it is squeezed out laterally its radius of gyration, and therefore its moment of inertia, increase, and so its rate of rotation must slow. If it slows, however, the rotation no longer "fits" the rotation of the underlying earth. There are two possible consequences: either the water can rotate with respect to the earth or it can move to a different latitude where its newly acquired rotation will fit. It usually does the latter. Hence a body of water whose rotation has been slowed by being squashed vertically will usually move toward the Equator, where the vertical component of the earth's rotation is smaller; on the other hand, a body whose rotation has been speeded by being bulged up to replace water that has been swept away from the surface will usually move toward the poles.

In the band of water a couple of thousand miles wide along latitude 30 degrees this indirectly wind-driven flow moves water toward the Equator. Of course the regions of the ocean closer to the poles do not become empty of water; somewhere there must be a return flow. The returning water must also attain a rotation that fits the rotation of the underlying earth. If it flows north, it must gain counterclockwise rotation (or lose clockwise rotation). It does this by running in a strong current on the westward side of the ocean, changing its rotation

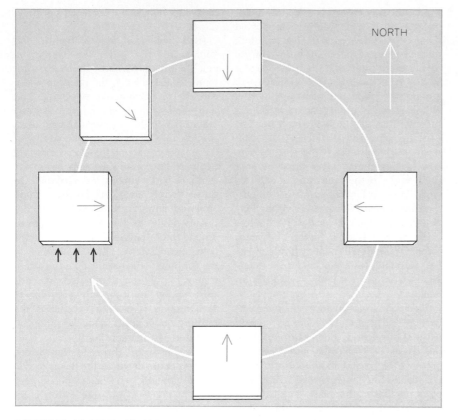

CORIOLIS ACCELERATION, caused by the earth's rotation, affects any object moving on the earth. It is directed at a right angle to the direction of motion (to the right in the Northern Hemisphere). If a frictionless slab is set in motion toward the north by a single impulse (*black arrows*), the Coriolis effect (*colored arrows*) moves the slab in a circle.

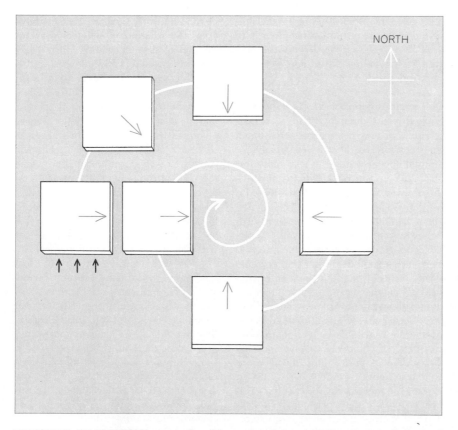

PRESENCE OF FRICTION causes the slab to slow down, spiraling in toward the center of the circle in the top illustration. A push to the north causes a spiral to the east.

by "rubbing its left shoulder" against the shore. The Gulf Stream is such a current; it is the return flow of water that was squeezed south by the wind-driven convergence of surface waters throughout the entire central North Atlantic. Most great ocean currents seem to be indirectly driven in this way.

It is worth noting that these return currents must be on the western side of the oceans (that is, off the eastern coasts of the land) in both hemispheres and regardless of whether the flow is northward or southward. The reason is that the earth's angular velocity of rotation is maximum counterclockwise to an observer looking down at the North Pole and maximum clockwise at the South Pole. Any south-flowing return current in either hemisphere must gain clockwise rotation (or lose counterclockwise rotation) if it is to fit when it arrives. It gains this rotation by friction on its right side, and so it must keep to the right—that is, to the west—of the ocean. On the other hand, a north-flowing return current must keep to the left—again the west!

This description of the general wind-driven circulation accords reasonably well with observations of the long-term characteristics of the ocean circulation. What happens on a shorter term, in response to changes of the atmospheric circulation and the wind-force pattern that results? The characteristic time constant of the Ekman layer is half a pendulum day, and there is every reason to believe this layer adjusts itself within a day or so to changes in the wind field. The indirectly driven flow is much harder to deal with. Its time constant is of the order of years, and we have no clear understanding of how it adjusts; the indirectly driven circulation may still be responding, in ways that are not clear, for years after an atmospheric change.

So far the discussion has been qualitative. To make it quantitative we need to know two things: the nature of the wind over the ocean at each time and place and the amount of force the wind exerts on the surface. Meteorologists are getting better at the first question, although there are some important gaps in our detailed information, notably in the Southern Ocean and in the South Pacific.

Investigation of the second problem, that of the quantitative relation between the wind flow and the force on the surface, is becoming a scientific discipline in its own right. Turbulent flow over a boundary is a complex phenomenon for which there is no really complete theory even in simple laboratory cases. Nevertheless, a great deal of experimental data has been collected on flows over solid surfaces, both in the laboratory and in nature, so that from an engineering point of view at least the situation is fairly well understood. The force exerted on a surface varies with the roughness of that surface and approximately with the square of the wind speed at some fixed height above it. A wind of 10 meters per second (about 20 knots, or 22 miles per hour) measured at a height of 10 meters will produce a force of some 30 tons per square kilometer on a field of mown grass or of about 70 tons per square kilometer on a ripe wheat field. On a really smooth surface such as glass the force is only about 10 tons per square kilometer.

When the wind blows over water, the whole thing is much more complicated. The roughness of the water is not a given characteristic of the surface but depends on the wind itself. Not only that, the elements that constitute the roughness—the waves—themselves move more or less in the direction of the wind. Recent evidence indicates that a large portion of the momentum transferred from the air into the water goes into waves rather than directly into making currents in the water; only as the waves break, or otherwise lose energy, does their momentum

become available to generate currents or produce Ekman layers. Waves carry a substantial amount of both energy and momentum (typically about as much as is carried by the wind in a layer about one wavelength thick), and so the wave-generation process is far from negligible. So far we have no theory that accounts in detail for what we observe.

A violently wavy surface belies its appearance by acting, as far as the wind is concerned, as though it were very smooth. At 10 meters per second, recent measurements seem to agree, the force on the surface is quite a lot less than the force over mown grass and scarcely more than it is over glass; some observations in light winds of two or three meters per second indicate that the force on the wavy surface is less than it is on a surface as smooth as glass. In some way the motion of the waves seems to modify the airflow so that air slips over the surface even more freely than it would without the waves. This seems not to be the case at higher wind speeds, above about five meters per second, but the force remains strikingly low compared with that over other natural surfaces.

One serious deficiency is the fact that there are no direct observations at all in those important cases in which the wind speed is greater than about 12 meters per second and has had time and

fetch (the distance over water) enough to raise substantial waves. (A wind of even 20 meters per second can raise waves eight or 10 meters high—as high as a three-story building. Making observations under such circumstances with the delicate instruments required is such a formidable task that it is little wonder none have been reported.) Some indirect studies have been made by measuring how water piles up against the shore when driven by the wind, but there are many difficulties and uncertainties in the interpretation of such measurements. Such as they are, they indicate that the apparent roughness of the surface increases somewhat under high-wind conditions, so that the force on the surface increases rather more rapidly than as the square of the wind speed.

Assuming that the force increases at least as the square of the wind speed, it is evident that high-wind conditions produce effects far more important than their frequency of occurrence would suggest. Five hours of 60-knot storm winds will put more momentum into the water than a week of 10-knot breezes. If it should be shown that for high winds the force on the surface increases appreciably more rapidly than as the square of the wind speed, then the transfer of momentum to the ocean will turn out to be dominated by what happens during

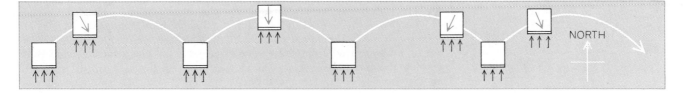

STEADY PUSH (*black arrows*), rather than a single impulse, is balanced, in the absence of friction, by the Coriolis effect (*colored arrows*), causing a series of loops. A steady force on a frictionless slab makes it move at a right angle to the direction of the force.

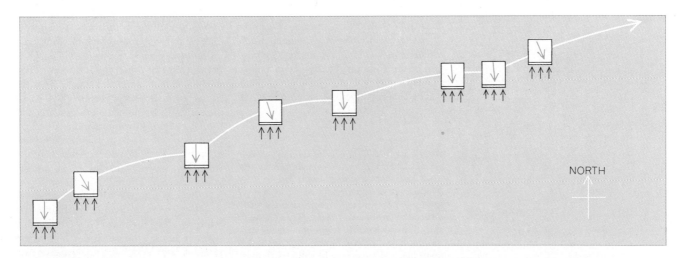

FRICTIONAL DRAG reduces the speed of the slab and thus of the Coriolis effect, which can no longer balance the driving force, and the amplitude of the loops is damped out gradually. A force toward the north therefore moves the slab toward the northeast.

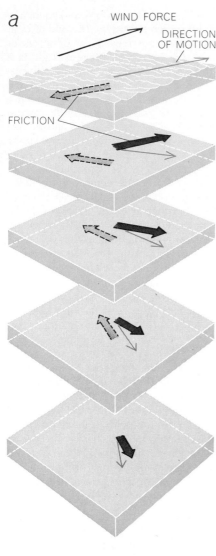

a

WIND FORCE

DIRECTION
OF MOTION

FRICTION

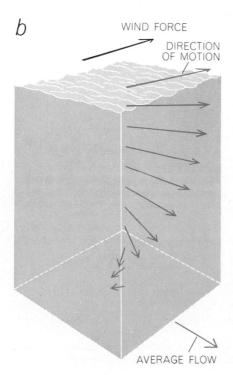

b

WIND FORCE

DIRECTION
OF MOTION

AVERAGE FLOW

BODY OF WATER can be thought of as a set of slabs (*a*), the top one driven by the wind and each driving the one below it by friction. At each stage the speed of flow is reduced and (in the Northern Hemisphere) directed more to the right. This "Ekman spiral" persists until friction becomes negligible. The "Ekman layer" in which this takes place (*b*) behaves like the frictionless slabs in the preceding illustrations. Its average flow is at right angle to wind driving it.

the occasional storm rather than by the long-term average winds.

It is tempting to try to infer high-wind behavior from what we know about lower wind speeds. Certainly the shapes of wavy surfaces appear nearly the same notwithstanding the size of the waves—as long as one disregards waves less than about five centimeters long, which are strongly affected by surface tension. Yet, curious as it may seem, the only thing that makes one wind-driven wave field different in any fundamental way from another is surface tension, even though it directly affects only these very short waves. Indeed, surface tension is the basis of the entire Beaufort wind scale, which depends on the number and nature of whitecaps; only the fact that the surface tension is better able to hold the surface together at low wind speeds than at high speeds enables us to see a qualitative difference in the nature of the sea surface at different wind speeds [*see illustration on page 38*]. Otherwise the waves would look just the same except for a difference in scale. If we were sure we could ignore surface-tension effects, then we could calculate the force the wind would exert at high wind speeds on the basis of data obtained at lower speeds, but one should be extremely cautious about such calculations, at least until some confirming measurements are available.

Whereas the ocean seems primarily to be driven by surface forces, the atmosphere is a heat engine that makes use of heat received from the sun to develop the mechanical energy of its motion. Any heat engine functions by accepting thermal energy at a comparatively high temperature, discharging some of this thermal energy at a lower temperature and transforming the rest into mechanical energy. The atmosphere does this by absorbing energy at or near its base and radiating it away from much cooler high levels. A substantial proportion of the required heating from below comes from the ocean.

This energy comes in two forms. If cooler air blows over warmer water, there is a direct heat flow into the air. What is usually more important, though, is the evaporation of water from the surface into the air. Evaporation causes cooling, that is, it removes heat, in this case from the surface of the water. When the moisture-laden air is carried to a high altitude, where expansion under reduced atmospheric pressure causes it to cool, the water vapor may recondense into water droplets and the heat that was given up by the surface of the water is transferred to the air. If the cloud that is formed evaporates again, as it sometimes does, the atmosphere gains no net thermal energy. If the water falls to the surface as rain or snow, however, then there has been a net gain and it is available to drive the atmosphere. Typically the heat gained by the atmosphere through this evaporation-condensation process is considerably more than the heat gained by direct thermal transfer through the surface.

Virtually everywhere on the surface of the ocean, averaged over a year, the ocean is a net source of heat to the atmosphere. In some areas the effect is much more marked than in others. For example, some of the most important return currents, such as the Gulf Stream in the western Atlantic and the Kuroshio Current in the western Pacific off Japan, contain very warm water and move so rapidly that the water has not cooled even when it arrives far north of the tropical and subtropical regions where it gained its high temperature. At these northern latitudes the characteristic wind direction is from the west, off the continent. In winter, when the continents are cold, air blowing from them onto this abnormally warm water receives great quantities of heat, both by direct thermal transfer and in the form of water vapor.

The transfer of heat and water vapor depends on a disequilibrium at the interface of the water and the air. Within a millimeter or so of the water the air temperature is not much different from that of the surface water, and the air is nearly saturated with water vapor. The small differences are nevertheless crucial, and the lack of equilibrium is maintained by the mixing of air near the surface with air at higher levels, which is typically appreciably cooler and lower in water-vapor content. The mixing mechanism is a turbulent one, the turbulence gaining its energy from the wind. The higher the wind speed is, the more vigorous the turbulence is and therefore the higher the rates of heat and moisture transfer are. These rates tend to increase linearly with the wind speed, but even less is known about the details of this phenomenon than about the wind force on water. One source of complication is the fact that, as I mentioned above, the wind-to-water transfer of momentum is effected partly by wave-generation mechanisms. When the wind makes waves, it must transfer not only momentum but also important amounts of energy—energy that is not available to provide the turbulence needed to produce the mixing that would effect the transfer of heat and water vapor.

At fairly high wind speeds another phenomenon arises that may be of considerable importance. I mentioned that when surface tension is no longer able to hold the water surface together at high wind speeds, spray droplets blow off the top of the waves. Some of these drops fall back to the surface, but others evaporate and in doing so supply water vapor to the air. They have another important role: The tiny residues of salt that are left over when the droplets of seawater evaporate are small enough and light enough to be carried upward by the turbulent air. They act as nuclei on which condensation may take place, and so they play a role in returning to the atmosphere the heat that is lost in the evaporation process.

The ocean's great effect on climate is illustrated by a comparison of the temperature ranges in three Canadian cities, all at about the same latitude but with very different climates [*see top illustration at right*]. Victoria is a port on the southern tip of Vancouver Island, on the eastern shore of the Pacific Ocean; Winnipeg is in the middle of the North American land mass; St. John's is on the island of Newfoundland, jutting into the western Atlantic. The most striking climatic difference among the three is the enormous temperature range at Winnipeg compared with the two coastal cities. The range at St. John's, although much less, is still greater than at Victoria, probably because at St. John's the air usually blows from the direction of the continent and the effect of the water is somewhat less dominant than at Victoria, which typically receives its air directly from the ocean. St. John's is colder than Victoria because it is surrounded by cold water of the Labrador Current.

The influence of the ocean is associated with its enormous thermal capacity. Every day, on the average, the earth absorbs from the sun and reradiates into space enough heat to raise the temperature of the entire atmosphere nearly two degrees Celsius (three degrees Fahrenheit). Yet the thermal ca-

	VICTORIA	WINNIPEG	ST. JOHN'S
MEAN JULY MAXIMUM	68	80.1	68.9
MEAN JANUARY MINIMUM	35.6	−8.1	18.5

MODERATING EFFECT of the ocean on climate is illustrated by a comparison of the temperature range (in degrees Fahrenheit) at three Canadian cities. The range between minimums and maximums is much greater at Winnipeg than at coastal Victoria or St. John's.

pacity of the atmosphere is equivalent to that of only the top three meters of the ocean, or only a few percent of the 100 meters or so of ocean water that is heated in summer and cooled in winter. (The great bulk of ocean water, more than 95 percent of it, is so deep that surface heating does not penetrate, and its temperature is independent of season.) If the ocean lost its entire heat supply for a day but continued to give up heat in a normal way, the temperature of the upper 100 meters would drop by only about a tenth of a degree.

Compared with the land, the ocean heats slowly in summer and cools slowly in winter, so that its temperature is much less variable. Moreover, because air has so much less thermal capacity, when it blows over water it tends to come to the water temperature rather than vice versa. For these reasons maritime climates are much more equable than continental ones.

Although the ocean affects the atmo-sphere's temperature more than the atmosphere affects the ocean's, the ocean is cooled when it gives up heat to the atmosphere. The density of ocean water is controlled by two factors, temperature and salinity, and evaporative cooling tends to make the water denser by affecting both factors: it lowers the temperature and, since evaporation removes water but comparatively little salt, it also increases the salinity. If surface water becomes denser than the water underlying it, vigorous vertical convective mixing sets in. In a few places in the ocean the cooling at the surface can be so intense that the water will sink and mix to great depths, sometimes right to the bottom. Such occurrences are rare both in space and in time, but once cold water has reached great depths it is heated from above very slowly, and so it tends to stay deep for a long time with little change in temperature; there is some evidence of water that has remained cold and deep in the ocean for more than

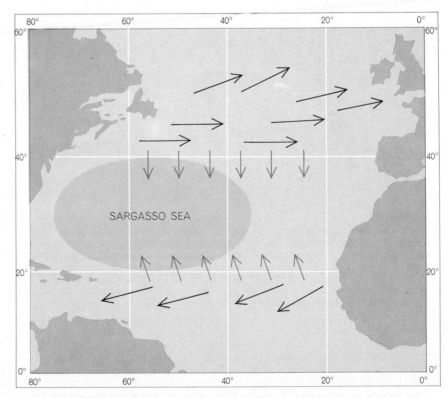

PREVAILING-WIND DIRECTIONS (*black arrows*) and the resulting Ekman-layer flows (*colored arrows*) in the North Atlantic drive water into the region of the Sargasso Sea.

1,000 years. With this length of residence not much of the heavy, cold water needs to be produced every year for it to constitute, as it does, the bulk of the ocean water.

The sinking of water cooled at the surface is one aspect of another important feature of the ocean: the flow induced by differences in density, which is to say the flow induced principally by temperature and salt content. This thermohaline circulation of the ocean is in addition to the wind-driven circulation discussed earlier.

In its thermohaline aspects the ocean itself acts as a heat engine, although it is far less efficient than the atmosphere. Roughly speaking, the ocean can be divided into two layers: a rather thin upper one whose density is comparatively low because it is warmed by the sun, and a thick lower one, a fraction of a percent denser and composed of water only a few degrees above the freezing point that has flowed in from those few areas where it is occasionally created. Somewhere—either distributed over the ocean or perhaps only locally near the shore and in other special places—there is mixing between these layers. The mixing is of such a nature that the cold deep water is mixed into the warm upper water rather than the other way around,

that is, the cold water is added to the warm from the bottom [*see upper illustration at right*]. Once the water is in the upper layer its motion is largely governed by the wind-driven circulation, although density differences still play a role. In one way or another some of this surface water arrives at a location and time at which it is cooled sufficiently to sink again and thus complete the circulation.

This picture can be rounded out by consideration of the effects of the earth's rotation, which are in some ways quite surprising. The deep water that mixes into the upper layer must have a net upward motion. (The motion is far too

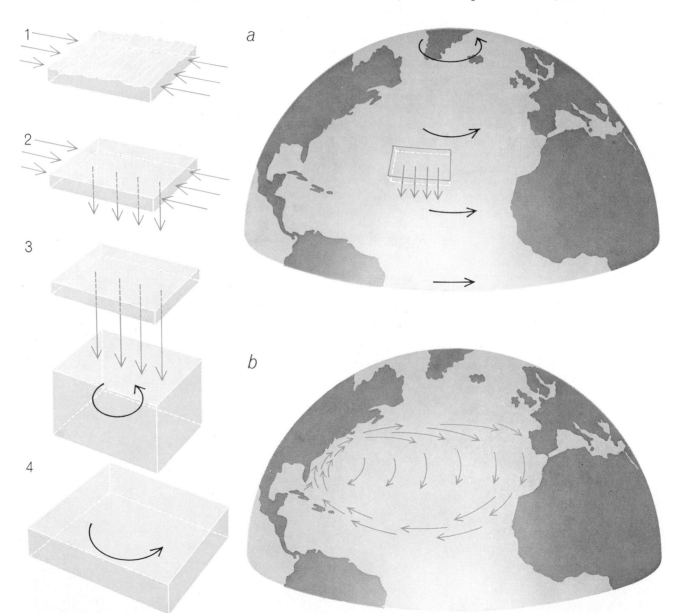

MAJOR CURRENTS are generated by a mechanism involving the Ekman-layer flow and the earth's rotation. The Ekman-layer inflow shown in the preceding illustration (*1*) produces a downflow (*2*) that presses on the underlying water (*3*), squeezing it outward (*4*) and thus reducing its rate of rotation (*curved black arrow*). There is a rate of rotation appropriate to each latitude, and when the rotation of a body of water is reduced, it must move (*colored arrows*) toward the Equator until its new rotation "fits" (*a*). For this reason there is a general movement of water from the mid-latitudes toward the Equator (*b*). That water must be replaced, and the water replacing it must have the proper rotation. This is accomplished by a return flow that runs along the western shore of the ocean, changing its rotation by "rubbing its shoulder" against the coast, as the Gulf Stream does in the Atlantic Ocean.

small to measure, only a few meters per year, but we infer its existence indirectly.) To make possible this upward flow there must be a compensating lateral inflow. Remember that on the rotating earth this lateral inflow results in an increase in speed of rotation, and so for it to continue to fit the rotation of the underlying earth the water must move toward the nearest pole; it must flow away from the equatorial regions. Yet

the source of this cold deep water is at high latitudes! How does it get near the Equator to supply the demand?

The answer is similar to the one for the wind-driven circulation: The cold water must flow in a western boundary current, in order to gain the proper rotation as it moves [*see lower illustration on this page*]. There is some direct evidence of the inferred concentrated western boundary current in the North At-

lantic, and there are hints of it in the South Pacific, but most of the rest is based on inference. There seems to be no source of cold deep water in the North Pacific, so that the deep water there must come from Antarctic regions.

We have seen that the atmosphere drives the ocean and that heat supplied from the ocean is largely instrumental in releasing energy for the atmo-

THERMOHALINE CIRCULATION, the flow induced by density rather than wind action, begins with the creation of dense, cold water that sinks to great depths. Under certain conditions this deep water mixes upward into the warm surface layer (*color*) as shown here. As it moves up, this water increases its rotation and so it must move generally from the Equator toward the two poles.

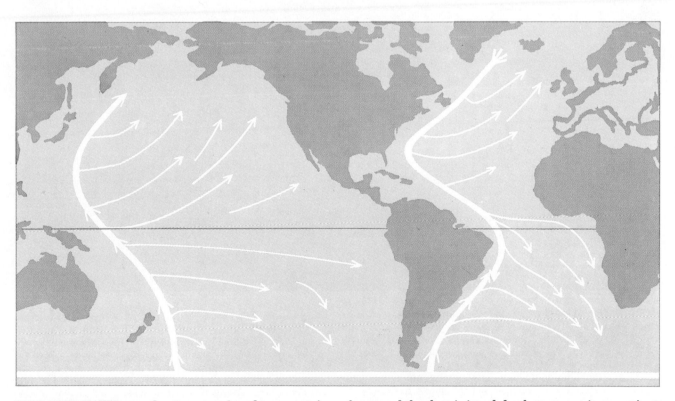

COLD DEEP WATER must flow in western boundary currents in order to arrive at the Equator and thus be able to move poleward as it mixes upward. Details of deep circulation are still almost unknown and the chart is intended only to suggest its approximate directions. There is some evidence of such boundary currents in the North Atlantic and there are some hints in the South Pacific.

OCEAN AND ATMOSPHERE, the two thin fluid films in which life is sustained and whose nature and motion determine the environment, dominate this color photograph of the watery hemisphere of the earth. The picture was made on January 21, 1968, by a spin-scan camera on NASA's Applications Technology Satellite 3, in synchronous orbit 22,300 miles above the Pacific Ocean. The camera experiment was proposed and developed by Verner E. Suomi and Robert J. Parent of the University of Wisconsin's Space Science and Engineering Center. As the camera spins it scans a 2.2-mile-wide strip across the earth, then steps down in latitude and scans another strip; in about 25 minutes a 2,400-strip picture is completed. What the camera transmits to the earth is an electrical signal representing the amount of green, red and blue light in each successive picture element, and from these signals a color negative is built up at the receiving station. Such photographs yield information on the interrelation of atmospheric and oceanic phenomena. In this picture, for example, the convective pattern over the southeastern Pacific indicates that air heated by the sea is rising.

sphere. There is a great deal of feedback between the two systems. The atmospheric patterns determine the oceanic flows, which in turn influence where—and how much—heat is released to the atmosphere. Further, the atmospheric flow systems determine how much cloud cover there will be over certain parts of the ocean and therefore how much—and where—the ocean will be heated. The system is not a particularly stable one. Every locality has its abnormally cold or mild winters and its abnormally wet or dry summers. The persistence of such anomalies over several months almost certainly involves the ocean, because the characteristic time constants of purely atmospheric phenomena are simply too short. Longer-term climatological variations such as the "little ice age" that lasted for about 40 years near the beginning of the 19th century are even more likely to have involved changes in the ocean's circulation. And then there are the more dramatic events of the great Pleistocene glaciations.

There are any number of theories for these events and, since experts disagree, it is incumbent on the rest of us to refrain from dogmatic statements. Nevertheless, it does not seem impossible that the ocean-atmosphere system has a number of more or less stable configurations. That is, there may be a number of different patterns in which the atmosphere can drive the ocean in such a way that the ocean releases heat to the atmosphere in the right quantity and at the right places to allow the pattern to continue. Of course the atmosphere is extremely turbulent, so that its equilibrium is constantly being disturbed. If the system is stable, then forces must come into

INFRARED IMAGERY delineates the temperature structure of bodies of water and is used to study currents and wave patterns. This image of the shoulder of the Gulf Stream is from the Antisubmarine Warfare Environmental Prediction Services Project of the Naval Oceanographic Office. It was made by an airborne scanner at low altitude and shows several hundred yards of the boundary between the warm current and cooler water off Cape Hatteras. The range is from about 13 to 21 degrees Celsius, with the warm water darker.

play that tend to restore conditions after each such disturbance. If there are a number of different stable patterns, however, it is possible that a particularly large disturbance might tip the system from one stable condition to another.

One can imagine a gambler's die lying on the floor of a truck running over a rough road; the die is stable on any of its six faces, so that in spite of bouncing and vibration the same face usually remains up—until a particularly big bump jars it so that it lands with a different face up, whereupon it is stable in its new position. It seems not at all impossible that the ocean-atmosphere system behaves something like this. Perhaps in recent years we have been bouncing along with, say, a four showing. Perhaps 200 years ago the die flipped over to three for a moment, then flipped back to four. It could one day jounce over to a snake eye and bring a new ice age!

5

The Control of the Water Cycle

INTRODUCTION

Since the Romans began their ambitious program of building aqueducts, engineering control of the distribution of water on the earth's surface has been a constant concern of civilizations. Today, the United States alone uses almost 1/200 of the total rainfall on all land areas of the Earth for purposes ranging from drinking water to irrigation. It has become imperative that we use our water resources wisely and interfere with the hydrological cycle with surgical precision rather than bulldozing. Because we are now quantifying our knowledge of the water cycle, we can better appreciate just where in the water balance in air, sea, and land we can have the most effect with the least disruption.

In their article, "The Control of the Water Cycle," José P. Peixoto and M. Ali Kettani review the water cycle and then, focusing on the atmospheric transport of water vapor, discuss the time and space relations of evaporation into the atmosphere, precipitation from it to land or sea surface, and the slow response of groundwater under arid lands to an excess of evaporation over precipitation. The authors propose the application of ingenious devices to increase precipitation over dry ground, to use the fresh water outflow from rivers to the oceans, and to recover fresh water from the oceans.

The point of these proposals is not that they are practical blueprints for immediate action but that an increasing quantitative knowledge of the natural water cycle stimulates engineering thought that may have practical value. After all, the major interference in the natural hydrological cycle has been the centuries-old practice of drilling water wells to extract deeply buried fresh water. So what we are discussing is not *whether* to interfere but how to do so intelligently. And what appear to be exotic ideas serve to sharpen our thinking about one of the world's urgent resource problems.

The Control of the Water Cycle

by José P. Peixoto and M. Ali Kettani
April 1973

The recent emphasis in hydrologic studies on the crucial role played by the general circulation of the atmosphere has led to schemes for altering this vast natural process

"We made from water every living thing."
—The Koran (Prophets 30)

"All the rivers run into the sea; yet the sea is not full."
—Ecclesiastes 1:7

The importance of water to the creation and sustenance of every living thing, including prosperous human communities, was profoundly appreciated by the civilization-builders of the ancient Middle East. So was the vital role of the hydrologic cycle, the wondrously fickle process by which the terrestrial supply of water is constantly replenished, thus ensuring the continued existence in a given place of man and his creations.

Indeed, the loss of an adequate water supply can be implicated in the demise of many human settlements in the ancient world, most notably in the disappearance of the original high civilizations in the formerly "Fertile Crescent" of the Tigris and the Euphrates river valleys. In more recent times the fall of the Roman Empire and the decline of the Arabian caliphates have been blamed on a more or less sudden decline in the amount of water available to their populations. In the modern world vast, potentially prosperous regions of Africa, Asia, Australia and South America—regions stretching over millions of square kilometers in the subtropical latitudes—are sparsely populated today because of a lack of water. Even in prosperous countries such as the U.S. the increasing danger of an imbalance between water production and water consumption is acutely felt.

Yet in spite of the obvious importance of water to the life and prosperity of humankind, societies have gone on for centuries using this precious commodity in a basically inefficient and wasteful way. It is only recently that some tentative efforts have been made to consider the possibilities of controlling the water cycle on a large scale to better meet human needs. In order to understand why this long overdue development has finally begun, it is necessary to review some recent trends in the study of hydrology.

During the past few decades the water balance at the earth's surface has been the subject of intensive research. Until the mid-1930's most of these in-

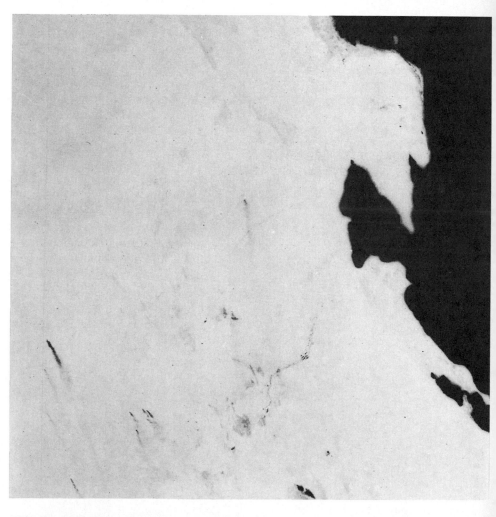

SITE OF PROPOSED PROJECT for controlling the water cycle over a section of the Persian Gulf is shown in this photograph, made from an altitude of almost 600 miles by a camera on board the National Aeronautics and Space Administration's Earth Resources Technology Satellite (ERTS). The proposed "heliohydroelectric" scheme, put forward in

vestigations dealt only with the terrestrial branch of the water cycle, examining the balance of precipitation, evaporation, transpiration, runoff, underground flow and changes in ground-water storage. This conventional approach to the water budget often yielded results that were not widely accepted by other hydrologists; at best the results were representative of the water budget only in limited regions or watershed systems. Actually the problem was that evaporation from the surface of the earth is not uniform; the water released is continuously removed by turbulent diffusion near the surface, by convection currents due to temperature differences in the air and by the wind. All these factors condition the phenomenon of evaporation. Moreover, there is no single instrument that can measure natural evaporation directly. Measuring precipitation by direct means in many parts of the world is possible, but precipitation tends to be spotty, and simultaneous widespread sampling is at times inadequate.

In view of these difficulties some recent studies have approached the problem of the water balance by considering the atmospheric branch of the hydrologic cycle instead of the terrestrial branch. This approach is based on the principles of continuity and conservation of mass in the atmosphere rather than at the earth's surface. It has become feasible in recent years as a result of the rapid expansion of the upper-air sensor network and the resulting improvement in the quality of aerological observations, which are now adequate to deal quantitatively with the large-scale processes of the atmospheric branch of the water cycle. This information in turn has become available for evaluating the water balance at the earth's surface. For example, the Planetary Circulations Project at the Massachusetts Institute of Technology, with which we have both been associated, is engaged in studying the general circulation of the atmosphere. This work has given impetus to a large number of investigations of the relations between the flux of water vapor in the atmosphere and the hydrologic cycle. It is out of investigations such as these that specific schemes for the control of the water cycle are beginning to emerge.

Why Water Is Where It Is

In primeval days our planet was devoid of any atmosphere, since elements such as helium and hydrogen were too light to be trapped by the earth's gravitational field. The earth had no oceans, and its entire surface was covered with active volcanoes from which poured lava, gases (mainly hydrogen and hydrogen-rich compounds) and vapors (mostly water vapor). Water molecules in the volcanic emissions were promptly photodissociated, or split by sunlight, into their constituent atoms: hydrogen and oxygen. The hydrogen escaped and the oxygen reacted with ammonia (NH_3) and methane (CH_4) to form nitrogen (N) and carbon dioxide (CO_2). Thus the earth's original atmosphere is believed

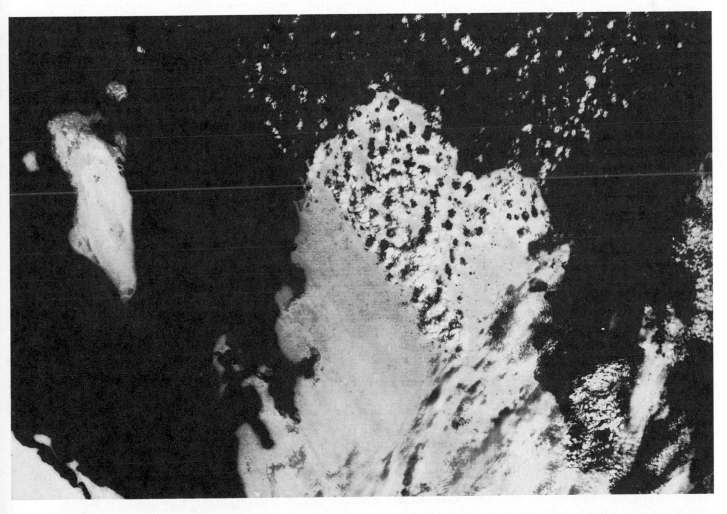

1970 by one of the authors (Kettani) in collaboration with L. M. Gonsalves, would utilize solar energy to generate through evaporation an inflow of water from the open sea into an artificial reservoir called Dawhat Salwah (*see illustration on pages 62 and 63*). The reservoir would be formed by building causeways between Saudi Arabia (*landmass at left*) and Bahrain (*island at center*) and between Bahrain and Qatar (*peninsula at right*). The inflow of water through a conduit could be used to generate electricity.

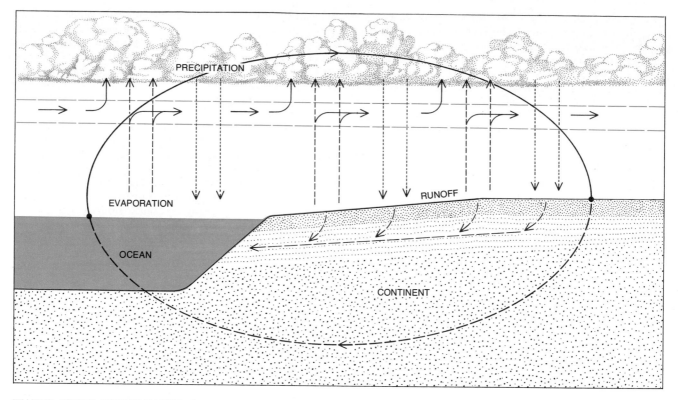

WATER CYCLE OF THE EARTH, depicted in this highly schematic diagram, can be regarded as consisting of two parts: an atmospheric branch (*solid arc*) and a terrestrial branch (*broken arc*). The earth yields water to the atmosphere mainly through evaporation from the oceans. This water is transported by the general circulation of the atmosphere in both the gaseous phase (water vapor) and the condensed phases (clouds). Eventually the water precipitates back to the earth, either as a solid or as a liquid. If it falls on land, it can infiltrate underground, run off in rivers and streams into the ocean or evaporate directly back to the atmosphere.

to have been a mixture of the last two gases. Its composition later changed to its present state when plants appeared and started absorbing the carbon dioxide and liberating oxygen. After the first process of atmospheric formation had been stabilized, the excess water began accumulating in depressions, forming the oceans.

The amount of water on the earth can be considered constant during the evolution of man. True, some new water is being produced by the remaining few volcanoes and hot springs. Most steam ejected by volcanoes, however, is actually either rainwater that had saturated the upper layers of rock or seawater that was trapped at the time marine sediments were deposited. A certain amount of water may be destroyed in the upper atmosphere by the photodissociation of water vapor by solar radiation. Both effects are negligible on the geological time scale.

At present the water existing on the earth is distributed in three separate reservoirs. In order of importance they are the oceans, the continents and the atmosphere. The interior of the earth contains an appreciable amount of water, either dissolved or chemically combined with solid or molten rock, but

there is no satisfactory estimate of the amount of water thus locked up.

About 97.3 percent of all the water in the hydrosphere is in the oceans; the remaining 2.7 percent is on the continents, mostly in the glaciers of the Arctic and the Antarctic. The atmosphere contains a mere hundred-thousandth of the total water present on the earth. In terms of volume the total water in the oceans amounts to $1,350 \times 10^{15}$ cubic meters; the entire atmosphere normally contains a constant quantity of water on the order of $.013 \times 10^{15}$ cubic meters.

The water on the continents is in turn distributed in several reservoirs, namely glaciers (29×10^{15} cubic meters), ground water (8.4×10^{15} cubic meters), lakes and rivers ($.2 \times 10^{15}$ cubic meters) and the living matter of the biosphere ($.0006 \times 10^{15}$ cubic meters). The amount of water locked in the polar ice is impressively large, totaling some 1.8 percent of all the water in the hydrosphere. Of the total amount of underground water, vadose water (water present in soils) accounts for only $.066 \times 10^{15}$ cubic meters. The remainder is about evenly divided between reservoirs deeper than 800 meters and reservoirs shallower than that level.

This distribution of terrestrial water

changes with time. For instance, continental ice caps have periodically developed and melted again during the past two million years. If they were to melt altogether, the sea level would rise by as much as 60 meters and submerge a large area of the continents. During periods of maximum glaciation the sea level may have been lowered by as much as 140 meters, uncovering large areas of the continental shelves. These fluctuations are accompanied by a disturbance in the balance of the water cycle.

The Water Cycle

The earth is constantly releasing water vapor into the atmosphere, both through evaporation from the surface (seas, lakes, rivers, glaciers, snowfields, man-made reservoirs, soil and so on) and through the transpiration of living matter. This natural transfer of water from the earth's surface to the atmosphere is extremely important, since it conditions the characteristics of the air masses, the energetics of the atmosphere, and the establishment and evolution of the hydrologic cycle.

The water cycle is the fundamental concept of hydrology. It is a consequence of the principle of conservation of water in its three phases on the earth. It describes a closed sequence of natural phenomena by which water enters the atmosphere from the surface in the gaseous phase and returns to the surface in the liquid or solid phase. There it is partly retained or runs off, to evaporate again into the atmosphere. Energy is continuously exchanged between the atmosphere and the surface. It is this exchange that induces the circulation of water in a continuous cycle.

Except for a negligible outflow from the interior of the earth, all the energy that drives the water cycle comes from the sun. The solar radiation reaching the earth's orbit amounts to two calories per square centimeter per minute, a quantity termed the solar constant. The amount of solar energy actually received by the earth in one minute is equal to the product of the solar constant and the cross-sectional area of the earth: 2.55×10^{18} calories. This amount corresponds to an average energy of .5 calorie per minute per square centimeter, or a fourth of the solar constant. In terms of power that energy flux amounts to a total continuous input of about $.178 \times 10^{18}$ watts. As the solar radiation proceeds toward the surface of the earth, its ultraviolet component is strongly absorbed by the ozone in the upper atmosphere, heating that part of the atmosphere. More radiation is absorbed by air molecules, is reflected back into space by clouds or is simply scattered at lower altitudes. Nevertheless, on the average about 81 percent of the incoming radiation manages to reach the earth's surface. Roughly a third of this energy is immediately reflected back into space. The remaining two-thirds is absorbed by the continents and the oceans. Of the absorbed radiation 77 percent is ultimately reradiated as long-wave radiation through the atmosphere back into space. The remaining 23 percent goes into the evaporation of water.

The most important meteorological factor in the hydrologic cycle therefore is sunlight, the supply of energy that causes and maintains the circulation of water in the cycle. Air temperature, humidity and wind mainly condition the evapotranspiration process. Clouds, the visible deposits of water in the atmosphere, in turn condition the solar radiation that reaches the earth.

The hydrologic cycle has two distinct branches: the atmospheric branch, in which the horizontal flux of water is mainly in the gaseous phase, and the terrestrial branch, in which the flux of water in the liquid phase and the solid phase predominates. At the surface—the interface that separates the two branches of the cycle—complex processes arise owing to boundary conditions. Even in each branch of the cycle the physical processes are not simple, because neither the atmosphere nor the soil is homogeneous. The influence of regional and local factors in hydrologic phenomena must therefore be taken into account.

The Role of the Atmosphere

The amount of water vapor in the atmosphere varies widely, both in space

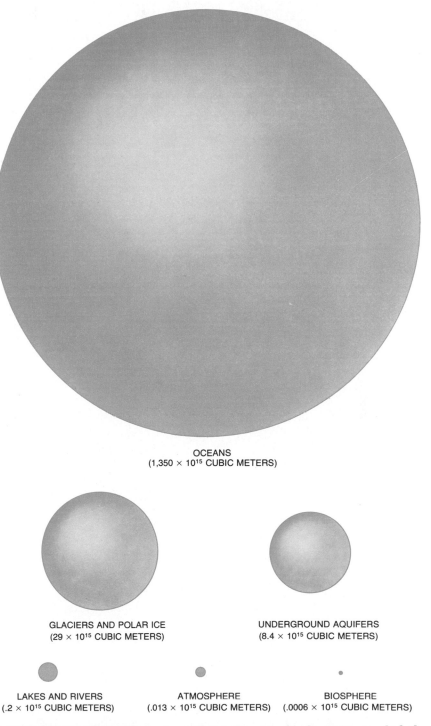

OCEANS
(1,350 × 10¹⁵ CUBIC METERS)

GLACIERS AND POLAR ICE
(29 × 10¹⁵ CUBIC METERS)

UNDERGROUND AQUIFERS
(8.4 × 10¹⁵ CUBIC METERS)

LAKES AND RIVERS
(.2 × 10¹⁵ CUBIC METERS)

ATMOSPHERE
(.013 × 10¹⁵ CUBIC METERS)

BIOSPHERE
(.0006 × 10¹⁵ CUBIC METERS)

DISTRIBUTION OF WATER on the earth is indicated in this illustration, in which the amount of water present in various natural reservoirs is represented in terms of comparative spherical volumes. The number under the name of each reservoir denotes the contents of that reservoir in cubic meters. Although the atmosphere contains a mere hundred-thousandth of all the water in the hydrosphere, the influence of this small amount on the climate of the earth and on the location of hydrologic resources is far out of proportion to its mass.

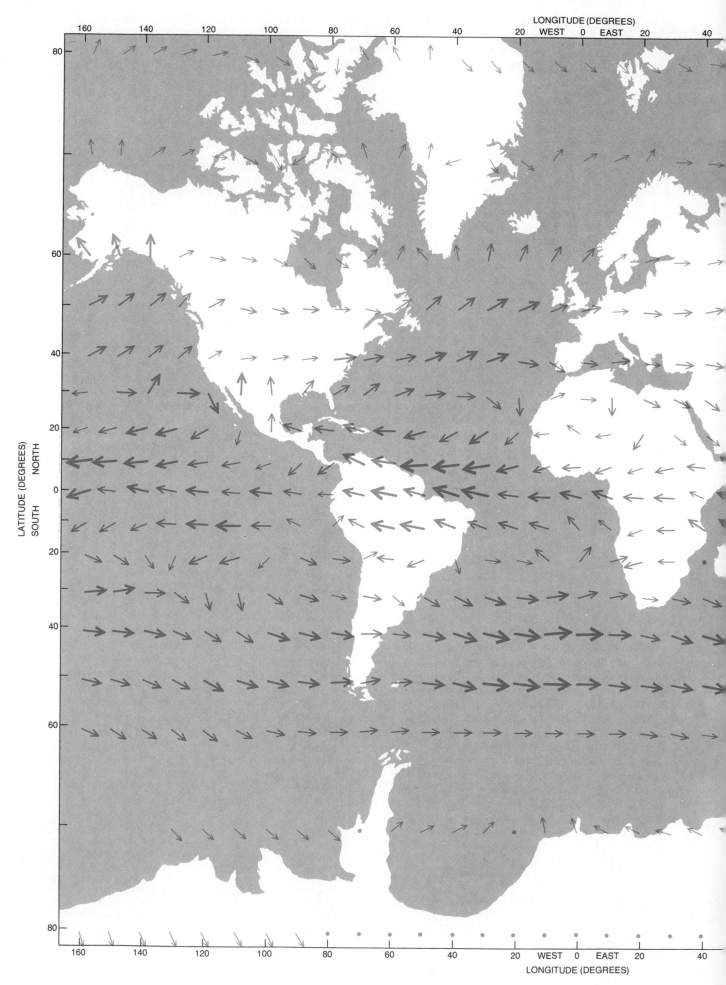

LATITUDE (DEGREES)
SOUTH   NORTH

LONGITUDE (DEGREES)

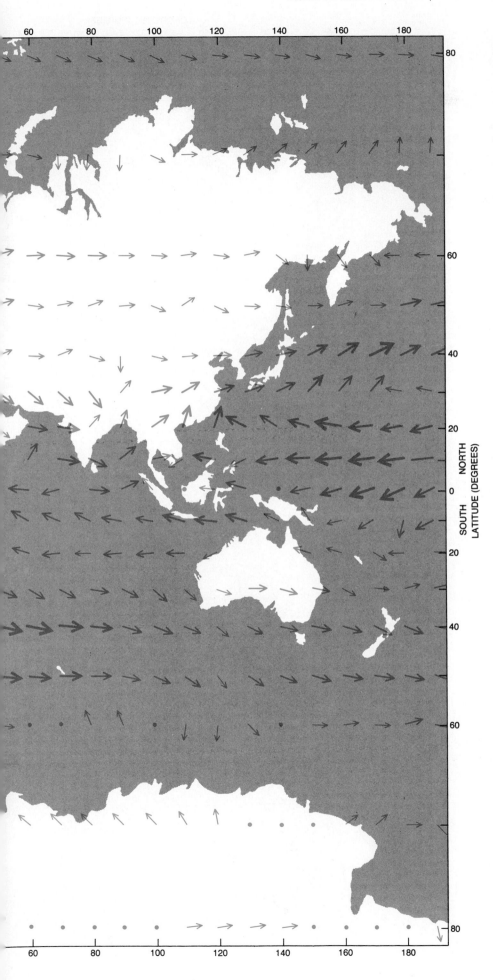

and in time, from a negligible minimum up to values of the order of .6 percent of the total atmospheric mass of a column of air. The annual mean water-vapor content averaged for the entire atmosphere is rather small, amounting to about .3 percent of the total mass of the atmosphere. That quantity is of the same order of magnitude as the amount of water contained in freshwater lakes. Hence the atmosphere is a comparatively small reservoir of water; at a given instant it holds only a tiny part of all the water that participates in the hydrologic cycle. In fact, if all the atmospheric water vapor were to condense at once, only a moderate amount of precipitation would result. Even though the total amount of water in the atmosphere is small, however, there is a huge transport of water vapor by atmospheric circulations of various scales in space and in time.

The influence of this small amount of water vapor on the climate of the earth and on hydrologic resources is far out of proportion to its mass. Water vapor plays a major role in the overall energetics of the earth and in the general circulation of the atmosphere. It is the most important factor in all radiative processes of the atmosphere in that it regulates the energy balance through the absorption and transmission of radiation. In addition, the processes of evaporation, condensation and sublimation make possible the observed energy balance and

MOVEMENT OF WATER VAPOR
(HUNDREDS OF GRAMS PER
CENTIMETER PER SECOND)

• LESS THAN 1
→ 1 TO 4
→ 5 TO 8
→ 9 TO 12
→ 13 TO 16
→ 17 TO 20
→ 21 TO 24

GLOBAL TRANSPORT of water vapor by the atmospheric circulation was calculated by one of the authors (Peixoto) on the basis of data gathered at numerous stations from pole to pole during the International Geophysical Year (1958). The data consisted of daily measurements of the wind and the water content of the atmosphere obtained with the aid of the IGY aerological radio-sensing network. The averaged results, which show the mean flow of moisture during the year as a field of vectors, represent the lower half of the atmosphere; the upper atmosphere contains only a negligible amount of moisture. The asymmetric pattern of the vector arrows reflects important irregularities in the global transport of atmospheric water.

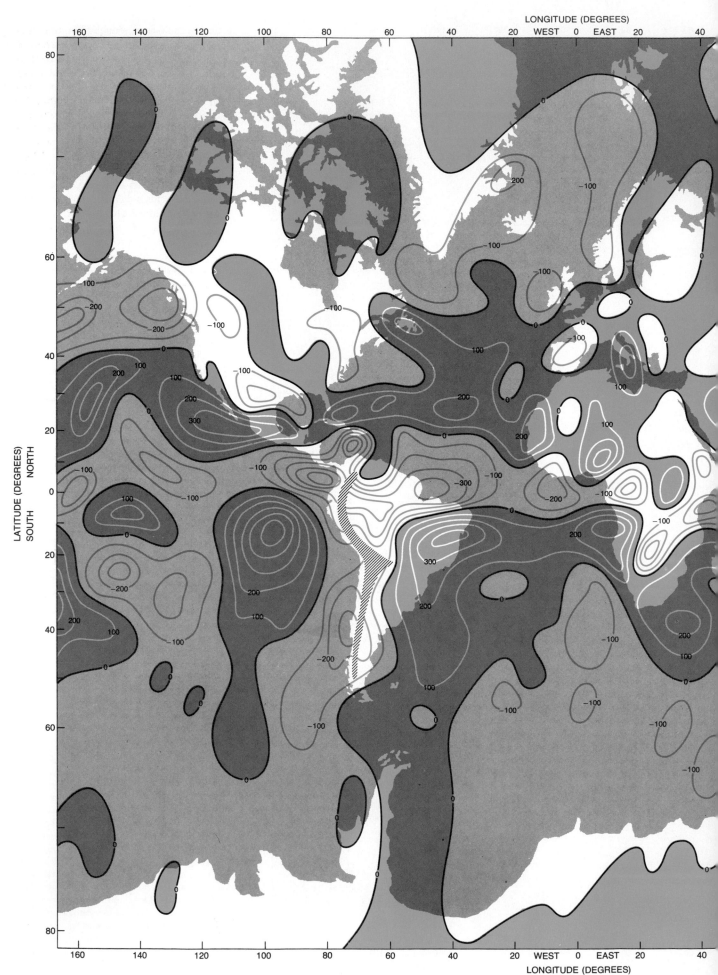

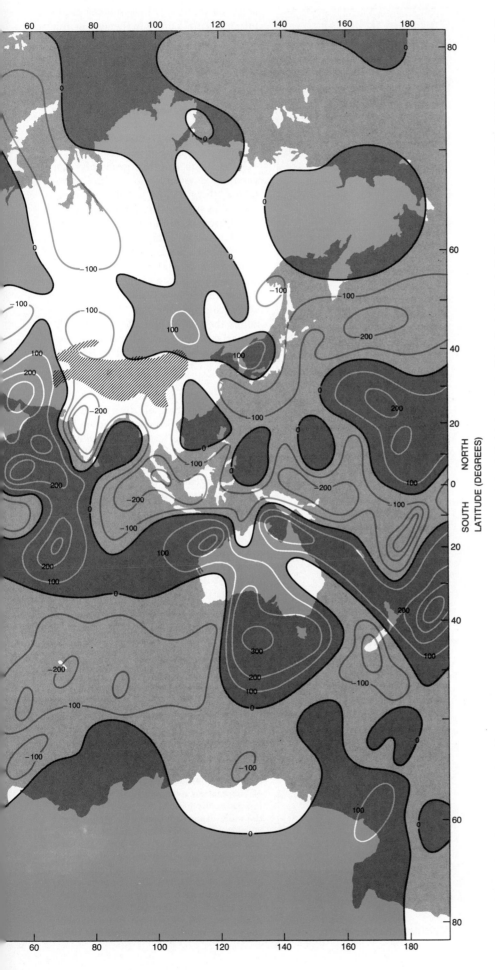

the water balance within the physical system of the earth and the atmosphere.

On the average, whenever there is an imbalance between precipitation and evaporation, there is a need for a net transport of water vapor to or from the locality by atmospheric circulations. Can such water transports in the atmosphere be assessed from the knowledge of the wind and the water content of the atmosphere? If the answer is affirmative, then an independent check on the precipitation-evaporation difference is available. That would be a most welcome development in view of the shortcomings of the existing methods of measuring these quantities separately. It must be understood, however, that for this purpose time averages of the wind and the humidity will not do, because it turns out that the effect of the deviations of these factors from their arithmetic mean cannot be neglected. The process must be evaluated instantaneously, say once each day; then an average of the results must be taken over a long enough interval. That must be done for several levels in the atmosphere up to an elevation of about five or six kilometers. Therefore, with the use of many stations at which soundings are taken, the volume of the calculations soon becomes enormous. Fortunately modern data-processing techniques are powerful enough to cope with the task.

Some years ago one of us (Peixoto) made an attempt to perform this new type of moisture-transport calculation over the entire globe on a daily basis from pole to pole for the International Geophysical Year (1958). The project was part of the comprehensive M.I.T. study of the general circulation of the earth's atmosphere and of its energetics. The results in one form showed the mean flow of moisture during the year as a field of vectors [see *illustration on*

MATHEMATICAL OPERATION was performed on the field of vectors in the illustration on pages 54 and 55 in order to compare the aerological data with conventional climatological estimates. The outcome is a contour map in which regions where there is a divergence in the field of vectors (*white contours on color background*) correspond to net sources of moisture and hence to places where there is on the average an excess of evaporation over precipitation, whereas regions where convergence prevails (*color contours on white background*) correspond to net sinks of water vapor and hence to places where precipitation exceeds evaporation. The numbers on the contours indicate the equivalent depth of liquid water in centimeters per year.

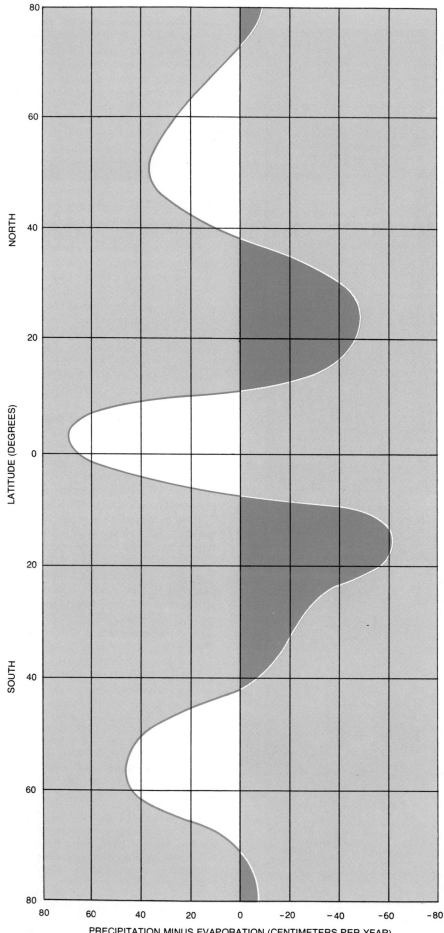

LATITUDE (DEGREES)

NORTH

SOUTH

PRECIPITATION MINUS EVAPORATION (CENTIMETERS PER YEAR)

pages 54 and 55]. These indicators are interesting to examine in themselves, but a second operation must be performed on them in order to compare the outcome with the conventional climatological estimates, which are arrived at by evaluating the excess of water that crosses the boundaries into any unit area of the earth's surface. In mathematical terms this operation is known as the convergence of a field of vectors.

When there is a divergence of water vapor in a given region of the atmosphere, there will be a source of moisture. This means that for the same interval there is on the average an excess of evaporation over precipitation. On the other hand, when convergence prevails, there is a sink of water vapor, and precipitation exceeds evaporation. Therefore the numerical value of the divergence equals the rate of evaporation minus the rate of precipitation. The divergence-convergence distribution corresponding to the vector picture is shown on the map on pages 56 and 57. This is the first such global chart ever to be drawn from actual aerological data.

The distribution of divergence and convergence can be averaged around the world for various latitudes [*see illustration at left*]. Inspection of these figures shows that on the average there is (1) a net convergence in the equatorial zone, where heavy precipitation occurs; (2) a net convergence in the middle and high latitudes, where the large excess precipitation over evaporation is associated with the migratory cyclones, the polar front and the alternation of air masses; (3) a net divergence in the subtropical

AVERAGE DISTRIBUTION of divergence zones (*dark color*) and convergence zones (*white*) around the world were computed for various latitudes from the contour map on the preceding two pages. The resulting curve is almost symmetric with respect to the Equator, where there is a comparatively narrow zone of net convergence (excess of precipitation over evaporation). Flanking this intertropical convergence zone are two broader zones of net divergence over the subtropical latitudes, where the large excess of evaporation over precipitation, particularly over the oceans, accounts for most of the water vapor supplied to the atmosphere. The net convergence in the middle and high latitudes is associated with the migratory cyclones, the polar front and the alternation of air masses. There is a small net divergence in the polar regions. The main effect of this alternation of zones on the hydrologic cycle is a net transport of water vapor northward and southward from primary sources of moisture in subtropical belts.

regions in both hemispheres, where there is strong evaporation, associated with the large semipermanent anticyclones. The results agree quite well with those obtained from independent earlier estimates of evaporation minus precipitation by climatological methods.

Thus the primary and most important sources of moisture for the entire atmosphere are found over the subtropical regions, mainly in the oceans, where evaporation proceeds continuously. The moisture supplied to the atmosphere is transported from the sources and carried by the atmospheric circulation into regions of prevailing convergence, where it condenses and falls as precipitation. Therefore the theory of the formation of precipitation from evaporation in a given place cannot be accepted. Judging from the present data, it is probable that the path length of the atmospheric branch of the water cycle varies from one locality to another. Taking as a criterion the average distance between divergence and convergence centers, it appears that the mean path length in temperate latitudes is of the order of 1,000 kilometers.

The Role of the Terrestrial Branch

Regions of convergence must have a means for disposing of the excess of precipitating water. Over land this means generally drainage by rivers, and hence convergence regions should coincide with the catchment basins of large streams. This conclusion agrees with findings on the divergence-convergence map, an outcome that lends support to the validity of the aerological method used.

The equatorial regions of the Atlantic and Pacific oceans show a general convergence. Marked centers of strong convergence are found just south of Panama, off the east coast of South America near the Equator and over Brazil (with high values in the headwaters of the Amazon River, as would be expected). This belt of convergence extends through the Atlantic Ocean with a center over the Gulf of Guinea. In equatorial Africa the regions of convergence are found in the general vicinity of the headwaters and drainage basins of many large rivers. The areas of largest convergence are found in Ethiopia and the Somali highlands, from which the Blue Nile derives most of its water; in northern Zaire over the headwaters of the Ubangi and Congo rivers; in southern Angola, Rhodesia and the eastern part of southwest Africa, where the Zambezi, Orange and Limpopo rivers originate.

In Ghana and the Ivory Coast there is convergence, and much of the excess of water associated with it is probably fed into the Niger, the Volta and perhaps the Senegal River. An extensive area of convergence extends from India through Southeast Asia, Indonesia and the entire Central Pacific to the west coast of Central America, with several intense centers. Major river systems found in this belt include the Indus, the Ganges, the Brahmaputra, the Salween, the Mekong and the Yangtze. All these areas are known to have extremely high values of precipitation.

The mid-latitude regions around the Southern Hemisphere also show many areas of convergence. The convergence areas are related to the polar front that is associated with the tracks of extratropical storms across the South Atlantic, the South Indian Ocean and the South Pacific. The extensive area of convergence that covers most of South America corresponds to the basins and headwaters of a large system of rivers (including the Magdalena, the Orinoco, the Amazon, the Paraguay, the Paraná and the Uruguay). There is a convergence center that extends through the eastern South Pacific and along the Chilean coast, where strong thunderstorm activity predominates. In the neighboring regions the salt content of the ocean is low.

The subtropical regions of the South Atlantic, the Indian Ocean and the South Pacific show strong and extensive areas of divergence (excess of evaporation over precipitation). In the Atlantic region the divergence belt is elongated in an east-west direction without any interruption. The strong divergence off the coast of central Brazil is reflected by a correspondingly higher salt content in the waters of that region of the Atlantic. By and large all the divergence regions over the oceans are associated with high salinity. These areas and the corresponding ones in the Northern Hemisphere are areas where evaporation attains its maximum value. They are associated with the anticyclonic belt of the subtropical latitudes and are generally situated on the margins of the semipermanent high-pressure cells where clear skies predominate. On land, areas of large divergence are found over northeastern Brazil, a region noted for droughts; in Angola, Uganda, Kenya and South Africa and over most of Australia. In the Northern Hemisphere there are strong divergences over southern Mexico and Lower California in North America, over the Sahara in Africa and over Arabia and Iran in Asia. In the polar regions it seems that there is a small net divergence. With the presence of ice there is comparatively little evaporation. The divergence is higher over the North Pole than over the South Pole.

The regions of strongest divergence tend to be found over the subtropical regions, although there are exceptions such as the center over the eastern coast of Asia. When these regions are over ocean surfaces, there is of course no shortage of water to supply the net evaporation; it is always replenished by ocean currents. The situation is otherwise where the regions of strong divergence are over land, as they are, for example, over the Sahara.

Where does the water come from in such places? It is now known that under this seemingly arid region huge amounts of water are buried in large formations of porous rock. Evaporation from these aquifers has been found to amount in certain areas to as much as 200,000 cubic meters per year per square kilometer. Minerals formed by the deposition of salt in the course of evaporation are common in the Sahara, where such evaporites are found in the salt flats called chotts. Deposits of minerals in the form of "desert varnish" are the result of strong evaporation followed by the active oxidation of a thin layer that consists mainly of iron compounds that have accumulated on exposed rock surfaces.

The excess of evaporation over precipitation has to be made up by underground drainage from the less arid areas that supply the water. The study of the flow of water below the surface in desert areas is demanding, but it is an enterprise that has increasing practical significance. As early as 1940 Bo Hellström of the Royal Institute of Technology of Stockholm made such a study of the eastern Sahara near the Nile. It seems from historical and archaeological evidence that the Khârga Oasis in Egypt was more abundantly supplied with water from springs in ancient times than it is today. Apparently the decline was rather sudden and can be attributed to the puncturing through erosion and subsequent leakage of a subterranean water-bearing rock stratum in the bed of the Nile. This had the effect of decreasing the hydraulic pressure of the underground water over the entire region and thus reducing the discharge of the Khârga springs. Hellström suggested that the erection of a high dam at Aswan in Egypt would in effect repair the damage. The higher head of the impounded waters would then prevent the leakage from the break, which is located just upstream from Aswan.

In a more recent study Robert P. Ambroggi of the United Nations Food and Agriculture Organization also analyzed the problem of water under the Sahara. He reached the conclusion that there are several internal drainage areas below the surface, indicating the existence of large sources of underground water [see "Water under the Sahara," by Robert P. Ambroggi; SCIENTIFIC AMERICAN, May, 1966]. The existence of similar underground flows passing the oases said to be connected with the Nile River were reported by F. M. Ali of the Egyptian Meteorological Department at the Symposium on Tropical Meteorology held in Nairobi in 1960.

It has been suggested that there may be another example of underground drainage in the Chad basin. According to the French hydrologist G. Drouhin, although Lake Chad receives water from the Sahara and Lagone rivers, its area and salinity content remain constant. Without a sizable subterranean drainage, the salinity due to the combined effects of evaporation and the transport of salts by the rivers should increase. Hydrological observations indicate that there is a flow toward the north-northeast in an aquifer that extends under the

lake. In a more recent work based on hydrological and geological evidence Eugene M. Rasmusson of the National Oceanic and Atmospheric Administration found some indications of large sources of underground water in the desert areas of the U.S.

Control of the Water Cycle

In order to decide in what direction the water cycle should be controlled, it is first necessary to determine just what human needs for water are. Water is a necessary component of all industrial or agricultural development. In the Tropics, for example, to raise one metric ton of wheat requires some 8,000 metric tons of water. In industry 200 tons of water are needed to make one metric ton of steel, and 20 tons of water are consumed in refining one ton of gasoline.

The water needs of an "average" person range from 900 cubic meters per year in an agricultural society such as Iran to about 2,700 cubic meters per year in a highly industrial society such as the U.S. Of the 2,700 cubic meters of water consumed by the average

American, a mere .5 cubic meter is needed for drinking, 200 cubic meters is used in the home and the rest is spent in industry and agriculture. In the Tropics three cubic meters per year per square meter would be needed for agriculture. Assuming, then, that the water needs of an average man are about 1,500 cubic meters per year, the total world consumption of water would be 4.5 trillion cubic meters per year. Just to keep up with the increase in world population the water demand will increase by about 100 billion cubic meters per year. Since the amount of water that traverses the water cycle in a year is limited, a solution to the water problem should clearly be found soon.

What do we really mean by water need? Water molecules in the water cycle, just like electrons in an electric circuit, do work in traversing a system. For example, the water drunk by a man is not destroyed; it passes through his body, doing work in the process, and then rejoins the water cycle. In this system the body of the man acts as the load in the cycle, whereas the water acts as the flow of electrons. Similarly, the water applied in the irrigation of a crop

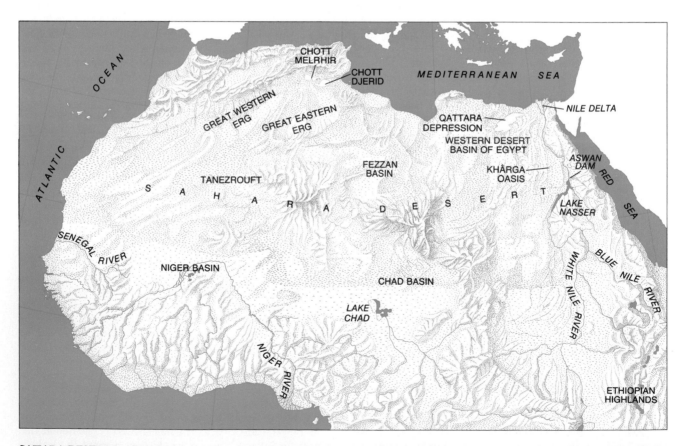

SAHARA DESERT has been found in recent years to conceal vast underground water resources concentrated in seven major storage reservoirs: the Great Western Erg, the Great Eastern Erg, the Tanezrouft, the Niger basin, the Fezzan basin, the Chad basin and the Western Desert basin of Egypt (the last being the largest). The total water-storage capacity of these seven basins is estimated at 15.3 trillion cubic meters. Other important hydrologic features of the Sahara region discussed in the text are designated on map.

does work by passing through the plants before rejoining the cycle through evapotranspiration. Basically the sun is the equivalent of the generator in the electric circuit.

Continuing with the electric-circuit analogy, if the work done by water is to be increased, there are three possibilities: (1) to increase the amount of water in each cycle by improving the efficiency of the cycle at the load and at the generator, (2) to increase the number of cycles possible and (3) to accelerate the cycle by increasing the rate of both evaporation and precipitation.

At this point some further considerations should be taken into account in discussing the control of the water cycle. First, water as a carrier is able to do work only when it is in its pure or quasi-pure state. Second, since man is a land dweller, he uses the water on the continents rather than the water in the seas; any water falling on the ocean is lost for human purposes. Third, water is mostly used in the liquid phase. Under these conditions one can establish three principles for the control of the water cycle: (1) Evaporation from the continents, from freshwater surfaces and from underground aquifers should be minimized as much as possible. (2) Since seawater cannot be used as such, evaporation from the ocean surfaces should be accelerated as much as possible. (3) No water should be allowed to proceed in the cycle, flowing into the oceans or evaporating into the atmosphere, until it has done the largest possible amount of useful work. The desalting of seawater is a way to obtain fresh water outside the natural cycle, through the application of nuclear energy, fossil-fuel energy or concentrated solar energy.

Control of the Atmospheric Branch

One way of attacking the problem of lack of water in the arid and semiarid regions of the globe would be to speed up the hydrologic cycle in the atmospheric branch by the artificial inducement of precipitation over the region to be benefited, at times and in amounts previously fixed and with the assurance that the water thus formed would not fall on adjacent regions or on the sea. In fact, the atmosphere can be regarded as a resource for the extraction of liquid water by artificial means. There have been some attempts in this direction through the seeding of clouds with silver iodide, frozen carbon dioxide ("dry ice") and other substances. In spite of the good results obtained under certain conditions, the technique cannot be used at present because artificially induced precipitation is not yet economically justified or even physically defined. Nonetheless, this field of research remains very active. One can be confident that the ultimate success of methods for the artificial release and control of precipitation will depend on the acquisition of a much deeper knowledge of cloud physics, namely the mechanisms of nucleation and the ways in which nature produces rain and snow. Present attempts at rainmaking must thus be undertaken cautiously.

The formation of dew is another way of extracting moisture from the atmosphere. Dew has been a source of water since the Middle Ages, but only recently has it been collected on a larger scale. For example, it is a source of water for human uses in Gibraltar. The capture of water through the interception of fog has also been attempted with some success.

A different line of attack on the problem of control of the atmospheric branch of the water cycle has been advanced by Victor P. Starr and David Anati of M.I.T. They have proposed that liquid water could be recovered from the atmosphere through a partial duplication of natural moist-convection processes, in which air warmed and loaded with water vapor by the heat of the sun rises until it reaches a point at which the water vapor condenses. They suggest an experimental device consisting essentially of a large chimney, perhaps 100 meters in diameter, erected to some height above the free condensation level of a parcel of ground [see *illustration at right*]. This device, called an aerological accelerator, would be self-energizing; no source of energy would be necessary other than the automatic release of the latent heat of the condensing water vapor. Starr and Anati consider a hypothetical experiment within a tube extending to a

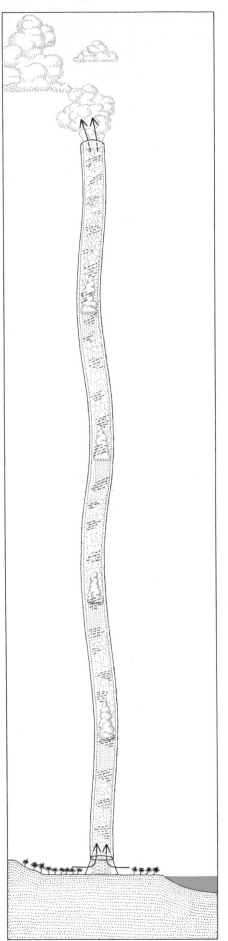

HYPOTHETICAL DEVICE for recovering liquid water from the atmosphere by partially duplicating natural moist-convection processes has been proposed by Victor P. Starr and David Anati of the Massachusetts Institute of Technology. Their device, called the aerological accelerator, would consist essentially of a long, double-walled plastic tube filled with helium and allowed to float freely in the atmosphere at a fixed height above a parcel of land in a normally arid coastal region. They suggest a tube with a height of about 3,000 meters and a cross-sectional area of 10,000 square meters. Moist air ascending within the tube should reach the condensation level, leading to precipitation. The scheme could act in part as a mechanism for the desalting of seawater.

height of 3,000 meters and having a cross-sectional area of 10,000 square meters. Assume that the vertical velocity of the moist air is of the order of 20 meters per second, and that one gram of water is released from each kilogram of air in the ascending column. A simple calculation shows that the water obtained would be sufficient to irrigate several square kilometers to a depth of 30 centimeters per year, if meteorological conditions are favorable during a fair fraction of this period.

Starr and Anati have considered Arabia and the Gulf of Mexico as possible locations for experiments with the aero-

logical accelerator. A computation based on aerological data for Aden on June 5, 1958, shows that about 1,000 metric tons of water per hour can conceivably be extracted from the gaseous hydrosphere with a device having about the same dimensions as the one described above. It should be mentioned that no precipitation has been recorded in Aden for the month of June during the 10-year period from 1951 through 1960, and that the mean annual precipitation barely exceeds four centimeters per year.

One can take advantage of the dependence of the convection efficiency on the moisture content of the intake air to appreciably reduce the necessary minimum height of the tube. That would require the evaporation of additional moisture from the sea surface in a proper manner at the intake. The scheme would then act, at least in part, as a mechanism for the desalting of ocean water. From a water-production point of view the "captive rainstorm" would for its size have some advantages over a natural one, since the lateral entrainment of drier air, which has a damping effect on the moist-convection dynamics, would be absent. Thus the presence of the tube should favor the desired process.

The influence of such a device might extend beyond its immediate location: natural convection at higher levels could develop as a result of its action, and individual cells of free convection could possibly detach themselves from the top of the tube and drift off as autonomous thunderstorms. Obstacles to the construction of an experimental unit include the need for much more study to determine the best theoretical design, the best location and so on, the engineering difficulties connected with the construction, and the high cost of the enterprise. The scientific and practical interest of such a project could nonetheless be great enough to warrant a considerable effort under the proper circumstances.

Real prototypes of the aerological accelerator are found in ventilation shafts of deep gold mines in South Africa. A single such shaft can produce as much as 33,000 cubic meters of water per year. Water obtained in this way has actually been put to agricultural use. The greatest obstacle to such schemes is the need for a mechanism to remove the raindrops from the rising air.

Control of Surface Runoff

The arid regions around the world include most of the African continent outside its equatorial zone, the entire Mid-

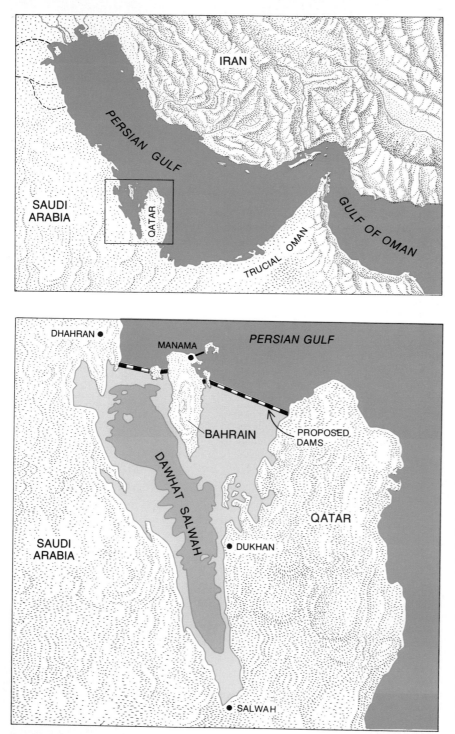

SCALE of the heliohydroelectric project advanced by Kettani and Gonsalves is suggested by the map of the entire Persian Gulf area at top and by the more detailed map at bottom. The total surface area of the proposed reservoir (*lighter shades of color*) would be 6,460 kilometers. The marginal lands that could be reclaimed from the sea in Dawhat Salwah, amounting to some 4,100 square kilometers, are indicated in the lightest shade of color.

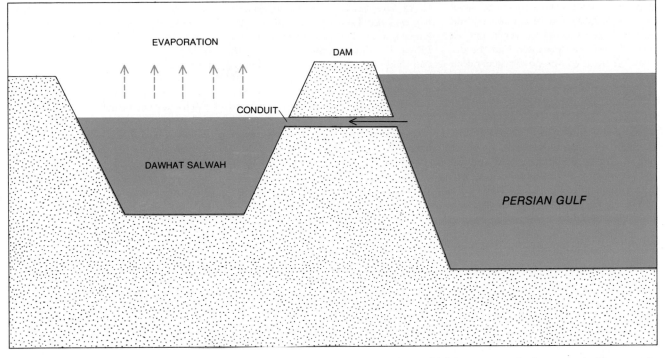

CONDUIT would link the Dawhat Salwah reservoir with the open sea. The rapid loss of water due to evaporation in the confined reservoir would create a hydraulic head across the conduit, which could be exploited to produce electric power by controlling the flow of water in such a way that water is admitted to the conduit at the same rate at which it evaporates in the reservoir. For example, with a net evaporation rate of 3.5 meters per year and an optimum head of 12.4 meters, 300 million kilowatt-hours of electricity could be produced. In addition it has been suggested that magnesium could be obtained from the reservoir by precipitation methods.

dle East, all of central Asia, most of Australia and most of the western part of the Western Hemisphere. Many of these regions lie near the shore, where water is plentiful. One might therefore be tempted to seek a solution to the water problem in desalination projects. The largest desalting plant in the U.S., however, produces no more than 3.6 million cubic meters of fresh water per year, merely enough to satisfy the needs of a town of 3,600 people. The largest desalting plant in the U.S.S.R. is expected to yield 55 million cubic meters per year, enough for a city of 55,000 people.

These figures compare poorly with the discharges of many rivers that are allowed to dump huge quantities of fresh water into the sea, often in areas where these waters are most needed. For example, the water discharged by the Amazon (6.8 trillion cubic meters per year) is enough to satisfy the needs of almost twice the present population of the entire earth. The Congo River would satisfy the needs of four times the population of Africa (1.35 trillion cubic meters per year). The Mississippi alone discharges into the ocean every year as much water as the entire consumption of the U.S. (560 billion cubic meters per year). This waste goes on even in areas where water is most badly needed, such

as the Middle East. There the Tigris and the Euphrates dump into the Persian Gulf as much water as the entire consumption of Iran. If this water could be brought to Arabia by pipeline, the entire economy of the Arabian peninsula would be changed.

To cite just one example where the control of running water might boost the economic capacity of a country, consider the case of Egypt. That country is literally a gift of the Nile; it has a population of 34 million people, two-thirds of whom live in the 25,000 square kilometers of the Nile delta. The remaining third live in about 14,000 square kilometers along the banks of the river. Thus the entire population is concentrated in a mere 39,000 square kilometers in a country with a total area of one million square kilometers. The Nile itself has an average yearly discharge at Aswan of 92 billion cubic meters, enough for a population of 100 million people if utilized fully and efficiently. The waters of the Nile, which cross Egypt on a 1,550-kilometer path, come from the Ethiopian highlands (84 percent) and from the lakes plateau of central Africa (16 percent).

The construction of the high dam at Aswan certainly reduced the annual amount of water lost to the Mediter-

ranean. It also brought 28,000 square kilometers of desert land into cultivation (of which 8,000 square kilometers are in Egypt and the rest in the Sudan) and generates 10 billion kilowatt-hours of electric energy. The lake behind the dam, however, introduces a new loss, a loss due to evaporation. (Seepage is not considered a direct loss because it helps to refill the underground aquifers.) Assuming an evaporation rate of three meters per year, one can see that as much as 24 billion cubic meters of water would be evaporated yearly from the 7,800-square-kilometer surface of the lake. This amount is huge, representing about a fourth of the entire discharge of the Nile at Aswan. Either water reservoirs presenting such a large free surface to the sun should be avoided in arid zones or a solution to the evaporation problem should be found. It might be well to examine the possibility of covering the entire evaporating surface by a layer of some material that would eliminate or greatly reduce evaporation, such as a layer of liquid hexadecanol, plastic covers, wood plates or dyes to retard heat absorption.

The Qattara Depression, some 200 kilometers west of Cairo and 60 kilometers south of the Mediterranean, has long fascinated engineers and scientists. The depression has a surface area at sea level of about 20,000 square kilometers, and its lowest point is 134 meters below sea level. It has been proposed that the depression be used for power generation by admitting seawater from the Mediterranean. It would seem more beneficial, however, to devote this large area to cultivation and settlement. If only a ninth of the Nile were diverted into it, the depression could be made the home of as many as 10 million people, and the silt brought by the Nile could transform large areas into farmland. The waste water would reenter the water cycle through evaporation.

Control of Underground Water

In sloping soils about 20 percent of precipitation infiltrates the ground. In flat, permeable soils the figure can be as high as 80 percent. These waters feed underground aquifers and make up for their loss through natural springs and evapotranspiration.

Underground seas have been discovered in arid regions such as Iran, Arabia and North Africa. Ancient peoples were already acquainted with these aquifers; for example, some 3,000 years ago a system of qanats, or underground channels, was built in Iran for tapping underground water [see "The Qanats of Iran," by H. E. Wulff; SCIENTIFIC AMERICAN, April, 1968]. The system was expanded under the Islamic caliphates to the rest of the Middle East and to North Africa. The qanat system consists of a simple network of underground channels bringing water by gravity from the highland aquifers to the lowlands. Even today this system supplies 75 percent of the water needs of Iran.

The underground water in Arabia was discovered as a result of drilling for oil. At present a great effort is being made to develop its potentialities. The eastern region of Arabia alone has seven major aquifers in Cenozoic and Mesozoic strata, with hydraulic gradients decreasing from west to east. Near the Persian Gulf the Mesozoic strata reach a depth of 1,000 meters. The strata are not too deep to be mined efficiently.

The discovery of the vast underground water potential of the Sahara by Ambroggi and others confirmed a prediction made by Starr and one of us (Peixoto) in 1958 on the basis of aerological findings. There are seven major storage basins under the Sahara, having a total area of 4.5 million square kilometers and a total water-storage capacity of 15.3 trillion cubic meters. They are the Great Western Erg, the Great Eastern Erg and the Tanezrouft in Algeria; the Niger basin in Mali; the Fezzan basin in Libya; the Chad basin in Chad, Niger and Nigeria, and the Western Desert basin of Egypt [see illustration on page 60]. The last is the largest, with a total area of 1.8 million square kilometers and a total water-storage capacity of six trillion cubic meters. It extends into neighboring African countries.

Consider the Great Eastern Erg of Algeria, with an area of 375,000 square kilometers and a water-storage capacity of 1.7 trillion cubic meters. Most of this water is fossil water gathered after the last ice age, when the Sahara must have been a tropical region with heavy rainfall. At present the aquifer is being recharged by the flood rains from the northern Tell mountains. The total natural recharge is estimated to be about 900 million cubic meters per year. Since the total capacity of the aquifer does not change appreciably, there has to be some discharge. That discharge is evaporation from a large depression called Chott Melrhir-Djerid. The chott is dry except during rainy seasons, and its lowest point is 31 meters below sea level. It extends to Tunisia, where only about 20 kilometers of sandy soil separates it from the Mediterranean. Chott Melrhir proper is entirely below sea level and extends over an area of about 10,000 square kilometers.

If Chott Melrhir were flooded entirely by water from the Mediterranean, the discharge of the Great Eastern Erg through evaporation would be greatly reduced, thus limiting the loss of usable underground water. Moreover, the oil of eastern Algeria could be shipped directly from where it is produced, and great quantities of salt minerals could be extracted from the newly formed gulf. The annual supply of fresh water made available in this way would be enough to support the settlement of about a million people in a region that now has scarcely 200,000 inhabitants.

Control of the Oceans

Some attempts have been made to control the water cycle over the oceans. The desalting of seawater is only an example of such attempts. Schemes on a broader scale have been proposed, such as the "heliohydroelectric" project put forward in 1970 by one of us (Kettani) in collaboration with L. M. Gonsalves. The proposed scheme utilizes solar energy to generate through evaporation an inflow of water from the Persian Gulf to an artificially made reservoir called Dawhat Salwah. The reservoir, which would be formed by building causeways between Saudi Arabia and Bahrain and between Bahrain and Qatar, would have a total surface area of 6,460 square kilometers [see illustrations on page 62].

The system formed would consist of two reservoirs linked by a conduit [see illustration on preceding page]. The rapid loss of water due to the evaporation in the confined reservoir would generate a hydraulic head across the conduit, since the water level in the open sea would not be affected. Thus the flow of water across the conduit could be controlled in such a way that if water were admitted in the conduit at the same rate at which it evaporated in the enclosed reservoir, electric power could be generated. With a net evaporation rate of 3.5 meters per year and an optimum head of 12.4 meters, 300 million kilowatt-hours of electricity could be produced. Four other avenues of economic exploitation have been considered: mineral production, road communication, land reclamation and sea farming.

The marginal lands that would be gained from the sea in Dawhat Salwah would amount to about 4,100 square kilometers. The effect on Bahrain would be considerable: its area would increase almost four times. Because the amount of water in the atmosphere cannot be increased, the effect of creating such an artificial depression would be to raise the level of all the oceans of the world by .25 millimeter. Presumably nobody would object to this rise.

The project is actually a replica of a natural phenomenon in central Asia, where the Caspian Sea sends 8.4 billion cubic meters through a narrow strait into a 10,400-square-kilometer desert basin called the Kara Bogaz Gol [see "The Sea that Spills into a Desert," by Maurice A. Garbell; SCIENTIFIC AMERICAN, August, 1963]. The Caspian, however, is a closed sea, and whatever water it loses goes directly into the overall water cycle. As a result the surface level of the Caspian has been going down for the past 90 years. The flow into the Kara Bogaz Gol has not been large enough to keep up with the evaporation rate, and the area of that desert lake is still decreasing.

Antarctic ice is the most important source of fresh water waiting to be tapped. The feasibility of towing Antarctic icebergs to arid areas in America, Africa, Asia and Australia has been considered. It has been suggested that it would be possible to tow a train of 10 icebergs containing a total of about 10 billion cubic meters of fresh water to southern California in about 10 months, with a loss of water content of only 10 percent. The icebergs could be moored at their destination and then quarried for human use.

Epilogue

The hydrologic cycle on a planetary scale can be regarded as the result of a gigantic distillation scheme extending over the entire earth. The heating of the subtropical regions by solar radiation leads to a continuous evaporation of water, which is released into the atmosphere and is removed by the wind circulation and transported, mostly in the lower half of the atmosphere, northward and southward into other latitudes. During the transfer process in the gaseous phase, because of cooling due to expansion, some of the water vapor eventually condenses to form clouds, which lead to precipitation over the equatorial regions and over the middle and high latitudes. Finally the combined action of oceans, rivers and underground flows provides the compensating return of water in the liquid or solid phases to the source regions, closing the sequence of natural phenomena involved in the transition phase of the water. Thereafter the water is once again available to be released into the atmosphere and to start the cycle anew.

Thus the hydrologic cycle is maintained by the general circulation of the atmosphere. We cannot understand the mechanisms of the hydrologic cycle unless we understand the mechanisms of the general circulation, and vice versa. This complex feedback process leads to a highly nonlinear coupling of the hydrologic cycle and the general circulation, which if well understood will make their control a foreseeable possibility.

ASWAN DAM has contributed significantly to the economic capacity of the Nile valley region by reducing the annual amount of water lost to the Mediterranean, by bringing thousands of square kilometers of desert land into cultivation and by generating prodigious amounts of electric energy. The loss of water due to evaporation from the lake behind the dam, however, is huge, representing about a fourth of the entire discharge of the Nile at Aswan. One solution to the evaporation problem that has been suggested is to cover the entire evaporating surface of the lake by a layer of some material that would retard heat absorption. The dam is at upper left in this ERTS photograph.

6

The Oxygen Cycle

INTRODUCTION

If there is one natural resource that most of us take for granted, it is the air we breathe. In the late 1960s a small scare wave was produced when some people started to wonder aloud whether pollution of the atmosphere would eventually lead to a loss of atmospheric oxygen and thus to death for all higher organisms. This unwarranted fear (unwarranted because our present supply of atmospheric oxygen is so enormous compared with any process that might reduce the supply) was voiced when many earth scientists and biologists were coming to a new appreciation of the oxygen cycle of the planet, its functioning, and its early evolution in the earth's history. Curiously enough, it was geological speculation on the evolution of oxygen that played a most important role in opening up knowledge of the present-day cycle.

In "The Oxygen Cycle," Preston Cloud, a geologist and paleontologist who has long been concerned with Precambrian geology, the earliest fossils, and the evolution of life, has teamed up with Aharon Gibor, a biologist, to explore in this chapter the nature of the oxygen cycle. Quickly we come to realize that oxygen cannot be divorced from carbon. The two are indissolubly linked in all the processes of life and earth. To understand the flow of oxygen from reduced (molecular oxygen) to oxidized states (oxygen in carbon dioxide and oxides), we must pair it with the associated flow of carbon from reduced (organic carbon) to oxidized forms (carbon dioxide and carbonates).

Oxygen is the by-product of one of the great biological inventions, photosynthesis, and the fuel for another biological innovation that took place early in the history of life, respiration. Though there are inorganic mechanisms in the outer atmosphere that can produce small amounts of molecular oxygen, it is photosynthesis that has produced our oxygen. One fascinating question is how early and how much? That question and others have assumed even more interest as we have begun to explore Mars, a planet that has a very small amount of oxygen in an extremely thin atmosphere. By understanding the oxygen cycle on earth we may learn how to interpret that cycle on a smaller, colder, planet on which the existence of life remains to be proven.

The Oxygen Cycle

by Preston Cloud and Aharon Gibor
September 1970

*The oxygen in the atmosphere was originally put there
by plants. Hence the early plants made possible the
evolution of the higher plants and animals that require
free oxygen for their metabolism*

The history of our planet, as recorded in its rocks and fossils, is reflected in the composition and the biochemical peculiarities of its present biosphere. With a little imagination one can reconstruct from that evidence the appearance and subsequent evolution of gaseous oxygen in the earth's air and water, and the changing pathways of oxygen in the metabolism of living things.

Differentiated multicellular life (consisting of tissues and organs) evolved only after free oxygen appeared in the atmosphere. The cells of animals that are truly multicellular in this sense, the Metazoa, obtain their energy by breaking down fuel (produced originally by photosynthesis) in the presence of oxygen in the process called respiration. The evolution of advanced forms of animal life would probably not have been possible without the high levels of energy release that are characteristic of oxidative metabolism. At the same time free oxygen is potentially destructive to all forms of carbon-based life (and we know no other kind of life). Most organisms have therefore had to "learn" to conduct their oxidations anaerobically, primarily by removing hydrogen from foodstuff rather than by adding oxygen. Indeed, the anaerobic process called fermentation is still the fundamental way of life, underlying other forms of metabolism.

Oxygen in the free state thus plays a role in the evolution and present functioning of the biosphere that is both pervasive and ambivalent. The origin of life

and its subsequent evolution was contingent on the development of systems that shielded it from, or provided chemical defenses against, ordinary molecular oxygen (O_2), ozone (O_3) and atomic oxygen (O). Yet the energy requirements of higher life forms can be met only by oxidative metabolism. The oxidation of the simple sugar glucose, for example, yields 686 kilocalories per mole; the fermentation of glucose yields only 50 kilocalories per mole.

Free oxygen not only supports life; it arises from life. The oxygen now in the atmosphere is probably mainly, if not wholly, of biological origin. Some of it is converted to ozone, causing certain high-energy wavelengths to be filtered out of the radiation that reaches the surface of the earth. Oxygen also combines with a wide range of other elements in the earth's crust. The result of these and other processes is an intimate evolutionary interaction among the biosphere, the atmosphere, the hydrosphere and the lithosphere.

Consider where the oxygen comes from to support the high rates of energy release observed in multicellular organisms and what happens to it and to the carbon dioxide that is respired [*see illustration on page 72*]. The oxygen, of course, comes from the air, of which it constitutes roughly 21 percent. Ultimately, however, it originates with the decomposition of water molecules by light energy in photosynthesis. The 1.5 billion cubic kilometers of water on the earth are split by photosynthesis and reconsti-

tuted by respiration once every two million years or so. Photosynthetically generated oxygen temporarily enters the atmospheric bank, whence it is itself recycled once every 2,000 years or so (at current rates). The carbon dioxide that is respired joins the small amount (.03 percent) already in the atmosphere, which is in balance with the carbon dioxide in the oceans and other parts of the hydrosphere. Through other interactions it may be removed from circulation as a part of the carbonate ion (CO_3^-) in calcium carbonate precipitated from solution. Carbon dioxide thus sequestered may eventually be returned to the atmosphere when limestone, formed by the consolidation of calcium carbonate sediments, emerges from under the sea and is dissolved by some future rainfall.

Thus do sea, air, rock and life interact and exchange components. Before taking up these interactions in somewhat greater detail let us examine the function oxygen serves within individual organisms.

Oxygen plays a fundamental role as a building block of practically all vital molecules, accounting for about a fourth of the atoms in living matter. Practically all organic matter in the present biosphere originates in the process of photosynthesis, whereby plants utilize light energy to react carbon dioxide with water and synthesize organic substances. Since carbohydrates (such as sugar), with the general formula $(CH_2O)_n$, are the common fuels that are stored by plants, the essential reaction of photosynthesis can be written as $CO_2 + H_2O + \text{light energy} \rightarrow CH_2O + O_2$. It is not immediately obvious from this formulation which of the reactants serves as the source of oxygen atoms in the carbohydrates and which is the source of free molecular oxygen. In 1941 Samuel Ruben and Mar-

RED BEDS rich in the oxidized (ferric) form of iron mark the advent of oxygen in the atmosphere. The earliest continental red beds are less than two billion years old; the red sandstones and shales of the Nankoweap Formation in the Grand Canyon (*opposite page*) are about 1.3 billion years old. The appearance of oxygen in the atmosphere, the result of photosynthesis, led in time to the evolution of cells that could survive its toxic effects and eventually to cells that could capitalize on the high energy levels of oxidative metabolism.

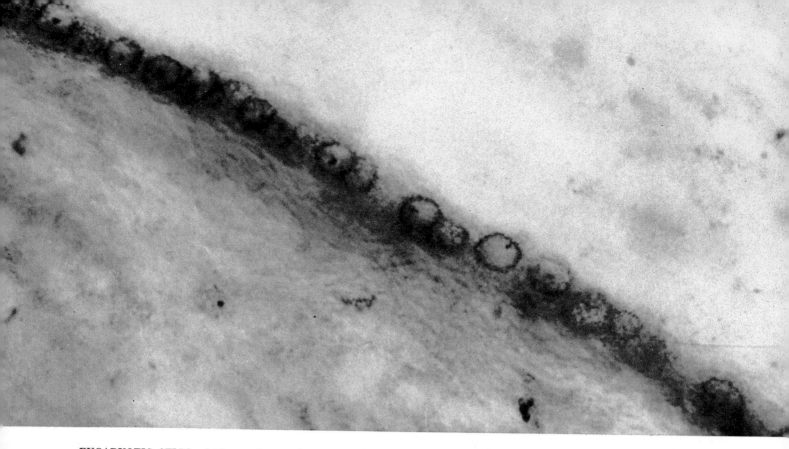

EUCARYOTIC CELLS, which contain a nucleus and divide by mitosis, were, like oxygen, a necessary precondition for the evolution of higher life forms. The oldest eucaryotes known were found in the Beck Spring Dolomite of eastern California by Cloud and his colleagues. The photomicrograph above shows eucaryotic cells with an average diameter of 14 microns, probably green algae. The regular occurrence and position of the dark spots suggest they may be remnants of nuclei or other organelles. Other cell forms, which do not appear in the picture, show branching and large filament diameters that also indicate the eucaryotic level of evolution.

PROCARYOTIC CELLS, which lack a nucleus and divide by simple fission, were a more primitive form of life than the eucaryotes and persist today in the bacteria and blue-green algae. Procaryotes were found in the Beck Spring Dolomite in association with the primitive eucaryotes such as those in the photograph at the top of the page. A mat of threadlike procaryotic blue-green algae, each thread of which is about 3.5 microns in diameter, is seen in the photomicrograph below. It was made, like the one at top of page, by Gerald R. Licari. Cells of this kind, among others, presumably produced photosynthetic oxygen before eucaryotes appeared.

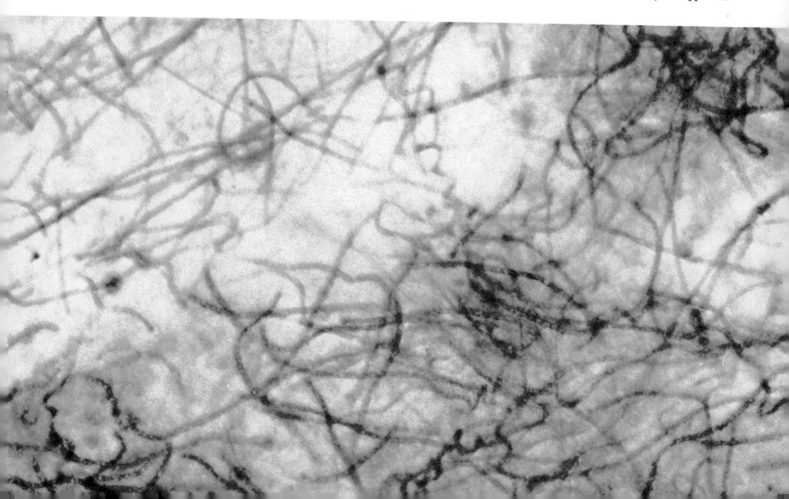

tin D. Kamen of the University of California at Berkeley used the heavy oxygen isotope oxygen 18 as a tracer to demonstrate that the molecular oxygen is derived from the splitting of the water molecule. This observation also suggested that carbon dioxide is the source of the oxygen atoms of the synthesized organic molecules.

The primary products of photosynthesis undergo a vast number of chemical transformations in plant cells and subsequently in the cells of the animals that feed on plants. During these processes changes of course take place in the atomic composition and energy content of the organic molecules. Such transformations can result in carbon compounds that are either more "reduced" or more "oxidized" than carbohydrates. The oxidation-reduction reactions between these compounds are the essence of biological energy supply and demand. A more reduced compound has more hydrogen atoms and fewer oxygen atoms per carbon atom; a more oxidized compound has fewer hydrogen atoms and more oxygen atoms per carbon atom. The combustion of a reduced compound liberates more energy than the combustion of a more oxidized one. An example of a molecule more reduced than a carbohydrate is the familiar alcohol ethanol (C_2H_6O); a more oxidized molecule is pyruvic acid ($C_3H_4O_3$).

Differences in the relative abundance of hydrogen and oxygen atoms in organic molecules result primarily from one of the following reactions: (1) the removal (dehydrogenation) or addition (hydrogenation) of hydrogen atoms, (2) the addition of water (hydration), followed by dehydrogenation; (3) the direct addition of oxygen (oxygenation). The second and third of these processes introduce into organic matter additional oxygen atoms either from water or from molecular oxygen. On decomposition the oxygen atoms of organic molecules are released as carbon dioxide and water. The biological oxidation of molecules such as carbohydrates can be written as the reverse of photosynthesis: $CH_2O + O_2 \rightarrow CO_2 + H_2O$ + energy. The oxygen atom of the organic molecule appears in the carbon dioxide and the molecular oxygen acts as the acceptor for the hydrogen atoms.

The three major nonliving sources of oxygen atoms are therefore carbon dioxide, water and molecular oxygen, and since these molecules exchange oxygen atoms, they can be considered as a common pool. Common mineral oxides such as nitrate ions and sulfate ions are also oxygen sources for living organisms,

which reduce them to ammonia (NH_3) and hydrogen sulfide (H_2S). They are subsequently reoxidized, and so as the oxides circulate through the biosphere their oxygen atoms are exchanged with water.

The dynamic role of molecular oxygen is as an electron sink, or hydrogen acceptor, in biological oxidations. The biological oxidation of organic molecules proceeds primarily by dehydrogenation: enzymes remove hydrogen atoms from the substrate molecule and transfer them to specialized molecules that function as hydrogen carriers [see top illustration on pages 74 and 75]. If these carriers become saturated with hydrogen, no further oxidation can take place until some other acceptor becomes available. In the anaerobic process of fermentation organic molecules serve as the hydrogen acceptor. Fermentation therefore results in the oxidation of some organic compounds and the simultaneous reduction of others, as in the fermentation of glucose by yeast: part of the sugar molecule is oxidized to carbon dioxide and other parts are reduced to ethanol.

In aerobic respiration oxygen serves as the hydrogen acceptor and water is produced. The transfer of hydrogen atoms (which is to say of electrons and protons) to oxygen is channeled through an array of catalysts and cofactors. Prominent among the cofactors are the iron-containing pigmented molecules called cytochromes, of which there are several kinds that differ in their affinity for electrons. This affinity is expressed as the oxidation-reduction, or "redox," potential of the molecule; the more positive the potential, the greater the affinity of the oxidized molecule for electrons. For example, the redox potential of cytochrome b is .12 volt, the potential of cytochrome c is .22 volt and the potential of cytochrome a is .29 volt. The redox potential for the reduction of oxygen to water is .8 volt. The passage of electrons from one cytochrome to another down a potential gradient, from cytochrome b to cytochrome c to the cytochrome a complex and on to oxygen, results in the alternate reduction and oxidation of these cofactors. Energy liberated in such oxidation-reduction reactions is coupled to the synthesis of high-energy phosphate compounds such as adenosine triphosphate (ATP). The special copper-containing enzyme cytochrome oxidase mediates the ultimate transfer of electrons from the cytochrome a complex to oxygen. This activation and binding of oxygen is seen as the fundamental step, and possibly

the original primitive step, in the evolution of oxidative metabolism.

In cells of higher organisms the oxidative system of enzymes and electron carriers is located in the special organelles called mitochondria. These organelles can be regarded as efficient low-temperature furnaces where organic molecules are burned with oxygen. Most of the released energy is converted into the high-energy bonds of ATP.

Molecular oxygen reacts spontaneously with organic compounds and other reduced substances. This reactivity explains the toxic effects of oxygen above tolerable concentrations. Louis Pasteur discovered that very sensitive organisms such as obligate anaerobes cannot tolerate oxygen concentrations above about 1 percent of the present atmospheric level. Recently the cells of higher organisms have been found to contain organelles called peroxisomes, whose major function is thought to be the protection of cells from oxygen. The peroxisomes contain enzymes that catalyze the direct reduction of oxygen molecules through the oxidation of metabolites such as amino acids and other organic acids. Hydrogen peroxide (H_2O_2) is one of the products of such oxidation. Another of the peroxisome enzymes, catalase, utilizes the hydrogen peroxide as a hydrogen acceptor in the oxidation of substrates such as ethanol or lactic acid. The rate of reduction of oxygen by the peroxisomes increases proportionately with an increase in oxygen concentration, so that an excessive amount of oxygen in the cell increases the rate of its reduction by peroxisomes.

Christian de Duve of Rockefeller University has suggested that the peroxisomes represent a primitive enzyme system that evolved to cope with oxygen when it first appeared in the atmosphere. The peroxisome enzymes enabled the first oxidatively metabolizing cells to use oxygen as a hydrogen acceptor and so reoxidize the reduced products of fermentation. In some respects this process is similar to the oxidative reactions of the mitochondria. Both make further dehydrogenation possible by liberating oxidized hydrogen carriers. The basic difference between the mitochondrial oxidation reactions and those of peroxisomes is that in peroxisomes the steps of oxidation are not coupled to the synthesis of ATP. The energy released in the peroxisomes is thus lost to the cell; the function of the organelle is primarily to protect against the destructive effects of free molecular oxygen.

Oxygen dissolved in water can diffuse

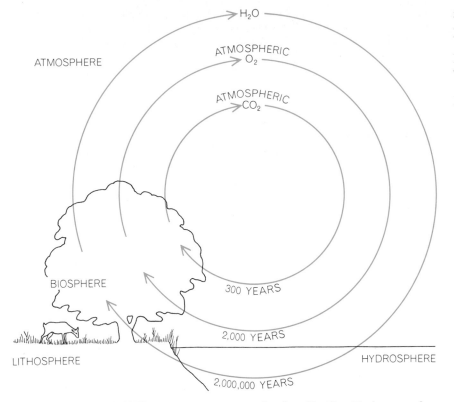

BIOSPHERE EXCHANGES water vapor, oxygen and carbon dioxide with the atmosphere and hydrosphere in a continuing cycle, shown here in simplified form. All the earth's water is split by plant cells and reconstituted by animal and plant cells about every two million years. Oxygen generated in the process enters the atmosphere and is recycled in about 2,000 years. Carbon dioxide respired by animal and plant cells enters the atmosphere and is fixed again by plant cells after an average atmospheric residence time of about 300 years.

across both the inner and the outer membranes of the cell, and the supply of oxygen by diffusion is adequate for single cells and for organisms consisting of small colonies of cells. Differentiated multicellular organisms, however, require more efficient modes of supplying oxygen to tissues and organs. Since all higher organisms depend primarily on mitochondrial aerobic oxidation to generate the energy that maintains their active mode of life, they have evolved elaborate systems to ensure their tissues an adequate supply of oxygen, the gas that once was lethal (and still is, in excess). Two basic devices serve this purpose: special chemical carriers that increase the oxygen capacity of body fluids, and anatomical structures that provide relatively large surfaces for the rapid exchange of gases. The typical properties of an oxygen carrier are exemplified by those of hemoglobin and of myoglobin, or muscle hemoglobin. Hemoglobin in blood readily absorbs oxygen to near-saturation at oxygen pressures such as those found in the lung. When the blood is exposed to lower oxygen pressures as it moves from the lungs to other tissues, the hemoglobin discharges most of its bound oxygen. Myoglobin, which acts as

a reservoir to meet the sharp demand for oxygen in muscle contraction, gives up its oxygen more rapidly. Such reversible bonding of oxygen in response to changes in oxygen pressure is an essential property of biochemical oxygen carriers.

Lungs and gills are examples of anatomical structures in which large wet areas of thin membranous tissue come in contact with oxygen. Body fluids are pumped over one side of these membranes and air, or water containing oxygen, over the other side. This ensures a rapid gas exchange between large volumes of body fluid and the environment.

How did the relations between organisms and gaseous oxygen happen to evolve in such a curiously complicated manner? The atmosphere under which life arose on the earth was almost certainly devoid of free oxygen. The low concentration of noble gases such as neon and krypton in the terrestrial atmosphere compared with their cosmic abundance, together with other geochemical evidence, indicates that the terrestrial atmosphere had a secondary origin in volcanic outgassing from the earth's interior. Oxygen is not known among the gases so released, nor is it

found as inclusions in igneous rocks. The chemistry of rocks older than about two billion years is also inconsistent with the presence of more than trivial quantities of free atmospheric oxygen before that time. Moreover, it would not have been possible for the essential chemical precursors of life—or life itself—to have originated and persisted in the presence of free oxygen before the evolution of suitable oxygen-mediating enzymes.

On such grounds we conclude that the first living organism must have depended on fermentation for its livelihood. Organic substances that originated in nonvital reactions served as substrates for these primordial fermentations. The first organism, therefore, was not only an anaerobe; it was also a heterotroph, dependent on a preexisting organic food supply and incapable of manufacturing its own food by photosynthesis or other autotrophic processes.

The emergence of an autotroph was an essential step in the onward march of biological evolution. This evolutionary step left its mark in the rocks as well as on all living forms. Some fated eobiont, as we may call these early life forms whose properties we can as yet only imagine, evolved and became an autotroph, an organism capable of manufacturing its own food. Biogeological evidence suggests that this critical event may have occurred more than three billion years ago.

If, as seems inescapable, the first autotrophic eobiont was also anaerobic, it would have encountered difficulty when it first learned to split water and release free oxygen. John M. Olson of the Brookhaven National Laboratory recently suggested biochemical arguments to support the idea that primitive photosynthesis may have obtained electrons from substances other than water. He argues that large-scale splitting of water and release of oxygen may have been delayed until the evolution of appropriate enzymes to detoxify this reactive substance.

We nevertheless find a long record of oxidized marine sediments of a peculiar type that precedes the first evidence of atmospheric oxygen in rocks about 1.8 billion years old; we do not find them in significant amounts in more recent strata. These oxidized marine sediments, known as banded iron formations, are alternately iron-rich and iron-poor chemical sediments that were laid down in open bodies of water. Much of the iron in them is ferric (the oxidized form, Fe^{+++}) rather than ferrous (the reduced form, Fe^{++}), implying that there was a source of oxygen in the column of water above them. Considering the

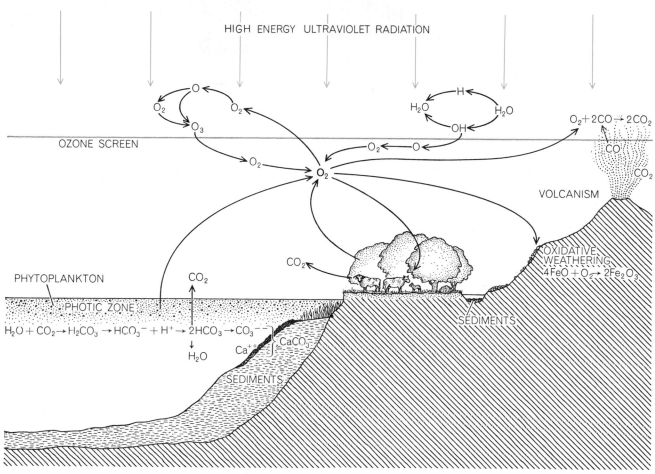

OXYGEN CYCLE is complicated because oxygen appears in so many chemical forms and combinations, primarily as molecular oxygen (O_2), in water and in organic and inorganic compounds. Some global pathways of oxygen are shown here in simplified form.

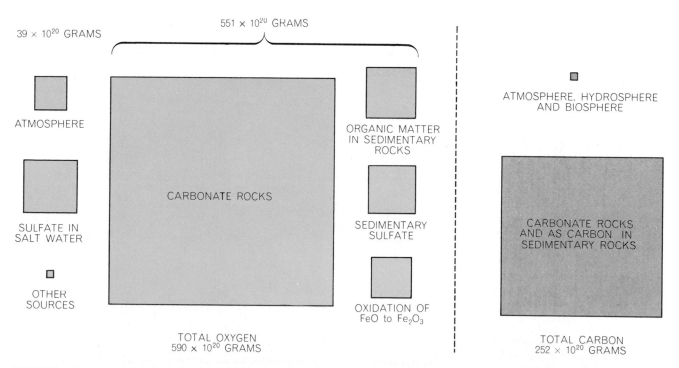

OXYGEN-CARBON BALANCE SHEET suggests that photosynthesis can account not only for all the oxygen in the atmosphere but also for the much larger amount of "fossil" oxygen, mostly in compounds in sediments. The diagram, based on estimates made by William W. Rubey, indicates that the elements are present in about the proportion, 12/32, that would account for their derivation through photosynthesis from carbon dioxide (one atom of carbon, molecular weight 12, to two of oxygen, molecular weight 16).

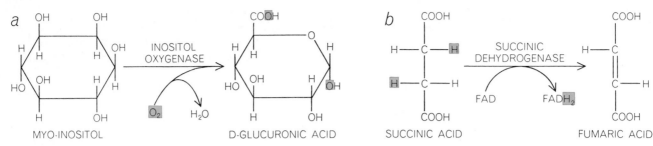

OXIDATION involves a decrease in the number of hydrogen atoms in a molecule or an increase in the number of oxygen atoms. It may be accomplished in several ways. In oxygenation (*a*) oxygen is added directly. In dehydrogenation (*b*) hydrogen is re-

problems that would face a water-splitting photosynthesizer before the evolution of advanced oxygen-mediating enzymes such as oxidases and catalases, one can visualize how the biological oxygen cycle may have interacted with ions in solution in bodies of water during that time. The first oxygen-releasing photoautotrophs may have used ferrous compounds in solution as oxygen acceptors—oxygen for them being merely a toxic waste product. This would have precipitated iron in the ferric form ($4FeO + O_2 \rightarrow 2Fe_2O_3$) or in the ferro-ferric form (Fe_3O_4). A recurrent imbalance of supply and demand might then account for the cyclic nature and differing types of the banded iron formations.

Once advanced oxygen-mediating enzymes arose, oxygen generated by increasing populations of photoautotrophs containing these enzymes would build up in the oceans and begin to escape into the atmosphere. There the ultraviolet component of the sun's radiation would dissociate some of the molecular oxygen into highly reactive atomic oxygen and also give rise to equally reactive ozone. Atmospheric oxygen and its reactive derivatives (even in small quantities) would lead to the oxidation of iron in sediments produced by the weathering of rocks, to the greatly reduced solubility of iron in surface waters (now oxygenated), to the termination of the banded iron formations as an important sedimentary type and to the extensive formation of continental red beds rich in ferric iron [*see illustration on page 68*]. The record of the rocks supports this succession of events: red beds are essentially restricted to rocks younger than about 1.8 billion years, whereas banded iron formation is found only in older rocks.

So far we have assumed that oxygen accumulated in the atmosphere as a consequence of photosynthesis by green plants. How could this happen if the entire process of photosynthesis and respiration is cyclic, representable by the reversible equation $CO_2 + H_2O + energy$

$\rightleftharpoons CH_2O + O_2$? Except to the extent that carbon or its compounds are somehow sequestered, carbohydrates produced by photosynthesis will be reoxidized back to carbon dioxide and water, and no significant quantity of free oxygen will accumulate. The carbon that is sequestered in the earth as graphite in the oldest rocks and as coal, oil, gas and other carbonaceous compounds in the younger ones, and in the living and dead bodies of plants and animals, is the

equivalent of the oxygen in oxidized sediments and in the earth's atmosphere! In attempting to strike a carbon-oxygen balance we must find enough carbon to account not only for the oxygen in the present atmosphere but also for the "fossil" oxygen that went into the conversion of ferrous oxides to ferric oxides, sulfides to sulfates, carbon monoxide to carbon dioxide and so on.

Interestingly, rough estimates made some years ago by William W. Rubey,

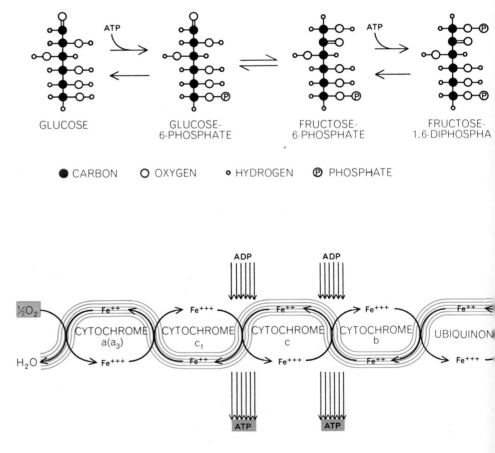

OXIDATIVE METABOLISM provides the energy that powers all higher forms of life. It proceeds in two phases: glycolysis (*top*), an anaerobic phase that does not require oxygen, and aerobic respiration (*bottom*), which requires oxygen. In glycolysis (or fermentation, the anaerobic process by which organisms such as yeast derive their energy) a molecule of the six-carbon sugar glucose is broken down into two molecules of the three-carbon compound pyruvic acid with a net gain of two molecules of adenosine triphosphate, the cellular

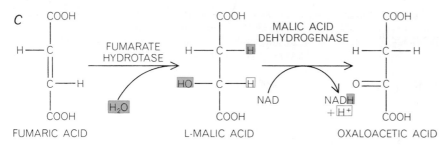

FUMARIC ACID   L-MALIC ACID   OXALOACETIC ACID

moved. In hydration-dehydrogenation (c) water is added and hydrogen is removed. Oxygenation does not occur in respiration, in which oxygen serves only as a hydrogen acceptor.

now of the University of California at Los Angeles, do imply an approximate balance between the chemical combining equivalents of carbon and oxygen in sediments, the atmosphere, the hydrosphere and the biosphere [see bottom illustration on page 73]. The relatively small excess of carbon in Rubey's estimates could be accounted for by the oxygen used in converting carbon monoxide to carbon dioxide. Or it might be due to an underestimate of the quantities of sulfate ion or ferric oxide in sediments. (Rubey's estimates could not include large iron formations recently discovered in western Australia and elsewhere.) The carbon dioxide in carbonate rocks does not need to be accounted for, but the oxygen involved in converting it to carbonate ion does. The recycling of sediments through metamorphism, mountain-building and the movement of ocean-floor plates under the continents is a variable of unknown dimensions, but

it probably does not affect the approximate balance observed in view of the fact that the overwhelmingly large pools to be balanced are all in the lithosphere and that carbon and oxygen losses would be roughly equivalent. The small amounts of oxygen dissolved in water are not included in this balance.

Nonetheless, water does enter the picture. Another possible source of oxygen in our atmosphere is photolysis, the ultraviolet dissociation of water vapor in the outer atmosphere followed by the escape of the hydrogen from the earth's gravitational field. This has usually been regarded as a trivial source, however. Although R. T. Brinkmann of the California Institute of Technology has recently argued that nonbiological photolysis may be a major source of atmospheric oxygen, the carbon-oxygen balance sheet does not support that belief, which also runs into other difficulties.

When free oxygen began to accumulate in the atmosphere some 1.8 billion years ago, life was still restricted to sites

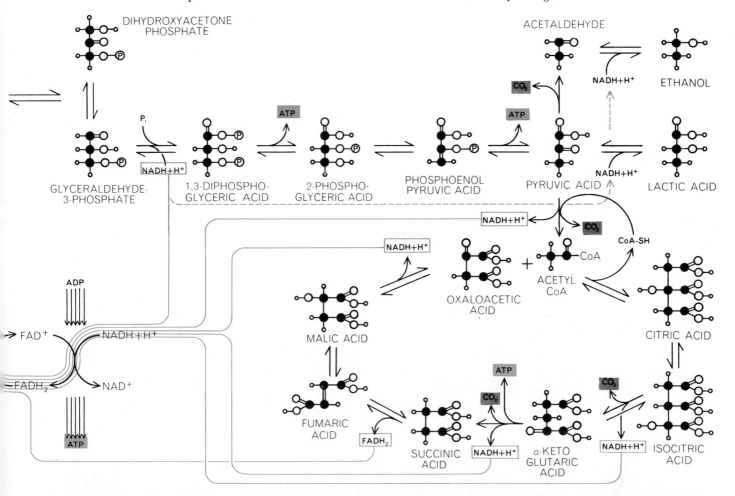

energy carrier. The pyruvic acid is converted into lactic acid in animal cells deprived of oxygen and into some other compound, such as ethanol, in fermentation. In aerobic cells in the presence of oxygen, however, pyruvic acid is completely oxidized to produce carbon dioxide and water. In the process hydrogen ions are removed. The electrons of these hydrogens (and of two removed in glycolysis) are passed along by two electron carriers, nicotinamide adenine dinucleotide (NAD) and flavin adenine dinucleotide (FAD), to a chain of respiratory enzymes, ubiquinone and the cytochromes, which are alternately reduced and oxidized. Energy released in the reactions is coupled to synthesis of ATP, 38 molecules of which are produced for every molecule of glucose consumed.

shielded from destructive ultraviolet radiation by sufficient depths of water or by screens of sediment. In time enough oxygen built up in the atmosphere for ozone, a strong absorber in the ultraviolet, to form a shield against incoming ultraviolet radiation. The late Lloyd V. Berkner and Lauriston C. Marshall of the Graduate Research Center of the Southwest in Dallas calculated that only 1 percent of the present atmospheric level of oxygen would give rise to a sufficient level of ozone to screen out the most deleterious wavelengths of the ultraviolet radiation. This also happens to be the level of oxygen at which Pasteur found that certain microorganisms switch over from a fermentative type of metabolism to an oxidative one. Berkner and Marshall therefore jumped to the conclusion (reasonably enough on the evidence they considered) that this was the stage at which oxidative metabolism arose. They related this stage to the first appearance of metazoan life somewhat more than 600 million years ago.

The geological record has long made it plain, however, that free molecular oxygen existed in the atmosphere well before that relatively late date in geo-logic time. Moreover, recent evidence is consistent with the origin of oxidative metabolism at least twice as long ago. Eucaryotic cells—cells with organized nuclei and other organelles—have been identified in rocks in eastern California that are believed to be about 1.3 billion years old [*see top illustration on page 70*]. Since all living eucaryotes depend on oxidative metabolism, it seems likely that these ancestral forms did too. The oxygen level may nonetheless have still been quite low at this stage. Simple diffusion would suffice to move enough oxygen across cell boundaries and within the cell, even at very low concentrations, to supply the early oxidative metabolizers. A higher order of organization and of atmospheric oxygen was required, however, for advanced oxidative metabolism. Perhaps that is why, although the eucaryotic cell existed at least 1.2 billion years ago, we have no unequivocal fossils of metazoan organisms from rocks older than perhaps 640 million years.

In other words, perhaps Berkner and Marshall were mistaken only in trying to make the appearance of the Metazoa coincide with the onset of oxidative metabolism. Once the level of atmospheric oxygen was high enough to generate an effective ozone screen, photosynthetic organisms would have been able to spread throughout the surface waters of the sea, greatly accelerating the rate of oxygen production. The plausible episodes in geological history to correlate with this development are the secondary oxidation of the banded iron formations and the appearance of sedimentary calcium sulfate (gypsum and anhydrite) on a large scale. These events occurred just as or just before the Metazoa first appeared in early Paleozoic time. The attainment of a suitable level of atmospheric oxygen may thus be correlated with the emergence of metazoan root stocks from premetazoan ancestors beginning about 640 million years ago. The fact that oxygen could accumulate no faster than carbon (or hydrogen) was removed argues against the likelihood of a rapid early buildup of oxygen.

That subsequent biospheric and atmospheric evolution were closely interlinked can now be taken for granted. What is not known are the details. Did oxygen levels in the atmosphere increase steadily throughout geologic time, marking regular stages of biological evolution such as the emergence of land plants, of

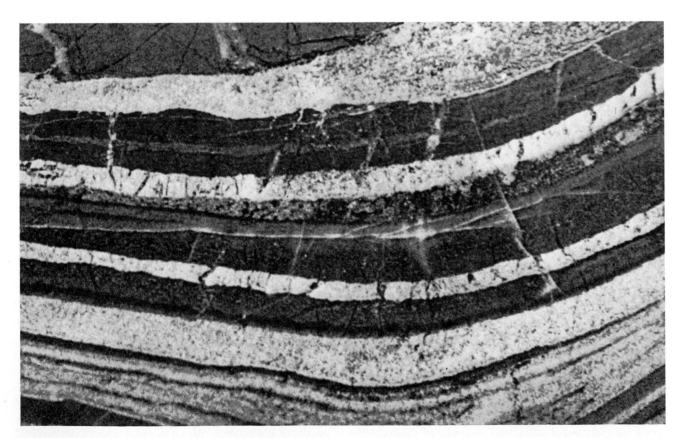

BANDED IRON FORMATION provides the first geological evidence of free oxygen in the hydrosphere. The layers in this polished cross section result from an alternation of iron-rich and iron-poor depositions. This sample from the Soudan Iron Formation in Minnesota is more than 2.7 billion years old. The layers, originally horizontal, were deformed while soft and later metamorphosed.

insects, of the various vertebrate groups and of flowering plants, as Berkner and Marshall suggested? Or were there wide swings in the oxygen level? Did oxygen decrease during great volcanic episodes, as a result of the oxidation of newly emitted carbon monoxide to carbon dioxide, or during times of sedimentary sulfate precipitation? Did oxygen increase when carbon was being sequestered during times of coal and petroleum formation? May there have been fluctuations in both directions as a result of plant and animal evolution, of phytoplankton eruptions and extinctions and of the extent and type of terrestrial plant cover? Such processes and events are now being seriously studied, but the answers are as yet far from clear.

What one can say with confidence is that success in understanding the oxy-

YEARS BEFORE PRESENT	LITHOSPHERE	BIOSPHERE	HYDROSPHERE	ATMOSPHERE
20 MILLION	GLACIATION	MAMMALS DIVERSIFY GRASSES APPEAR		OXYGEN APPROACHES PRESENT LEVEL
50 MILLION				
	COAL FORMATION VOLCANISM			
100 MILLION		SOCIAL INSECTS, FLOWERING PLANTS MAMMALS		ATMOSPHERIC OXYGEN INCREASES AT FLUCTUATING RATE
200 MILLION	GREAT VOLCANISM COAL FORMATION		OCEANS CONTINUE TO INCREASE IN VOLUME	
		INSECTS APPEAR LAND PLANTS APPEAR		
500 MILLION		METAZOA APPEAR RAPID INCREASE IN PHYTOPLANKTON		OXYGEN AT 3-10 PERCENT OF PRESENT ATMOSPHERIC LEVEL
	GLACIATION SEDIMENTARY CALCIUM SULFATE		SURFACE WATERS OPENED TO PHYTOPLANKTON	OXYGEN AT 1 PERCENT OF PRESENT ATMOSPHERIC LEVEL, OZONE SCREEN EFFECTIVE
1 BILLION	VOLCANISM			
		EUCARYOTES		OXYGEN INCREASING, CARBON DIOXIDE DECREASING
2 BILLION	RED BEDS	ADVANCED OXYGEN-MEDIATING ENZYMES	OXYGEN DIFFUSES INTO ATMOSPHERE	OXYGEN IN ATMOSPHERE
	GLACIATION BANDED IRON FORMATIONS OLDEST SEDIMENTS OLDEST EARTH ROCKS	FIRST OXYGEN-GENERATING PHOTOSYNTHETIC CELLS PROCARYOTES ABIOGENIC EVOLUTION	START OF OXYGEN GENERATION WITH FERROUS IRON AS OXYGEN SINK	
5 BILLION	(ORIGIN OF SOLAR SYSTEM)			NO FREE OXYGEN

CHRONOLOGY that interrelates the evolutions of atmosphere and biosphere is gradually being established from evidence in the geological record and in fossils. According to calculations by Lloyd V. Berkner and Lauriston C. Marshall, when oxygen in the atmosphere reached 1 percent of the present atmospheric level, it provided enough ozone to filter out the most damaging high-energy ultraviolet radiation so that phytoplankton could survive everywhere in the upper, sunlit layers of the seas. The result may have been a geometric increase in the amount of photosynthesis in the oceans that, if accompanied by equivalent sequestration of carbon, might have resulted in a rapid buildup of atmospheric oxygen, leading in time to the evolution of differentiated multicelled animals.

gen cycle of the biosphere in truly broad terms will depend on how good we are at weaving together the related strands of biospheric, atmospheric, hydrospheric and lithospheric evolution throughout geologic time. Whatever we may conjecture about any one of these processes must be consistent with what is known about the others. Whereas any one line of evidence may be weak in itself, a number of lines of evidence, taken together and found to be consistent, reinforce one another exponentially. This synergistic effect enhances our confidence in the proposed time scale linking the evolution of oxygen in the atmosphere and the management of the gaseous oxygen budget within the biosphere [see illustration on page 77].

The most recent factor affecting the oxygen cycle of the biosphere and the oxygen budget of the earth is man himself. In addition to inhaling oxygen and exhaling carbon dioxide as a well-behaved animal does, man decreases the oxygen level and increases the carbon dioxide level by burning fossil fuels and paving formerly green land. He is also engaged in a vast but unplanned experiment to see what effects oil spills and an array of pesticides will have on the world's phytoplankton. The increase in the albedo, or reflectivity, of the earth as a result of covering its waters with a molecule-thick film of oil could also affect plant growth by lowering the temperature and in other unforeseen ways. Reductions in the length of growing seasons and in green areas would limit terrestrial plant growth in the middle latitudes. (This might normally be counterbalanced by increased rainfall in the lower latitudes, but a film of oil would also reduce evaporation and therefore rainfall.) Counteracting such effects, man moves the earth's fresh water around to increase plant growth and photosynthesis in arid and semiarid regions. Some of this activity, however, involves the mining of ground water, thereby favoring processes that cause water to be returned to the sea at a faster rate than evaporation brings it to the land.

He who is willing to say what the final effects of such processes will be is wiser or braver than we are. Perhaps the effects will be self-limiting and self-correcting, although experience should warn us not to gamble on that. Oxygen in the atmosphere might be reduced several percent below the present level without adverse effects. A modest increase in the carbon dioxide level might enhance plant growth and lead to a corresponding increase in the amount of oxygen. Will a further increase in carbon dioxide also have (or renew) a "greenhouse effect," leading to an increase in temperature (and thus to a rising sea level)? Or will such effects be counterbalanced or swamped by the cooling effects of particulate matter in the air or by increased albedo due to oil films? It is anyone's guess. (Perhaps we should be more alarmed about a possible decrease of atmospheric carbon dioxide, on which all forms of life ultimately depend, but the sea contains such vast amounts that it can presumably keep carbon dioxide in the atmosphere balanced at about the present level for a long time to come.) The net effect of the burning of fossil fuels may in the long run be nothing more than a slight increase (or decrease?) in the amount of limestone deposited. In any event the recoverable fossil fuels whose combustion releases carbon dioxide are headed for depletion in a few more centuries, and then man will have other problems to contend with.

What we want to stress is the indivisibility and complexity of the environment. For example, the earth's atmosphere is so thoroughly mixed and so rapidly recycled through the biosphere that the next breath you inhale will contain atoms exhaled by Jesus at Gethsemane and by Adolf Hitler at Munich. It will also contain atoms of radioactive strontium 90 and iodine 131 from atomic explosions and gases from the chimneys and exhaust pipes of the world. Present environmental problems stand as a grim monument to the cumulatively adverse effects of actions that in themselves were reasonable enough but that were taken without sufficient thought to their consequences. If we want to ensure that the biosphere continues to exist over the long term and to have an oxygen cycle, each new action must be matched with an effort to foresee its consequences throughout the ecosystem and to determine how they can be managed favorably or avoided. Understanding also is needed, and we are woefully short on that commodity. This means that we must continue to probe all aspects of the indivisible global ecosystem and its past, present and potential interactions. That is called basic research, and basic research at this critical point in history is gravely endangered by new crosscurrents of anti-intellectualism.

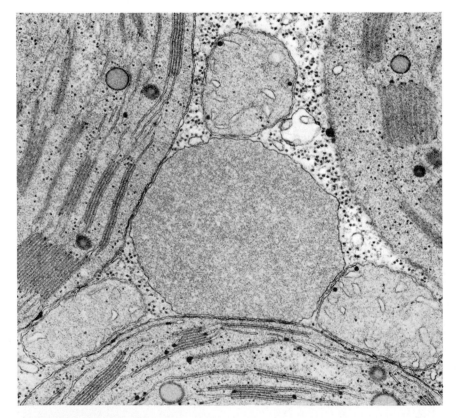

THREE ORGANELLES that are involved in oxygen metabolism in the living cell are enlarged 40,000 diameters in an electron micrograph of a tobacco leaf cell made by Sue Ellen Frederick in the laboratory of Eldon H. Newcomb at the University of Wisconsin. A peroxisome (center) is surrounded by three mitochondria and three chloroplasts. Oxygen is produced in the grana (layered objects) in the chloroplasts and is utilized in aerobic respiration in the mitochondria. Peroxisomes contain enzymes involved in oxygen metabolism.

7

The Carbon Cycle

INTRODUCTION

The counterpart to the oxygen cycle is the carbon cycle, with which it is inextricably bound, the major difference between the two being that oxygen is abundant, comprising 1/5 of the atmosphere, while carbon dioxide makes up only about 320 parts per million. Despite this small amount in the atmospheric reservoir, carbon dioxide exerts a strong influence on the surface environment of the earth. Absorption of infrared radiation by carbon dioxide affects the heat balance of the earth and so modifies climate by the greenhouse effect. Hence the concern expressed by Bert Bolin, in this article, "The Carbon Cycle," about the levels of atmospheric carbon dioxide that will be reached as we continue to burn fossil fuels. By burning fossil carbon, we are pumping carbon dioxide into the atmosphere at an enormously faster rate than natural processes did before the Industrial Revolution. How much this excess will contribute to a future warming of the globe, which might trigger the melting of polar ice caps, remains an important subject of discussion for scientists who will be monitoring the surface environment of the earth for the next generation.

To understand the carbon cycle and how it is affected by man, we must know the varied forms of carbon, their reservoirs on the surface and buried beneath it, the flow of material between reservoirs, and the geological, biological, and chemical processes that operate that flow. The relationships among plants and animals, organic matter, and soil constitute a part of the cycle that Bolin uses to show how movements of carbon change from day to night and during the year. In this and the other large reservoir in the whole process, the sea, the utilization of carbon is shown to be limited by other biological nutrients, mainly phosphorus and nitrogen. This effect has been a central part of discussions about control of algae in lakes that have been polluted by excess phosphorus. The carbon cycle has so many ramifications that only a few major points could be discussed in a short article. There is much that we still need to know: more accurate numbers that will enable us to make better predictions; greater knowledge of subcycles; awareness of the implications of the carbon cycle for the oxygen, nitrogen, and phosphorus cycles. The carbon cycle will undoubtedly continue to be a focus of scientific attention.

The Carbon Cycle

by Bert Bolin
September 1970

*The main cycle is from carbon dioxide to living matter
and back to carbon dioxide. Some of the carbon,
however, is removed by a slow epicycle that stores
huge inventories in sedimentary rocks*

The biosphere contains a complex mixture of carbon compounds in a continuous state of creation, transformation and decomposition. This dynamic state is maintained through the ability of phytoplankton in the sea and plants on land to capture the energy of sunlight and utilize it to transform carbon dioxide (and water) into organic molecules of precise architecture and rich diversity. Chemists and molecular biologists have unraveled many of the intricate processes needed to create the microworld of the living cell. Equally fundamental and no less interesting is the effort to grasp the overall balance and flow of material in the worldwide community of plants and animals that has developed in the few billion years since life began. This is ecology in the broadest sense of the word: the complex interplay between, on the one hand, communities of plants and animals and, on the other, both kinds of community and their nonliving environment.

We now know that the biosphere has not developed in a static inorganic environment. Rather the living world has profoundly altered the primitive lifeless earth, gradually changing the composition of the atmosphere, the sea and the top layers of the solid crust, both on land and under the ocean. Thus a study of the carbon cycle in the biosphere is fundamentally a study of the overall global interactions of living organisms and their

physical and chemical environment. To bring order into this world of complex interactions biologists must combine their knowledge with the information available to students of geology, oceanography and meteorology.

The engine for the organic processes that reconstructed the primitive earth is photosynthesis. Regardless of whether it takes place on land or in the sea, it can be summarized by a single reaction: $CO_2 + 2H_2A + light \rightarrow CH_2O + H_2O + 2A + energy$. The formaldehyde molecule CH_2O symbolizes the simplest organic compound; the term "energy" indicates that the reaction stores energy in chemical form. H_2A is commonly water (H_2O), in which case 2A symbolizes the release of free oxygen (O_2). There are, however, bacteria that can use compounds in which A stands for sulfur, for some organic radical or for nothing at all.

Organisms that are able to use carbon dioxide as their sole source of carbon are known as autotrophs. Those that use light energy for reducing carbon dioxide are called phototrophic, and those that use the energy stored in inorganic chemical bonds (for example the bonds of nitrates and sulfates) are called chemolithotrophic. Most organisms, however, require preformed organic molecules for growth; hence they are known as heterotrophs. The nonsulfur bacteria are an unusual group that is both photosyn-

thetic and heterotrophic. Chemoheterotrophic organisms, for example animals, obtain their energy from organic compounds without need for light. An organism may be either aerobic or anaerobic regardless of its source of carbon or energy. Thus some anaerobic chemoheterotrophs can survive in the deep ocean and deep lakes in the total absence of light or free oxygen.

There is more to plant life than the creation of organic compounds by photosynthesis. Plant growth involves a series of chemical processes and transformations that require energy. This energy is obtained by reactions that use the oxygen in the surrounding water and air to unlock the energy that has been stored by photosynthesis. The process, which releases carbon dioxide, is termed respiration. It is a continuous process and is therefore dominant at night, when photosynthesis is shut down.

If one measures the carbon dioxide at various levels above the ground in a forest, one can observe pronounced changes in concentration over a 24-hour period [*see top illustration on page 83*]. The average concentration of carbon dioxide in the atmosphere is about 320 parts per million. When the sun rises, photosynthesis begins and leads to a rapid decrease in the carbon dioxide concentration as leaves (and the needles of conifers) convert carbon dioxide into organic compounds. Toward noon, as the temperature increases and the humidity decreases, the rate of respiration rises and the net consumption of carbon dioxide slowly declines. Minimum values of carbon dioxide 10 to 15 parts per million below the daily average are reached around noon at treetop level. At sunset photosynthesis ceases while respiration continues, with the result that the carbon dioxide concentration close to the

CARBON LOCKED IN COAL and oil exceeds by a factor of about 50 the amount of carbon in all living organisms. The estimated world reserves of coal alone are on the order of 7,500 billion tons. The photograph on the opposite page shows a sequence of lignite coal seams being strip-mined in Stanton, N.D., by the Western Division of the Consolidation Coal Company. The seam, about two feet thick, is of low quality and is discarded. The second seam from the top, about three feet thick, is marketable, as is the third seam, 10 feet farther down. This seam is really two seams separated by about 10 inches of gray clay. The upper is some 3½ feet thick; the lower is about two feet thick. Twenty-four feet below the bottom of this seam is still another seam (*not shown*) eight feet thick, which is also mined.

ground may exceed 400 parts per million. This high value reflects partly the release of carbon dioxide from the decomposition of organic matter in the soil and partly the tendency of air to stagnate near the ground at night, when there is no solar heating to produce convection currents.

The net productivity, or net rate of fixation, of carbon dioxide varies greatly from one type of vegetation to another. Rapidly growing tropical rain forests annually fix between one kilogram and two kilograms of carbon (in the form of carbon dioxide) per square meter of land surface, which is roughly equal to the amount of carbon dioxide in a column of air extending from the same area of the earth's surface to the top of the atmosphere. The arctic tundra and the nearly barren regions of the desert may fix as

little as 1 percent of that amount. The forests and cultivated fields of the middle latitudes assimilate between .2 and .4 kilogram per square meter. For the earth as a whole the areas of high productivity are small. A fair estimate is that the land areas of the earth fix into organic compounds 20 to 30 billion net metric tons of carbon per year. There is considerable uncertainty in this figure; published estimates range from 10 to 100 billion tons.

The amount of carbon in the form of carbon dioxide consumed annually by phytoplankton in the oceans is perhaps 40 billion tons, or roughly the same as the gross assimilation of carbon dioxide by land vegetation. Both the carbon dioxide consumed and the oxygen released are largely in the form of gas dissolved near the ocean surface. Therefore most of the carbon cycle in the sea is self-con-

tained: the released oxygen is consumed by sea animals, and their ultimate decomposition releases carbon dioxide back into solution. As we shall see, however, there is a dynamic exchange of carbon dioxide (and oxygen) between the atmosphere and the sea, brought about by the action of the wind and waves. At any given moment the amount of carbon dioxide dissolved in the surface layers of the sea is in close equilibrium with the concentration of carbon dioxide in the atmosphere as a whole.

The carbon fixed by photosynthesis on land is sooner or later returned to the atmosphere by the decomposition of dead organic matter. Leaves and litter fall to the ground and are oxidized by a series of complicated processes in the soil. We can get an approximate idea of the rate at which organic matter in the soil is

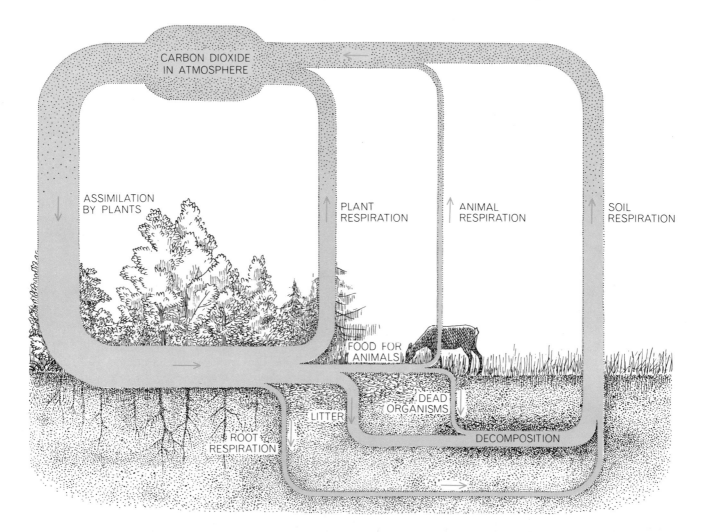

CARBON CYCLE begins with the fixation of atmospheric carbon dioxide by the process of photosynthesis, conducted by plants and certain microorganisms. In this process carbon dioxide and water react to form carbohydrates, with the simultaneous release of free oxygen, which enters the atmosphere. Some of the carbohydrate is directly consumed to supply the plant with energy; the carbon dioxide so generated is released either through the plant's leaves or through its roots. Part of the carbon fixed by plants is consumed by animals, which also respire and release carbon dioxide. Plants and animals die and are ultimately decomposed by microorganisms in the soil; the carbon in their tissues is oxidized to carbon dioxide and returns to the atmosphere. The widths of the pathways are roughly proportional to the quantities involved. A similar carbon cycle takes place within the sea. There is still no general agreement as to which of the two cycles is larger. The author's estimates of the quantities involved appear in the flow chart on page 86.

being transformed by measuring its content of the radioactive isotope carbon 14. At the time carbon is fixed by photosynthesis its ratio of carbon 14 to the non-radioactive isotope carbon 12 is the same as the ratio in the atmosphere (except for a constant fractionation factor), but thereafter the carbon 14 decays and becomes less abundant with respect to the carbon 12. Measurements of this ratio yield rates for the oxidation of organic matter in the soil ranging from decades in tropical soils to several hundred years in boreal forests.

In addition to the daily variations of carbon dioxide in the air there is a marked annual variation, at least in the Northern Hemisphere. As spring comes to northern regions the consumption of carbon dioxide by plants greatly exceeds the return from the soil. The increased withdrawal of carbon dioxide can be measured all the way up to the lower stratosphere. A marked decrease in the atmospheric content of carbon dioxide occurs during the spring. From April to September the atmosphere north of 30 degrees north latitude loses nearly 3 percent of its carbon dioxide content, which is equivalent to about four billion tons of carbon [see bottom illustration at right]. Since the decay processes in the soil go on simultaneously, the net withdrawal of four billion tons implies an annual gross fixation of carbon in these latitudes of at least five or six billion tons. This amounts to about a fourth of the annual terrestrial productivity referred to above (20 to 30 billion tons), which was based on a survey of carbon fixation. In this global survey the estimated contribution from the Northern Hemisphere, where plant growth shows a marked seasonal variation, constituted about 25 percent of the total tonnage. Thus two independent estimates of worldwide carbon fixation on land show a quite satisfactory agreement.

The forests of the world not only are the main carbon dioxide consumers on land; they also represent the main reservoir of biologically fixed carbon (except for fossil fuels, which have been largely removed from the carbon cycle save for the amount reintroduced by man's burning of it). The forests contain between 400 and 500 billion tons of carbon, or roughly two-thirds of the amount present as carbon dioxide in the atmosphere (700 billion tons). The figure for forests can be estimated only approximately. The average age of a tree can be assumed to be about 30 years, which implies that about 15 billion tons of carbon

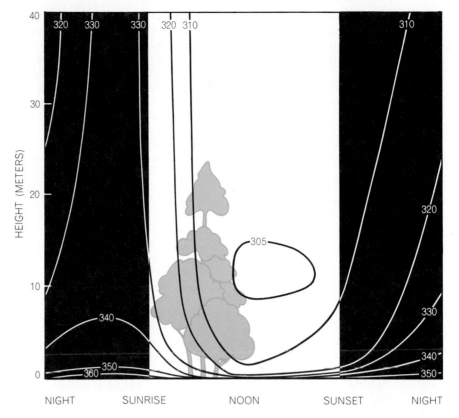

VERTICAL DISTRIBUTION OF CARBON DIOXIDE in the air around a forest varies with time of day. At night, when photosynthesis is shut off, respiration from the soil can raise the carbon dioxide at ground level to as much as 400 parts per million (ppm). By noon, owing to photosynthetic uptake, the concentration at treetop level can drop to 305 ppm.

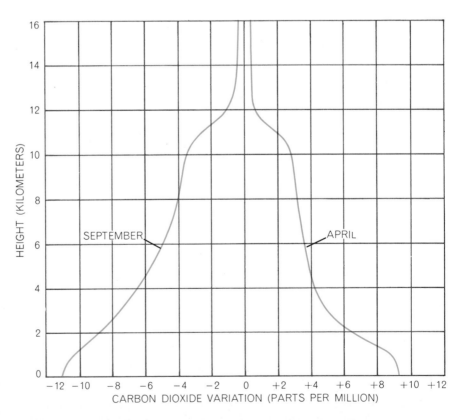

SEASONAL VARIATIONS in the carbon dioxide content of the atmosphere reach a maximum in September and April for the region north of 30 degrees north latitude. The departure from a mean value of about 320 ppm varies with altitude as shown by these two curves.

in the form of carbon dioxide is annually transformed into wood, which seems reasonable in comparison with a total annual assimilation of 20 to 30 billion tons.

The pattern of carbon circulation in the sea is quite different from the pattern on land. The productivity of the soil is mostly limited by the availability of fresh water and phosphorus, and only to a degree by the availability of other nutrients in the soil. In the oceans the overriding limitation is the availability of inorganic substances. The phytoplankton require not only plentiful supplies of phosphorus and nitrogen but also trace amounts of various metals, notably iron.

The competition for food in the sea is so keen that organisms have gradually developed the ability to absorb essential minerals even when these nutrients are available only in very low concentration. As a result high concentrations of nutrients are rarely found in surface waters, where solar radiation makes it possible for photosynthetic organisms to exist. If an ocean area is uncommonly productive, one can be sure that nutrients are supplied from deeper layers. (In limited areas they are supplied by the wastes of human activities.) The most productive waters in the world are therefore near the Antarctic continent, where the deep waters of the surrounding oceans well up and mix with the surface layers. There are similar upwellings along the coast of Chile, in the vicinity of Japan and in the Gulf Stream. In such regions fish are abundant and the maximum annual fixation of carbon approaches .3 kilogram per square meter. In the "desert" areas of the oceans, such

as the open seas of subtropical latitudes, the fixation rate may be less than a tenth of that value. In the Tropics warm surface layers are usually effective in blocking the vertical water exchange needed to carry nutrients up from below.

Phytoplankton, the primary fixers of carbon dioxide in the sea, are eaten by the zooplankton and other tiny animals. These organisms in turn provide food for the larger animals. The major part of the oceanic biomass, however, consists of microorganisms. Since the lifetime of such organisms is measured in weeks, or at most in months, their total mass can never accumulate appreciably. When microorganisms die, they quickly disintegrate as they sink to deeper layers. Soon most of what was once living tissue has become dissolved organic matter.

A small fraction of the organic particulate matter escapes oxidation and settles into the ocean depths. There it profoundly influences the abundance of chemical substances because (except in special regions) the deep layers exchange water with the surface layers very slowly. The enrichment of the deep layers goes hand in hand with a depletion of oxygen. There also appears to be an increase in carbon dioxide (in the form of carbonate and bicarbonate ions) in the ocean depths. The overall distribution of carbon dioxide, oxygen and various minor constituents in the sea reflects a balance between the marine life and its chemical milieu in the surface layers and the slow transport of substances by the general circulation of the ocean. The net effect is to prevent the ocean from becoming saturated with oxygen and to enrich the deeper strata with carbonate and bicarbonate ions.

The particular state in which we find the oceans today could well be quite different if the mechanisms for the exchange of water between the surface layers and the deep ones were either more intense or less so. The present state is determined primarily by the sinking of cold water in the polar regions, particularly the Antarctic. In these regions the water is also slightly saltier, and therefore still denser, because some of it has been frozen out in floating ice. If the climate of the earth were different, the distribution of carbon dioxide, oxygen and minerals might also be quite different. If the difference were large enough, oxygen might completely vanish from the ocean depths, leaving them to be populated only by chemibarotrophic bacteria. (This is now the case in the depths of the Black Sea.)

The time required to establish a new equilibrium in the ocean is determined by the slowest link in the chain of processes that has been described. This link is the oceanic circulation; it seems to take at least 1,000 years for the water in the deepest basins to be completely replaced. One can imagine other conditions of circulation in which the oceans would interact differently with sediments and rocks, producing a balance of substances that one can only guess at.

So far we have been concerned only with the basic biological and ecological processes that provide the mechanisms for circulating carbon through living organisms. Plants on land, with lifetimes measured in years, and phytoplankton in the sea, with lifetimes measured in weeks, are merely the innermost wheels in a biogeochemical machine that embraces the entire earth and that retains important characteristics over much longer time periods. In order to understand such interactions we shall need some rough estimates of the size of the various carbon reservoirs involved and the nature of their contents [see illustration on page 86]. In the context of the present argument the large uncertainties in such estimates are of little significance.

Only a few tenths of a percent of the immense mass of carbon at or near the surface of the earth (on the order of 20×10^{15} tons) is in rapid circulation in the biosphere, which includes the atmosphere, the hydrosphere, the upper portions of the earth's crust and the biomass itself. The overwhelming bulk of near-surface carbon consists of inorganic deposits (chiefly carbonates) and organic fossil deposits (chiefly oil shale, coal and petroleum) that required hundreds of

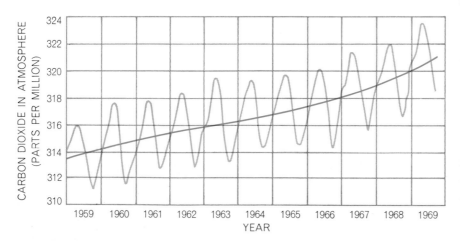

LONG-TERM VARIATIONS in the carbon dioxide content of the atmosphere have been followed at the Mauna Loa Observatory in Hawaii by the Scripps Institution of Oceanography. The sawtooth curve indicates the month-to-month change in concentration since January, 1959. The oscillations reflect seasonal variations in the rate of photosynthesis, as depicted in the bottom illustration on the preceding page. The smooth curve shows the trend.

OIL SHALE is one of the principal sedimentary forms in which carbon has been deposited over geologic time. This photograph, taken at Anvil Points, Colo., shows a section of the Green River Formation, which extends through Colorado, Utah and Wyoming. The formation is estimated to contain the equivalent of more than a trillion barrels of oil in seams containing more than 10 barrels of oil per ton of rock. Of this some 80 billion barrels is considered recoverable. The shale seams are up to 130 feet thick.

WHITE CLIFFS OF DOVER consist of almost pure calcium carbonate, representing the skeletons of phytoplankton that settled to the bottom of the sea over a period of millions of years more than 70 million years ago. The worldwide deposits of limestone, oil shale and other carbon-containing sediments are by far the largest repository of carbon: an estimated 20 quadrillion (10^{15}) tons.

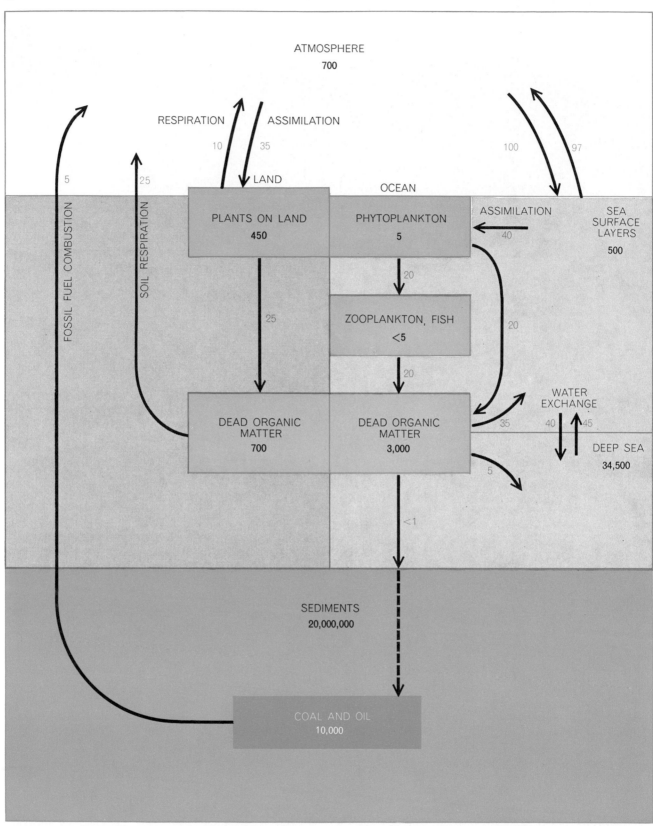

CARBON CIRCULATION IN BIOSPHERE involves two quite distinct cycles, one on land and one in the sea, that are dynamically connected at the interface between the ocean and the atmosphere. The carbon cycle in the sea is essentially self-contained in that phytoplankton assimilate the carbon dioxide dissolved in seawater and release oxygen back into solution. Zooplankton and fish consume the carbon fixed by the phytoplankton, using the dissolved oxygen for respiration. Eventually the decomposition of organic matter replaces the carbon dioxide assimilated by the phytoplank-

ton. All quantities are in billions of metric tons. It will be seen that the combustion of fossil fuels at the rate of about five billion tons per year is sufficient to increase the carbon dioxide in the atmosphere by about .7 percent, equivalent to adding some two parts per million to the existing 320 ppm. Since the observed annual increase is only about .7 ppm, it appears that two-thirds of the carbon dioxide released from fossil fuels is quickly removed from the atmosphere, going either into the oceans or adding to the total mass of terrestrial plants. The estimated tonnages are the author's.

millions of years to reach their present magnitude. Over time intervals as brief as those of which we have been speaking—up to 1,000 years for the deepocean circulation—the accretion of such deposits is negligible. We may therefore consider the life processes on land and in the sea as the inner wheels that spin at comparatively high velocity in the carbon-circulating machine. They are coupled by a very low gear to more majestic processes that account for the overall circulation of carbon in its various geologic and oceanic forms.

We now know that the two great systems, the atmosphere and the ocean, are closely coupled to each other through the transfer of carbon dioxide across the surface of the oceans. The rate of exchange has recently been estimated by measuring the rate at which the radioactive isotope carbon 14 produced by the testing of nuclear weapons has disappeared from the atmosphere. The neutrons released in such tests form carbon 14 by reacting with the nitrogen 14 of the atmosphere. In this reaction a nitrogen atom ($_7N^{14}$) captures a neutron and subsequently releases a proton, yielding $_6C^{14}$. (The subscript before the letter represents the number of protons in the nucleus; the superscript after the letter indicates the sum of protons and neutrons.)

The last major atmospheric tests were conducted in 1963. Sampling at various altitudes and latitudes shows that the constituents of the atmosphere became rather well mixed over a period of a few years. The decline of carbon 14, however, was found to be rapid; it can be explained only by assuming an exchange of atmospheric carbon dioxide, enriched in carbon 14, with the reservoir of much less radioactive carbon dioxide in the sea. The measurements indicate that the characteristic time for the residence of carbon dioxide in the atmosphere before the gas is dissolved in the sea is between five and 10 years. In other words, every year something like 100 billion tons of atmospheric carbon dioxide dissolves in the sea and is replaced by a nearly equivalent amount of oceanic carbon dioxide.

Since around 1850 man has inadvertently been conducting a global geochemical experiment by burning large amounts of fossil fuel and thereby returning to the atmosphere carbon that was fixed by photosynthesis millions of years ago. Currently between five and six billion tons of fossil carbon per year are being released into the atmosphere. This would be enough to increase the amount of carbon dioxide in the air by 2.3 parts per million per year if the carbon dioxide were uniformly distributed and not removed. Within the past century the carbon dioxide content of the atmosphere has risen from some 290 parts per million to 320, with more than a fifth of the rise occurring in just the past decade [*see illustration on page 84*]. The total increase accounts for only slightly more than a third of the carbon dioxide (some 200 billion tons in all) released from fossil fuels. Although most of the remaining two-thirds has presumably gone into the oceans, a significant fraction may well have increased the total amount of vegetation on land. Laboratory studies show that plants grow faster when the surrounding air is enriched in carbon dioxide. Thus it is possible that man is fertilizing fields and forests by burning coal, oil and natural gas. The biomass on land may have increased by as much as 15 billion tons in the past century. There is, however, little concrete evidence for such an increase.

Man has of course been changing his environment in other ways. Over the past century large areas covered with forest have been cleared and turned to agriculture. In such areas the character of soil respiration has undoubtedly changed, producing effects that might have been detectable in the atmospheric content of carbon dioxide if it had not been for the simultaneous increase in the burning of fossil fuels. In any case the dynamic equilibrium among the major carbon dioxide reservoirs in the biomass, the atmosphere, the hydrosphere and the soil has been disturbed, and it can be said that they are in a period of transition. Since even the most rapid processes of adjustment among the reservoirs take decades, new equilibriums are far from being established. Gradually the deep oceans become involved; their turnover time of about 1,000 years and their rate of exchange with bottom sediments control the ultimate partitioning of carbon.

Meanwhile human activities continue to change explosively. The acceleration in the consumption of fossil fuels implies that the amount of carbon dioxide in the atmosphere will keep climbing from its present value of 320 parts per million to between 375 and 400 parts per million by the year 2000, in spite of anticipated large removals of carbon dioxide by land vegetation and the ocean reservoir [*see illustrations on next page*]. A fundamental question is: What will happen over the next 100 or 1,000 years? Clearly the exponential changes cannot continue.

If we extend the time scale with which we are viewing the carbon cycle by several orders of magnitude, to hundreds of thousands or millions of years, we can anticipate large-scale exchanges between organic carbon on land and carbonates of biological origin in the sea. We do know that there have been massive exchanges in the remote past. Any discussion of these past events and their implications for the future, however, must necessarily be qualitative and uncertain.

Although the plants on land have probably played an important role in the deposition of organic compounds in the soil, the oceans have undoubtedly acted as the main regulator. The amount of carbon dioxide in the atmosphere is essentially determined by the partial pressure of carbon dioxide dissolved in the

GIANT FERN of the genus *Pecopteris*, which fixed atmospheric carbon dioxide 300 million years ago, left the imprint of this frond in a thin layer of shale just above a coal seam in Illinois. The specimen is in the collection of the Smithsonian Institution.

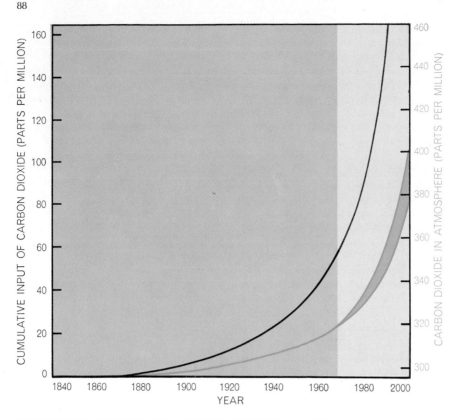

INCREASE IN ATMOSPHERIC CARBON DIOXIDE since 1860 is shown by the lower curve, with a projection to the year 2000. The upper curve shows the cumulative input of carbon dioxide. The difference between the two curves represents the amount of carbon dioxide removed by the ocean or by additions to the total biomass of vegetation on land.

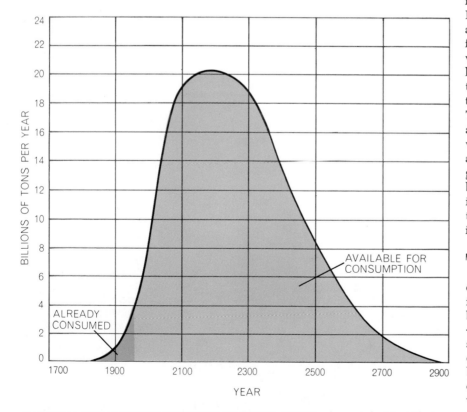

POSSIBLE CONSUMPTION PATTERN OF FOSSIL FUELS was projected by Harrison Brown in the mid-1950's. Here the fuel consumed is updated to 1960. If a third of the carbon dioxide produced by burning it all were to remain in the atmosphere, the carbon dioxide level would rise from 320 ppm today to about 1,500 ppm over the next several centuries.

sea. Over a period of, say, 100,000 years the leaching of calcium carbonates from land areas tends to increase the amount of carbon dioxide in the sea, but at the same time a converse mechanism—the precipitation and deposition of oceanic carbonates—tends to reduce the amount of carbon dioxide in solution. Thus the two mechanisms tend to cancel each other.

Over still longer periods of time—millions or tens of millions of years—the concentrations of carbonate and bicarbonate ions in the sea are probably buffered still further by reactions involving potassium, silicon and aluminum, which are slowly weathered from rocks and carried into the sea. The net effect is to stabilize the carbon dioxide content of the oceans and hence the carbon dioxide content of the atmosphere. Therefore it appears that the carbon dioxide environment, on which the biosphere fundamentally depends, may have been fairly constant right up to the time, barely a moment ago geologically speaking, when man's consumption of fossil fuels began to change the carbon dioxide content of the atmosphere.

The illustration on page 86 represents an attempt to synthesize into a single picture the circulation of carbon in nature, particularly in the biosphere. In addition to the values for inventories and transfers already mentioned, the flow chart contains other quantities for which the evidence is still meager. They have been included not only to balance the books but also to suggest where further investigation might be profitable. This may be the principal value of such an exercise. Such a flow chart also provides a semiquantitative model that enables one to begin to discuss how the global carbon system reacts to disturbances. A good model should of course include inventories and pathways for all the elements that play a significant role in biological processes.

The greatest disturbances of which we are aware are those now being introduced by man himself. Since his tampering with the biological and geochemical balances may ultimately prove injurious —even fatal—to himself, he must understand them much better than he does today. The story of the circulation of carbon in nature teaches us that we cannot control the global balances. Therefore we had better leave them close to the natural state that existed until the beginning of the Industrial Revolution. Out of a simple realization of this necessity may come a new industrial revolution.

8

The Nitrogen Cycle

INTRODUCTION

Nitrogen is chosen by many teachers to illustrate how biology, rather than inorganic chemistry, can play the dominant role in the cycle of an element. If life processes did not operate on the earth's surface, the nitrogen in the atmosphere would eventually be converted to nitrate by the available oxygen. Its actual transformations are almost all biologically mediated. Only by understanding the biology can one understand why nitrogen is present as the gaseous elemental molecule in the atmosphere.

In this article, "The Nitrogen Cycle," C. C. Delwiche observes that nitrogen is one of the minor constituents of rocks of the earth's crust, usually neglected for practical purposes in the study of all but a few quantitatively unimportant sedimentary rock types. In the atmosphere, one of the major reservoirs, it is stored as the relatively inert gas, nitrogen (N_2). But in the biological populations, the soils, and the oceans, the element is constantly cycling through the reduced ammonia, to oxidized states, nitrite and nitrate. All these fixed forms of nitrogen are necessary to bacteria, algae, and higher organisms for it is a key element in amino acids, proteins, and other nitrogenous components of cellular matter.

Recognition of the vital role of nitrogen in man's extensive alteration of his planetary environment has been growing as the demand for agricultural nitrate fertilizers has been steadily rising as they power the "green revolution." At the supply end of this cycle there is concern about the quantities of energy and other natural resources required to maintain the ever-increasing flow of nitrogen fertilizers to agriculture. There is also at the disposal end concern about the pollution of natural waters by unused nitrates shed by the agricultural soils. Bacterial transformation back to nitrogen on a large scale may become necessary for the public health. Yet, as Delwiche points out, there is much still to be learned about nitrogen fixation and denitrification, particularly in the surface waters of the oceans, which may play an important role in the stability of the global nitrogen cycle.

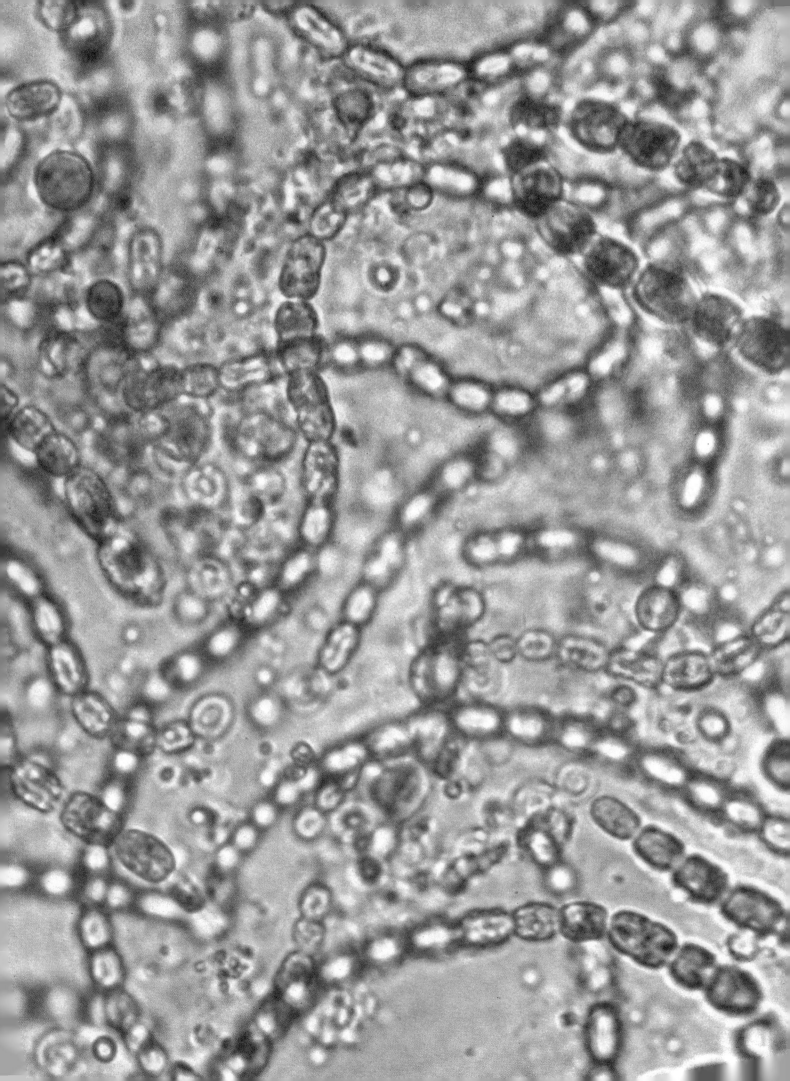

The Nitrogen Cycle

by C. C. Delwiche
September 1970

Nitrogen is 79 percent of the atmosphere, but it cannot be used directly by the large majority of living things. It must first be "fixed" by specialized organisms or by industrial processes

Although men and other land animals live in an ocean of air that is 79 percent nitrogen, their supply of food is limited more by the availability of fixed nitrogen than by that of any other plant nutrient. By "fixed" is meant nitrogen incorporated in a chemical compound that can be utilized by plants and animals. As it exists in the atmosphere nitrogen is an inert gas except to the comparatively few organisms that have the ability to convert the element to a combined form. A smaller but still significant amount of atmospheric nitrogen is fixed by ionizing phenomena such as cosmic radiation, meteor trails and lightning, which momentarily provide the high energy needed for nitrogen to react with oxygen or the hydrogen of water. Nitrogen is also fixed by marine organisms, but the largest single natural source of fixed nitrogen is probably terrestrial microorganisms and associations between such microorganisms and plants.

Of all man's recent interventions in the cycles of nature the industrial fixation of nitrogen far exceeds all the others in magnitude. Since 1950 the amount of nitrogen annually fixed for the production of fertilizer has increased approximately fivefold, until it now equals the amount that was fixed by all terrestrial ecosystems before the advent of modern agriculture. In 1968 the world's annual output of industrially fixed nitrogen amounted to about 30 million tons of ni-

trogen; by the year 2000 the industrial fixation of nitrogen may well exceed 100 million tons.

Before the large-scale manufacture of synthetic fertilizers and the wide cultivation of the nitrogen-fixing legumes one could say with some confidence that the amount of nitrogen removed from the atmosphere by natural fixation processes was closely balanced by the amount returned to the atmosphere by organisms that convert organic nitrates to gaseous nitrogen. Now one cannot be sure that the denitrifying processes are keeping pace with the fixation processes. Nor can one predict all the consequences if nitrogen fixation were to exceed denitrification over an extended period. We do know that excessive runoff of nitrogen compounds in streams and rivers can result in "blooms" of algae and intensified biological activity that deplete the available oxygen and destroy fish and other oxygen-dependent organisms. The rapid eutrophication of Lake Erie is perhaps the most familiar example.

To appreciate the intricate web of nitrogen flow in the biosphere let us trace the course of nitrogen atoms from the atmosphere into the cells of microorganisms, and then into the soil as fixed nitrogen, where it is available to higher plants and ultimately to animals. Plants and animals die and return the fixed nitrogen to the soil, at which point the nitrogen may simply be recycled through a new generation of plants and animals

or it may be broken down into elemental nitrogen and returned to the atmosphere [*see illustration on next two pages*].

Because much of the terminology used to describe steps in the nitrogen cycle evolved in previous centuries it has an archaic quality. Antoine Laurent Lavoisier, who clarified the composition of air, gave nitrogen the name azote, meaning without life. The term is still found in the family name of an important nitrogen-fixing bacterium: the Azotobacteraceae. One might think that fixation would merely be termed nitrification, to indicate the addition of nitrogen to some other substance, but nitrification is reserved for a specialized series of reactions in which a few species of microorganisms oxidize the ammonium ion (NH_4^+) to nitrite (NO_2^-) or nitrite to nitrate (NO_3^-). When nitrites or nitrates are reduced to gaseous compounds such as molecular nitrogen (N_2) or nitrous oxide (N_2O), the process is termed denitrification. "Ammonification" describes the process by which the nitrogen of organic compounds (chiefly amino acids) is converted to ammonium ion. The process operates when microorganisms decompose the remains of dead plants and animals. Finally, a word should be said about the terms oxidation and reduction, which have come to mean more than just the addition of oxygen or its removal. Oxidation is any process that removes electrons from a substance. Reduction is the reverse process: the addition of electrons. Since electrons can neither be created nor destroyed in a chemical reaction, the oxidation of one substance always implies the reduction of another.

One may wonder how it is that some organisms find it profitable to oxidize

BLUE-GREEN ALGAE, magnified 4,200 diameters on the opposite page, are among the few free-living organisms capable of combining nitrogen with hydrogen. Until this primary fixation process is accomplished, the nitrogen in the air (or dissolved in water) cannot be assimilated by the overwhelming majority of plants or by any animal. A few bacteria are also free-living nitrogen fixers. The remaining nitrogen-fixing microorganisms live symbiotically with higher plants. This micrograph, which shows blue-green algae of the genus *Nostoc*, was made by Herman S. Forest of the State University of New York at Geneseo.

nitrogen compounds whereas other organisms—even organisms in the same environment—owe their survival to their ability to reduce nitrogen compounds. Apart from photosynthetic organisms, which obtain their energy from radiation, all living forms depend for their energy on chemical transformations.

These transformations normally involve the oxidation of one compound and the reduction of another, although in some cases the compound being oxidized and the compound being reduced are different molecules of the same substance, and in other cases the reactants are fragments of a single molecular species. Ni-

trogen can be cycled because the reduced inorganic compounds of nitrogen can be oxidized by atmospheric oxygen with a yield of useful energy. Under anaerobic conditions the oxidized compounds of nitrogen can act as oxidizing agents for the burning of organic compounds (and a few inorganic com-

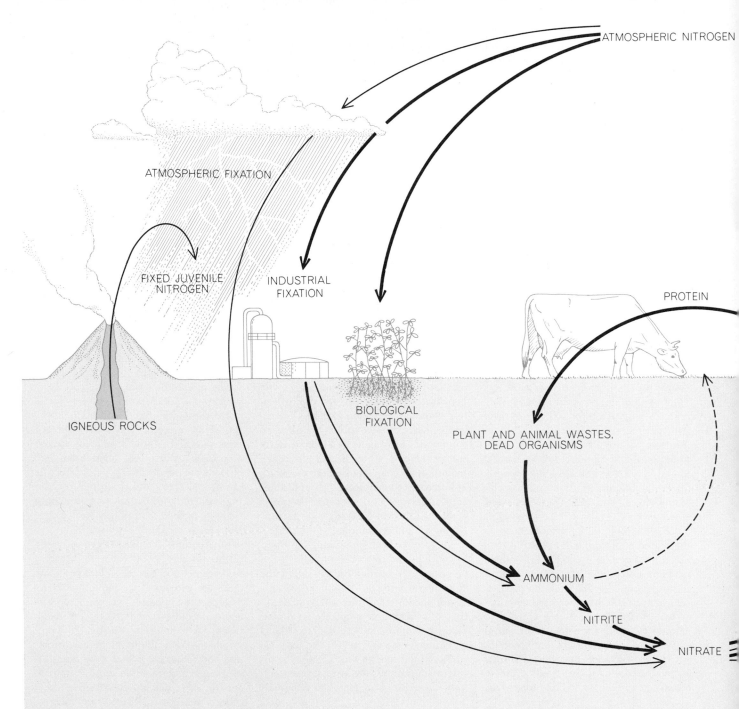

ATMOSPHERIC NITROGEN

ATMOSPHERIC FIXATION

FIXED JUVENILE NITROGEN

INDUSTRIAL FIXATION

PROTEIN

IGNEOUS ROCKS

BIOLOGICAL FIXATION

PLANT AND ANIMAL WASTES, DEAD ORGANISMS

AMMONIUM

NITRITE

NITRATE

NITROGEN CYCLE, like the water, oxygen and carbon cycles, involves all regions of the biosphere. Although the supply of nitrogen in the atmosphere is virtually inexhaustible, it must be combined with hydrogen or oxygen before it can be assimilated by higher plants, which in turn are consumed by animals. Man has intervened in the historical nitrogen cycle by the large-scale cultivation of nitrogen-fixing legumes and by the industrial fixation of nitrogen. The amount of nitrogen fixed annually by these two expedients now exceeds by perhaps 10 percent the amount of nitrogen fixed by terrestrial ecosystems before the advent of agriculture.

pounds), again with a yield of useful energy.

Nitrogen is able to play its complicated role in life processes because it has an unusual number of oxidation levels, or valences [*see illustration on page 95*]. An oxidation level indicates the number of electrons that an atom in a

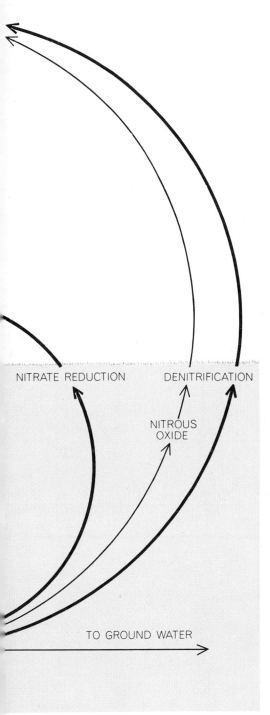

NITRATE REDUCTION   DENITRIFICATION

NITROUS
OXIDE

TO GROUND WATER

A cycle similar to the one illustrated also operates in the ocean, but its characteristics and transfer rates are less well understood. A global nitrogen flow chart, using the author's estimates, appears on the next page.

particular compound has "accepted" or "donated." In plants and animals most nitrogen exists either in the form of the ammonium ion or of amino ($-NH_2$) compounds. In either case it is highly reduced; it has acquired three electrons by its association with three other atoms and thus is said to have a valence of minus 3. At the other extreme, when nitrogen is in the highly oxidized form of the nitrate ion (the principal form it takes in the soil), it shares five of its electrons with oxygen atoms and so has a valence of plus 5. To convert nitrogen as it is found in the ammonium ion or amino acids to nitrogen as it exists in soil nitrates involves a total valence change of eight, or the removal of eight electrons. Conversely, to convert nitrate nitrogen into amino nitrogen requires the addition of eight electrons.

By and large the soil reactions that reduce nitrogen, or add electrons to it, release considerably more energy than the reactions that oxidize nitrogen, or remove electrons from it. The illustration on page 96 lists some of the principal reactions involved in the nitrogen cycle, together with the energy released (or required) by each. As a generalization one can say that for almost every reaction in nature where the conversion of one compound to another yields an energy of at least 15 kilocalories per mole (the equivalent in grams of a compound's molecular weight), some organism or group of organisms has arisen that can exploit this energy to survive.

The fixation of nitrogen requires an investment of energy. Before nitrogen can be fixed it must be "activated," which means that molecular nitrogen must be split into two atoms of free nitrogen. This step requires at least 160 kilocalories for each mole of nitrogen (equivalent to 28 grams). The actual fixation step, in which two atoms of nitrogen combine with three molecules of hydrogen to form two molecules of ammonia (NH_3), releases about 13 kilocalories. Thus the two steps together require a net input of at least 147 kilocalories. Whether nitrogen-fixing organisms actually invest this much energy, however, is not known. Reactions catalyzed by enzymes involve the penetration of activation barriers and not a simple change in energy between a set of initial reactants and their end products.

Once ammonia or the ammonium ion has appeared in the soil, it can be absorbed by the roots of plants and the nitrogen can be incorporated into amino acids and then into proteins. If the plant is subsequently eaten by an animal, the

nitrogen may be incorporated into a new protein. In either case the protein ultimately returns to the soil, where it is decomposed (usually with bacterial help) into its component amino acids. Assuming that conditions are aerobic, meaning that an adequate supply of oxygen is present, the soil will contain many microorganisms capable of oxidizing amino acids to carbon dioxide, water and ammonia. If the amino acid happens to be glycine, the reaction will yield 176 kilocalories per mole.

A few microorganisms represented by the genus *Nitrosomonas* employ nitrification of the ammonium ion as their sole source of energy. In the presence of oxygen, ammonia is converted to nitrite ion (NO_2^-) plus water, with an energy yield of about 65 kilocalories per mole, which is quite adequate for a comfortable existence. *Nitrosomonas* belongs to the group of microorganisms termed autotrophs, which get along without an organic source of energy. Photoautotrophs obtain their energy from light; chemoautotrophs (such as *Nitrosomonas*) obtain energy from inorganic compounds.

There is another specialized group of microorganisms, represented by *Nitrobacter,* that are capable of extracting additional energy from the nitrite generated by *Nitrosomonas*. The result is the oxidation of a nitrite ion to a nitrate ion with the release of about 17 kilocalories per mole, which is just enough to support the existence of *Nitrobacter*.

In the soil there are numerous kinds of denitrifying bacteria (for example *Pseudomonas denitrificans*) that, if obliged to exist in the absence of oxygen, are able to use the nitrate or nitrite ion as electron acceptors for the oxidation of organic compounds. In these reactions the energy yield is nearly as large as it would be if pure oxygen were the oxidizing agent. When glucose reacts with oxygen, the energy yield is 686 kilocalories per mole of glucose. In microorganisms living under anaerobic conditions the reaction of glucose with nitrate ion yields about 545 kilocalories per mole of glucose if the nitrogen is reduced to nitrous oxide, and 570 kilocalories if the nitrogen is reduced all the way to its elemental gaseous state.

The comparative value of ammonium and nitrate ions as a source of nitrogen for plants has been the subject of a number of investigations. One might think that the question would be readily resolved in favor of the ammonium ion: its valence is minus 3, the same as the valence of nitrogen in amino acids, whereas the valence of the nitrate ion is plus 5.

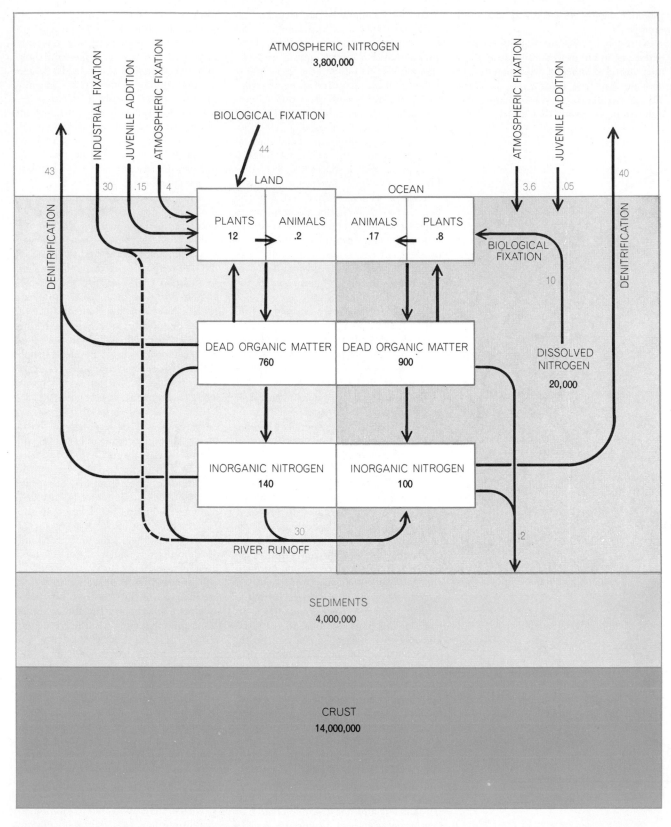

DISTRIBUTION OF NITROGEN in the biosphere and annual transfer rates can be estimated only within broad limits. The two quantities known with high confidence are the amount of nitrogen in the atmosphere and the rate of industrial fixation. The apparent precision in the other figures shown here reflects chiefly an effort to preserve indicated or probable ratios among different inventories. Thus the figures for atmospheric fixation and biological fixation in the oceans could well be off by a factor of 10. The figures for inventories are given in billions of metric tons; the figures for transfer rates (*color*) are given in millions of metric tons. Because of the extensive use of industrially fixed nitrogen the amount of nitrogen available to land plants may significantly exceed the nitrogen returned to the atmosphere by denitrifying bacteria in the soil. A portion of this excess fixed nitrogen is ultimately washed into the sea but it is not included in the figure shown for river runoff. Similarly, the value for oceanic denitrification is no more than a rough estimate that is based on the assumption that the nitrogen cycle was in overall balance before man's intervention.

On this basis plants must expend energy to reduce nitrogen from a valence of plus 5 to one of minus 3. The fact is, however, that there are complicating factors; the preferred form of nitrogen depends on other variables. Because the ammonium ion has a positive charge it tends to be trapped on clay particles near the point where it is formed (or where it is introduced artificially) until it has been oxidized. The nitrate ion, being negatively charged, moves freely through the soil and thus is more readily carried downward into the root zone. Although the demand for fertilizer in solid form (such as ammonium nitrate and urea) remains high, anhydrous ammonia and liquid ammoniacal fertilizers are now widely applied. The quantity of nitrogen per unit weight of ammonia is much greater than it is per unit of nitrate; moreover, liquids are easier to handle than solids.

Until the end of the 19th century little was known about the soil organisms that fix nitrogen. In fact, at that time there was some concern among scientists that the denitrifying bacteria, which had just been discovered, would eventually deplete the reserve of fixed nitrogen in the soil and cripple farm productivity. In an address before the Royal Society of London, Sir William Crookes painted a bleak picture for world food production unless artificial means of fixing nitrogen were soon developed. This was a period when Chilean nitrate reserves were the main source of fixed nitrogen for both fertilizer and explosives. As it turned out, the demand for explosives provided the chief incentive for the invention of the catalytic fixation process by Fritz Haber and Karl Bosch of Germany in 1914. In this process atmospheric nitrogen and hydrogen are passed over a catalyst (usually nickel) at a temperature of about 500 degrees Celsius and a pressure of several hundred atmospheres. In a French version of the process, developed by Georges Claude, nitrogen was obtained by the fractional liquefaction of air. In current versions of the Haber process the source of hydrogen is often the methane in natural gas [see illustration on page 97].

As the biological fixation of nitrogen and the entire nitrogen cycle became better understood, the role of the denitrifying bacteria fell into place. Without such bacteria to return nitrogen to the atmosphere most of the atmospheric nitrogen would now be in the oceans or locked up in sediments. Actually, of course, there is not enough oxygen in the

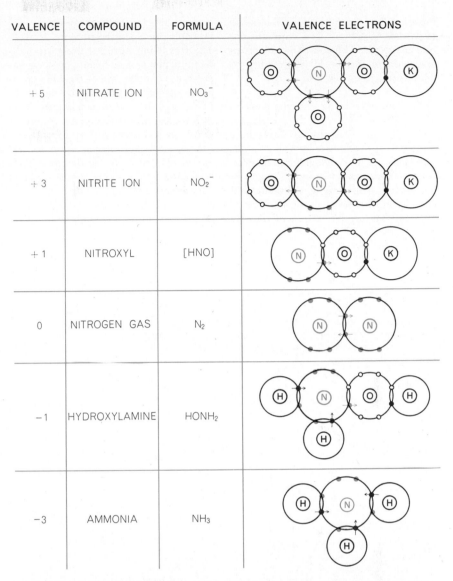

VALENCE	COMPOUND	FORMULA	VALENCE ELECTRONS
+ 5	NITRATE ION	NO_3^-	
+ 3	NITRITE ION	NO_2^-	
+ 1	NITROXYL	[HNO]	
0	NITROGEN GAS	N_2	
− 1	HYDROXYLAMINE	$HONH_2$	
− 3	AMMONIA	NH_3	

NITROGEN'S VARIETY OF OXIDATION LEVELS, or valence states, explains its ability to combine with hydrogen, oxygen and other atoms to form a great variety of biological compounds. Six of its valence states are listed with schematic diagrams (*right*) showing the disposition of electrons in the atom's outer (valence) shell. The ions are shown combined with potassium (K). In the oxidized (+) states nitrogen's outer electrons complete the outer shells of other atoms. In the reduced (−) states the two electrons needed to complete the outer shell of nitrogen are supplied by other atoms. Actually the outer electrons of two bound atoms spend some time in the shells of both atoms, contributing to the electrostatic attraction between them. Electrons of nitrogen (N) are in color; those of other atoms are black dots or open circles. The nitroxyl radical, HNO, is placed in brackets because it is not stable. It can exist in its dimeric form, hyponitrous acid (HONNOH).

atmosphere today to convert all the free nitrogen into nitrates. One can imagine, however, that if a one-way process were to develop in the absence of denitrifying bacteria, the addition of nitrates to the ocean would make seawater slightly more acidic and start the release of carbon dioxide from carbonate rocks. Eventually the carbon dioxide would be taken up by plants, and if the carbon were then deposited as coal or other hydrocarbons, the remaining oxygen would be available in the atmosphere to be com-

bined with nitrogen. Because of the large number of variables involved it is difficult to predict how the world would look without the denitrification reaction, but it would certainly not be the world we know.

The full story of the biological fixation of nitrogen has not yet been written. One would like to know how the activating enzyme (nitrogenase) used by nitrogen-fixing bacteria can accomplish at ordinary temperatures and pressures what takes hundreds of degrees and thousands of pounds of pressure in a synthetic-ammonia reactor. The total amount of nitrogenase in the world is probably no more than a few kilograms.

The nitrogen-fixing microorganisms are divided into two broad classes: those that are "free-living" and those that live in symbiotic association with higher plants. This distinction, however, is not as sharp as it was once thought to be, because the interaction of plants and microorganisms has varying degrees of intimacy. The symbionts depend directly on the plants for their energy supply and probably for special nutrients as well. The free-living nitrogen fixers are indirectly dependent on plants for their energy or, as in the case of the blue-green algae and photosynthetic bacteria, obtain energy directly from sunlight.

Although the nitrogen-fixation reaction is associated with only a few dozen species of higher plants, these species are widely distributed in the plant kingdom. Among the more primitive plants whose symbionts can fix nitrogen are the cycads and the ginkgos, which can be traced back to the Carboniferous period of some 300 million years ago [see bottom illustration on page 99]. It is probable that the primitive atmosphere of the earth contained ammonia, in which case the necessity for nitrogen fixation did not arise for hundreds of millions of years.

Various kinds of bacteria, particularly the Azotobacteraceae, are evidently the chief suppliers of fixed nitrogen in grasslands and other ecosystems where plants with nitrogen-fixing symbionts are absent. Good quantitative information on the rate of nitrogen fixation in such ecosystems is hard to obtain. Most investigations indicate a nitrogen-fixation rate of only two or three kilograms per hectare per year, with a maximum of perhaps five or six kilograms. Blue-green algae seem to be an important source of fixed nitrogen under conditions that favor their development [see illustration on page 90]. They may be a significant source in rice paddies and other environments favoring their growth. In natural ecosystems with mixed vegetation the symbiotic associations involving such plant genera as *Alnus* (the alders) and *Ceanothus* (the buckthorns) are important suppliers of fixed nitrogen.

For the earth as a whole, however, the greatest natural source of fixed nitrogen is probably the legumes. They are certainly the most important from an agronomic standpoint and have therefore been the most closely studied. The input of nitrogen from the microbial symbionts of alfalfa and other leguminous crops can easily amount to 350 kilograms per hectare, or roughly 100 times the annual rate of fixation attainable by nonsymbiotic organisms in a natural ecosystem.

Recommendations for increasing the world's food supply usually emphasize increasing the cultivation of legumes not only to enrich the soil in nitrogen but also because legumes (for example peas and beans) are themselves a food crop containing a good nutritional balance of amino acids. There are, however, several obstacles to carrying out such recommendations. The first is custom and taste. Many societies with no tradition of growing and eating legumes are reluctant to adopt them as a basic food.

For the farmer legumes can create a more immediate problem: the increased yields made possible by the extra nitrogen lead to the increased consumption of other essential elements, notably potassium and phosphorus. As a consequence farmers often say that legumes are "hard on the soil." What this really means is that the large yield of such crops places

REACTION	ENERGY YIELD (KILOCALORIES)
DENITRIFICATION	
1 $C_6H_{12}O_6 + 6KNO_3 \longrightarrow 6CO_2 + 3H_2O + 6KOH + 3N_2O$ GLUCOSE POTASSIUM POTASSIUM NITROUS NITRATE HYDROXIDE OXIDE	545
2 $5C_6H_{12}O_6 + 24KNO_3 \longrightarrow 30CO_2 + 18H_2O + 24KOH + 12N_2$ NITROGEN	570 (PER MOLE OF GLUCOSE)
3 $5S + 6KNO_3 + 2CaCO_3 \longrightarrow 3K_2SO_4 + 2CaSO_4 + 2CO_2 + 3N_2$ SULFUR POTASSIUM CALCIUM SULFATE SULFATE	132 (PER MOLE OF SULFUR)
RESPIRATION	

4 $C_6H_{12}O_6 + 6O_2 \longrightarrow 6CO_2 + 6H_2O$
                      CARBON   WATER
                      DIOXIDE
	686

AMMONIFICATION	
5 $CH_2NH_2COOH + 1\frac{1}{2}O_2 \longrightarrow 2CO_2 + H_2O + NH_3$ GLYCINE OXYGEN AMMONIA	176
NITRIFICATION	

6 $NH_3 + 1\frac{1}{2}O_2 \longrightarrow HNO_2 + H_2O$
                          NITROUS
                          ACID
	66

7 $KNO_2 + \frac{1}{2}O_2 \longrightarrow KNO_3$ POTASSIUM NITRITE	17.5
NITROGEN FIXATION	
8 $N_2 \longrightarrow 2N$ "ACTIVATION" OF NITROGEN	-160
9 $2N + 3H_2 \longrightarrow 2NH_3$	12.8

ENERGY YIELDS OF REACTIONS important in the nitrogen cycle show the various means by which organisms can obtain energy and thereby keep the cycle going. The most profitable are the denitrification reactions, which add electrons to nitrate nitrogen, whose valence is plus 5, and shift it either to plus 1 (as in N_2O) or zero (as in N_2). In the process, glucose (or sulfur) is oxidized. Reactions No. 1 and No. 2 release nearly as much energy as conventional respiration (*No. 4*), in which the agent for oxidizing glucose is oxygen itself. The ammonification reaction (*No. 5*) is one of many that release ammonium for nitrification. The least energy of all, but still enough to provide the sole energetic support for certain bacteria, is released by the nitrification reactions (*No. 6 and No. 7*), which oxidize nitrogen. Only nitrogen fixation, which is accomplished in two steps, calls for an input of energy. The true energy cost of nitrogen fixation to an organism is unknown, however.

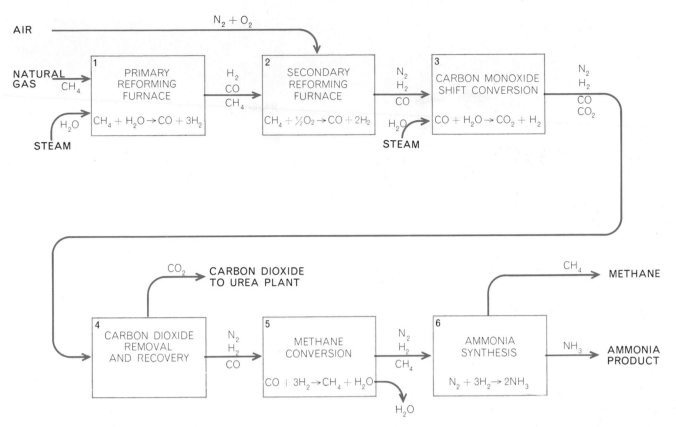

INDUSTRIAL AMMONIA PROCESS is based on the high-pressure catalytic fixation method invented in 1914 by Fritz Haber and Karl Bosch, which supplied Germany with nitrates for explosives in World War I. This flow diagram is based on the process developed by the M. W. Kellogg Company. As in most modern plants, the hydrogen for the basic reaction is obtained from methane, the chief constituent of natural gas, but any hydrocarbon source will do. In Step 1 methane and steam react to produce a gas rich in hydrogen. In Step 2 atmospheric nitrogen is introduced; the oxygen accompanying it is converted to carbon monoxide by partial combustion with methane. The carbon monoxide reacts with steam in Step 3. The carbon dioxide is removed in Step 4 and can be used elsewhere to convert some of the ammonia to urea, which has the formula $CO(NH_2)_2$. The last traces of carbon monoxide are converted to methane in Step 5. In Step 6 nitrogen and hydrogen combine at elevated temperature and pressure, in the presence of a catalyst, to form ammonia. A portion of the ammonia product can readily be converted to nitric acid by reacting it with oxygen. Nitric acid and ammonia can then be combined to produce ammonium nitrate, which, like urea, is another widely used fertilizer.

a high demand on all minerals, and unless the minerals are supplied the full benefit of the crop is not realized.

Symbiotic nitrogen fixers have a greater need for some micronutrients (for example molybdenum) than most plants do. It is now known that molybdenum is directly incorporated in the nitrogen-fixing enzyme nitrogenase. In Australia there were large areas where legumes refused to grow at all until it was discovered that the land could be made fertile by the addition of as little as two ounces of molybdenum per acre. Cobalt turns out to be another essential micronutrient for the fixation of nitrogen. The addition of only 10 parts per trillion of cobalt in a culture solution can make the difference between plants that are stunted and obviously in need of nitrogen and plants that are healthy and growing vigorously.

Although legumes and their symbionts are energetic fixers of nitrogen, there are indications that the yield of a legume crop can be increased still further by direct application of fertilizer instead of depending on the plant to supply all its own needs for fixed nitrogen. Additional experiments are needed to determine just how much the yield can be increased and how this increase compares with the industrial fixation of nitrogen in terms of energy investment. Industrial processes call for some 6,000 kilocalories per kilogram of nitrogen fixed, which is very little more than the theoretical minimum. The few controlled studies with which I am familiar suggest that the increase in crop yield achieved by the addition of a kilogram of nitrogen amounts to about the same number of calories. This comparison suggests that one can exchange the calories put into industrial fixation of nitrogen for the calories contained in food. In actuality this trade-off applies to the entire agricultural enterprise. The energy required for preparing, tilling and harvesting a field and for processing and distributing the product is only slightly less than the energy contained in the harvested crop.

Having examined the principal reactions that propel the nitrogen cycle, we are now in a position to view the process as a whole and to interpret some of its broad implications. One must be cautious in trying to present a worldwide inventory of a particular element in the biosphere and in indicating annual flows from one part of a cycle to another. The balance sheet for nitrogen [see top illustration on page 99] is particularly crude because we do not have enough information to assign accurate estimates to the amounts of nitrogen that are fixed and subsequently returned to the atmosphere by biological processes.

Another source of uncertainty involves the amount of nitrogen fixed by ionizing phenomena in the atmosphere. Although one can measure the amount of fixed nitrogen in rainfall, one is forced to guess how much is produced by ionization and how much represents nitrogen that has

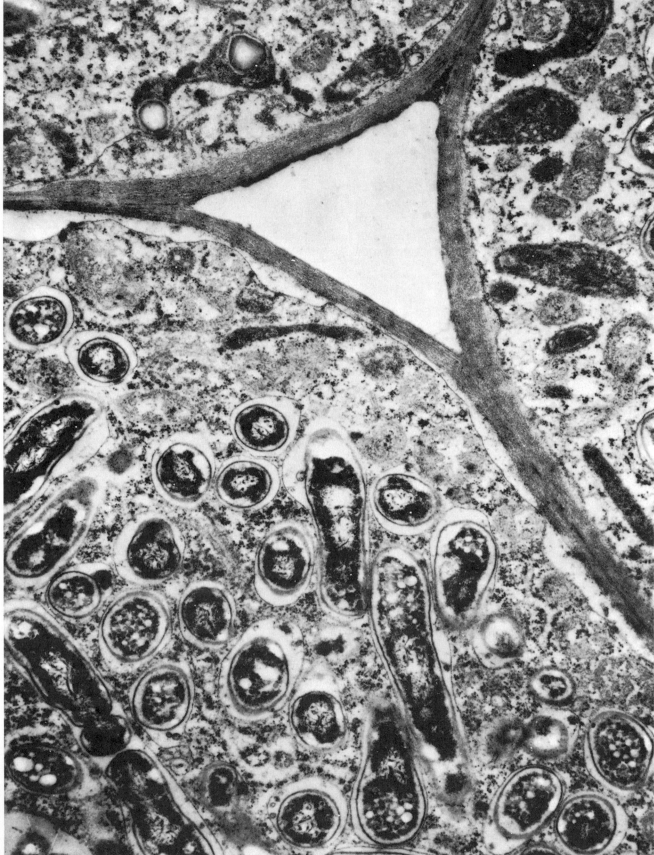

CROSS SECTION OF SOYBEAN ROOT NODULE, enlarged 22,-000 diameters, shows portions of three cells that have been infected by the nitrogen-fixing bacterium *Rhizobium japonicum*. More than two dozen bacteria are visible, each surrounded by a membrane. After the bacteria have divided, within a few days, each membrane will contain four to six "bacteroids." This electron micrograph was made by D. J. Goodchild and F. J. Bergersen of the Commonwealth Scientific and Industrial Research Organization in Australia.

entered the atmosphere from the land or the sea, either as ammonia or as oxides of nitrogen. Because the ocean is slightly alkaline it could release ammonia at a low rate, but that rate is almost impossible to estimate. Land areas are a more likely source of nitrogen oxides, and some reasonable estimates of the rate of loss are possible. One can say that the total amount of fixed nitrogen delivered to the earth by rainfall is of the order of 25 million metric tons per year. My own estimate is that 70 percent of this total is previously fixed nitrogen cycling through the biosphere, and that only 30 percent is freshly fixed by lightning and other atmospheric phenomena.

Another factor that is difficult to estimate is the small but steady loss of nitrogen from the biosphere to sedimentary rocks. Conversely, there is a continuous delivery of new nitrogen to the system by the weathering of igneous rocks in the crust of the earth. The average nitrogen content of igneous rocks, however, is considerably lower than that of sedimentary rocks, and since the quantities of the two kinds of rock are roughly equal, one would expect a net loss of nitrogen from the biosphere through geologic time. Conceivably this loss is just about balanced by the delivery of "juvenile" nitrogen to the atmosphere by volcanic action. The amount of fixed nitrogen reintroduced in this way probably does not exceed two or three million tons per year.

Whereas late-19th-century scientists worried that denitrifying bacteria were exhausting the nitrogen in the soil, we must be concerned today that denitrification may not be keeping pace with nitrogen fixation, considering the large amounts of fixed nitrogen that are being introduced in the biosphere by industrial fixation and the cultivation of legumes. It has become urgent to learn much more about exactly where and under what circumstances denitrification takes place.

We know first of all that denitrification does not normally proceed to any great extent under aerobic conditions. Whenever free oxygen is available, it is energetically advantageous for an organism to use it to oxidize organic compounds rather than to use the oxygen bound in nitrate salts. One can conclude that there must be large areas in the biosphere where conditions are sufficiently anaerobic to strongly favor the denitrification reaction. Such conditions exist wherever the input of organic materials exceeds the input of oxygen for their degradation. Typical areas where the deni-

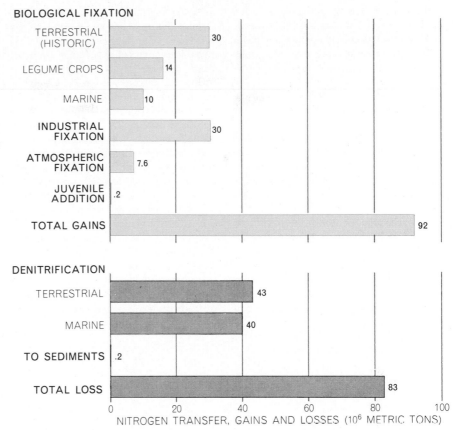

BALANCE SHEET FOR NITROGEN CYCLE, based on the author's estimates, indicates that nitrogen is now being introduced into the biosphere in fixed form at the rate of some 92 million metric tons per year (*colored bars*), whereas the total amount being denitrified and returned to the atmosphere is only about 83 million tons per year. The difference of some nine million tons may represent the rate at which fixed nitrogen is building up in the biosphere: in the soil, in ground-water reservoirs, in rivers and lakes and in the ocean.

ASSOCIATIONS OF TREES AND BACTERIA are important fixers of nitrogen in natural ecosystems. The ginkgo tree (*left*), a gymnosperm, has shown little outward change in millions of years. The alder (*right*), an angiosperm, is common in many parts of the world.

trification process operates close to the surface are the arctic tundra, swamps and similar places where oxygen input is limited. In many other areas where the input of organic material is sizable, however, denitrification is likely to be proceeding at some point below the surface, probably close to the level of the water table.

There are even greater uncertainties regarding the nitrogen cycle in the ocean. It is known that some marine organisms do fix nitrogen, but quantitative information is scanty. A minimum rate of denitrification can be deduced by estimating the amount of nitrate carried into the ocean by rivers. A reasonable estimate is 10 million metric tons per year in the form of nitrates and perhaps twice that amount in the form of organic material, a total of about 30 million tons. Since the transfer of nitrogen into sediments is slight, one can conclude that, at least before man's intervention in the nitrogen cycle, the ocean was probably capable of denitrifying that amount of fixed nitrogen.

The many blanks in our knowledge of the nitrogen cycle are disturbing when one considers that the amount of nitrogen fixed industrially has been doubling about every six years. If we add to this extra nitrogen the amounts fixed by the cultivation of legumes, it already exceeds (by perhaps 10 percent) the amount of nitrogen fixed in nature. Unless fertilizers and nitrogenous wastes are carefully managed, rivers and lakes can become loaded with the nitrogen carried in runoff waters. In such waterways and in neighboring ground-water systems the nitrogen concentration could, and in some cases already does, exceed the levels acceptable for human consumption. Under some circumstances bacterial denitrification can be exploited to control the buildup of fixed nitrogen, but much work has to be done to develop successful management techniques.

The problem of nitrogen disposal is aggravated by the nitrogen contained in the organic wastes of a steadily increasing human and domestic-animal population. Ideally this waste nitrogen should be recycled back to the soil, but efficient and acceptable means for doing so remain to be developed. At present it is economically sounder for the farmer to keep adding industrial fertilizers to his crops. The ingenuity that has been used to feed a growing world population will have to be matched quickly by an effort to keep the nitrogen cycle in reasonable balance.

9

The Global Circulation of Atmospheric Pollutants

INTRODUCTION

Watching a plume of dark smoke dissipate and be lost to sight within a short distance of the smoke stack, one might find it easy to believe that the pollutants that man produces are so mixed and diluted in the atmosphere that any harmful effects would be negligible. The trouble, says Reginald E. Newell, in the article "The Global Circulation of Atmospheric Pollutants," is that the testimony of our eyes—and noses—may be wrong. Very small quantities of pollutants, such as sulfur, carbon dioxide, and carbon monoxide, may be detected at various levels in the atmosphere by sensitive scientific instruments. Such measurements show that gases or particles injected into the atmosphere by man may not be homogeneously dispersed throughout the gaseous envelope surrounding the earth. This knowledge has come at a time when meteorologists have been making enormous gains in their understanding of the structure and dynamics of the atmosphere as well as how the products of atmospheric chemical reactions taking place in one part of the atmosphere may be transported to another. To study the fate of atmospheric pollutants, we must be aware of atmospheric dynamics, atmospheric chemical reactions, the links between the atmosphere and the oceans, and the release of particles and gases from the surface of the continents.

Air pollution is by definition an undesirable accumulation of substances added to the atmosphere in amounts greater than would ordinarily be present as a result of natural processes. Sulfur dioxide, and its breakdown products, for example, are not unnatural substances in the atmosphere; they are steadily contributed by a variety of normal earth processes and man-made sulfur in the atmosphere is only about half the amount normally injected by natural processes. Sulfur dioxide becomes a pollutant when it reaches abnormally high concentrations over local—usually urban—areas. Man is injecting a significant amount of carbon dioxide into the atmosphere, again only increasing the natural amount by an as yet small increment, but this increase may have some important effects on our climate.

Newell discusses theories of atmospheric circulation and the sensitive measurements of small amounts of pollutants and reaches some conclusions about how much the earth's climate may be affected by man's activities. He touches on many facets of air pollution, points out how little reliable data we have on the amounts and effects of pollution, and points up the possibility that even small amounts of pollutants may so affect the dynamics of the atmosphere as to affect our future seriously.

The Global Circulation of Atmospheric Pollutants

by Reginald E. Newell

January 1971

Worldwide wind and temperature patterns and the behavior of trace substances are studied in an effort to learn what effect changes in the atmosphere caused by man may have on the earth's climate

Pollution is more than a plume of smoke rising above a factory or a yellowish haze hanging over a city. The foreign substances man introduces into the air spread all over the globe and rise into the upper atmosphere. It is therefore important to learn how each of the major pollutants enters the atmosphere, the speed and extent of its spread and the ways in which it may alter the atmosphere and thus affect temperature and precipitation both locally and worldwide. In this article I shall discuss the movement of the various pollutants and review what is known about the effect on temperature of foreign substances in the atmosphere.

Of the total of 164 million metric tons of pollutants emitted each year in the U.S., about half comes from automobiles. Of the main component, carbon monoxide, 77 percent is from automobiles; so are most of the hydrocarbons and much of the oxides of nitrogen. The oxides of sulfur come mainly from electric power plants, small particles mainly from power plants and industry. In addition to these more obvious pollutants, vast quantities of water and carbon dioxide are produced by the burning of fossil fuels. Other pollutants are lead from automotive gasoline and ozone, which is produced by the action of sunlight on automobile exhaust. Radioactive substances introduced by man's activity include fission products from weapons tests, such as strontium 90, cesium 137 and iodine 131, and neutron-activated casing materials, such as tungsten 185, manganese 54, iron 55, rhodium 102 and cadmium 109. A satellite carrying a portable power plant using plutonium 238 as fuel accidentally burned up in the atmosphere over the southern Indian Ocean in April, 1964, at a height of from 40 to 50 kilometers instead of going into orbit, and the radioactivity from that point source has been tracked over the globe. The reprocessing of nuclear-fuel elements from power plants releases krypton 85, a radioactive gas with a half-life of about 10 years, which is gradually accumulating in the atmosphere.

Trace substances are, of course, present in nature. Carbon dioxide is taken up by plants during their growth cycle and released by the decay of plant material. Sulfur is also involved in plant processes and is abundant in the ocean, from which it is released by sea spray; it is sometimes injected high into the atmosphere in large amounts during volcanic eruptions. Ozone is produced in the upper atmosphere and carried downward toward the earth's surface, where it is destroyed [see "The Circulation of the Upper Atmosphere," by Reginald E. Newell; SCIENTIFIC AMERICAN, March, 1964]. As for radioactivity, radon and thoron gas emanate from the soil and decay in the atmosphere, giving rise to a chain of radioactive substances; some of the end products, such as lead 210, are transported up into the stratosphere. Cosmic rays entering the atmosphere collide with air molecules, usually in the stratosphere, to form radioactive nuclides such as beryllium 7, sodium 22 and carbon 14, some of which live long enough to find their way down to the surface.

Whether natural or man-made, some of these trace substances occur as gases, others as aerosols: finely divided liquid droplets or solid particles. Many are involved in phase transformations. For example, water vapor (gas) may be cooled to the point where it changes to water droplets or to ice crystals, forming clouds; sulfur dioxide gas may change in moist air to droplets of sulfuric acid. The gases diffuse and mix quite easily but the aerosols are governed by a number of factors that limit their spread. Large particles with radii of 10 microns (thou-

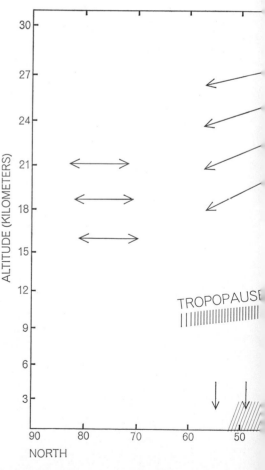

DISTRIBUTION AND MOVEMENTS of ozone (*black*), water (*gray*) and aerosols, or small solid and liquid particles (*color*), are

sandths of a millimeter) or more can be washed out of the air by raindrops or fall out directly. Very small particles can grow by coagulation until they too can be trapped in clouds and be washed out. Particles larger than about .3 micron cannot reach the upper atmosphere under normal circumstances because their fall velocities are greater than the average updraft speeds. Larger particles are nevertheless found in the upper atmosphere, some being introduced directly by volcanic eruptions, others growing there from smaller particles and gases introduced by the air motion. (Incidentally, the electrostatic precipitators on smokestacks are good for trapping particles with radii larger than about a micron; so is the human nose. Smaller particles can penetrate into the lungs, however, and so it is not always the larger particles one sees pouring from chimneys that are the most hazardous to health.)

Trace substances are moved over the globe by wind systems that fluctuate in strength and direction from day to day as cyclones and anticyclones move around the globe. The lowest 10 to 15

kilometers (six to nine miles) of the atmosphere, where the temperature decreases with height, is called the troposphere; above it to about 50 kilometers is the stratosphere, which in some respects resembles a series of stratified layers. Air parcels in both the troposphere and the stratosphere can be tabbed and tracked from one day to another [see illustration on next page]. The prevailing, or mean, wind, which is revealed by averaging over a long time period, blows from west to east over much of the middle latitudes in both hemispheres, but at low latitudes (and in the upper regions in general in summer) the prevailing wind is from the east. The mean wind transfers trace substances fairly rapidly round the globe; for example, the 35-meter-per-second west to east flow at 30 degrees north latitude gives a transit time of about 12 days. Clouds of debris from a nuclear explosion or volcanic eruption can often be identified as they make several circuits of the globe.

While the prevailing wind is along lines of latitude, north-south oscillations occur at the same time; the resulting north-south drift and an accompanying up-and-down motion give rise to an

overturning pattern at low latitudes called the Hadley-cell circulation [see illustration on page 106]. In the Tropics this large cell is a dominant feature of the circulation, whereas at middle latitudes north-south eddies in the prevailing-wind systems overshadow the mean north-south drifts. Air can be exchanged between the hemispheres by both the mean Hadley circulation and eddies in the upper tropical troposphere. As for vertical exchange, air passes into the stratosphere from the troposphere at low latitudes in the Hadley-cell circulation. Most of the transfer back into the troposphere is thought to occur close to the tropopause level near the middle-latitude jet stream [see illustration on these two pages], and some transfer back into the stratosphere may also occur there.

In the troposphere clouds, rain and thunderstorms are evidence of considerable vertical motion. Vertical velocities may reach 10 to 20 meters per second in thunderstorms but are generally no more than 10 centimeters per second in normal middle-latitude cyclones and anticyclones. In the stratosphere it is much harder to push an air parcel up or down

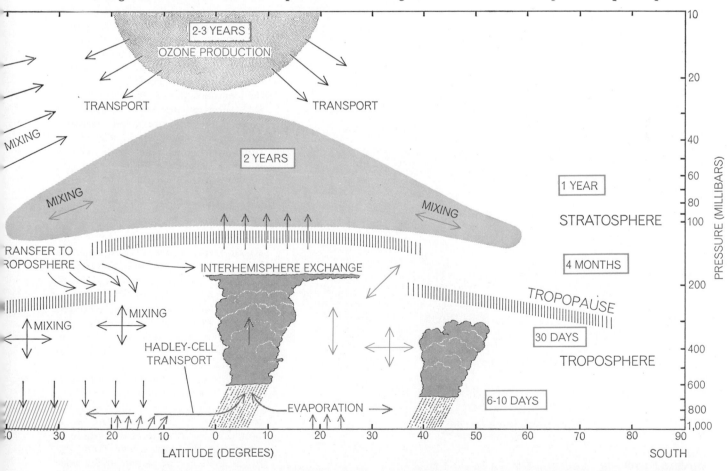

indicated on a schematic diagram drawn along a line of longitude. Effects illustrated for one hemisphere occur in both hemispheres. Boxed figures are residence times for aerosols. Pressure, measured in millibars, is often used by meteorologists as a measure of altitude. The tropopause is the boundary between the troposphere and the stratosphere; its altitude varies with latitude as indicated.

because the temperature increases with height, and so vertical motions rarely exceed a few centimeters per second and are often much smaller. The vertical spread of a trace substance in the stratosphere is therefore rather slow, like the downward migration of a card in a pack of cards that is being shuffled. There is much less shuffling at low latitudes than there is over the polar regions, so that trace constituents can stay in the equatorial stratosphere for several years.

Once the material reaches the bottom of the pack and enters the troposphere it can be mixed vertically rather rapidly. Small particles spend about 30 days in the troposphere before being washed out by rain. Gases spend varying periods there depending on the "sinks" by which each is removed from the atmosphere: incorporation into cloud droplets, reactions with other gases, loss to finely divided liquid or solid particles or the earth's surface and so on. Generally the tropospheric residence times of gases are from about two to four months—provided that there *is* a sink. (Krypton 85, for example, has no known sink and disappears only by radioactive decay.)

The global temperature pattern shows that the coldest air is over the Equator near the tropical tropopause at all seasons and over the winter pole in the stratosphere. The lowest temperatures over the Equator occur in January. Notice that temperature increases with latitude in the lower stratosphere. The temperature pattern is maintained by the liberation of latent heat, by radiation and by the motion of air masses, the sum of which is roughly in balance. When more water is rained out of a given air column than is evaporated into it from the surface, the column gains the latent heat that is liberated. The net effect of latent-heat liberation and radiative processes is that tropospheric air is heated at low latitudes and cooled at high latitudes, generating energy. In the troposphere the net effect of the radiative processes by themselves is to cool the air at all latitudes. Contributions come from absorption of incoming solar radiation, and from absorption and reemission of terrestrial long-wave radiation, by carbon dioxide, water vapor and ozone. Water vapor dominates in the lower layers, producing cooling of up to two degrees per day, with carbon dioxide of secondary importance; in the strato-

sphere carbon dioxide and ozone dominate [*see top illustration on page 109*]. Horizontal and vertical motions can also produce temperature changes. If air is forced to rise, it cools as it expands; if it is forced to sink, it contracts and becomes warmer, as in a bicycle pump. The former effect is thought to maintain the low temperatures over the tropical tropopause, the upward motion being forced from below, where the latent heat is liberated. On the other hand, compression is responsible for the inversion over the Los Angeles basin during a good fraction of the year—the increase of temperature with altitude that inhibits vertical mixing and traps pollutants near the surface.

Carbon dioxide is quite well mixed vertically, whereas water-vapor concentrations decrease with distance from the source (the earth's surface) and ozone decreases away from its source in the middle stratosphere [*see bottom illustration on page 109*]. Aerosols show two regions of high concentration. One is at the source (ground level). The other, in the lower stratosphere, is due to the direct injection of particles and the injection of gases from which particles form, together with a very slow removal rate.

It is fairly clear that if the present atmospheric temperature structure is maintained by a combination of air motions and effects that involve trace substances, it can be altered by changes in the concentration of trace substances. One therefore needs to know the natural cycles of the atmospheric trace constituents that are important in this context and the changes in the cycles that are being or may in the future be brought about by man.

Ozone (O_3) is produced by the photodissociation of molecular oxygen and the recombination of molecular with atomic oxygen, primarily above about 22 kilometers and at low latitudes. Ozone is transported toward the poles and downward by atmospheric motions between the subtropics and the high latitudes. From the lower stratosphere in the middle latitudes the ozone seeps into the troposphere, and it is eventually destroyed at the earth's surface or in reactions with aerosols in the surface layer. There is a maximum of ozone in the lower stratosphere in the spring and in the troposphere in the late spring; it is caused by an increase in the supply of energy to the stratosphere from the troposphere in this season and a concomitant increase in the large-scale mixing motions. Ozone stays for about four months in the middle-latitude lower

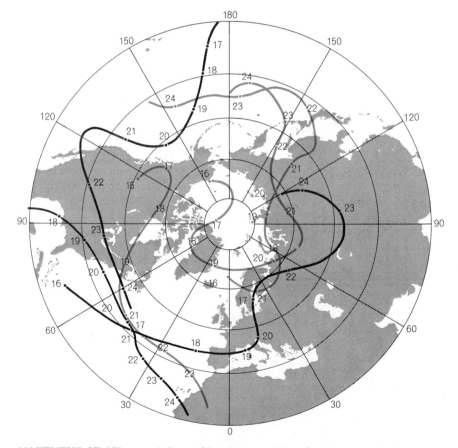

MOVEMENT OF AIR around the world is determined by identifying specific air parcels and tracking them. These tracks were worked out by Edwin Danielsen of the National Center for Atmospheric Research, who identified parcels in the troposphere (**black**) and stratosphere (**color**) and tracked them. The numbers represent successive days in April, 1964.

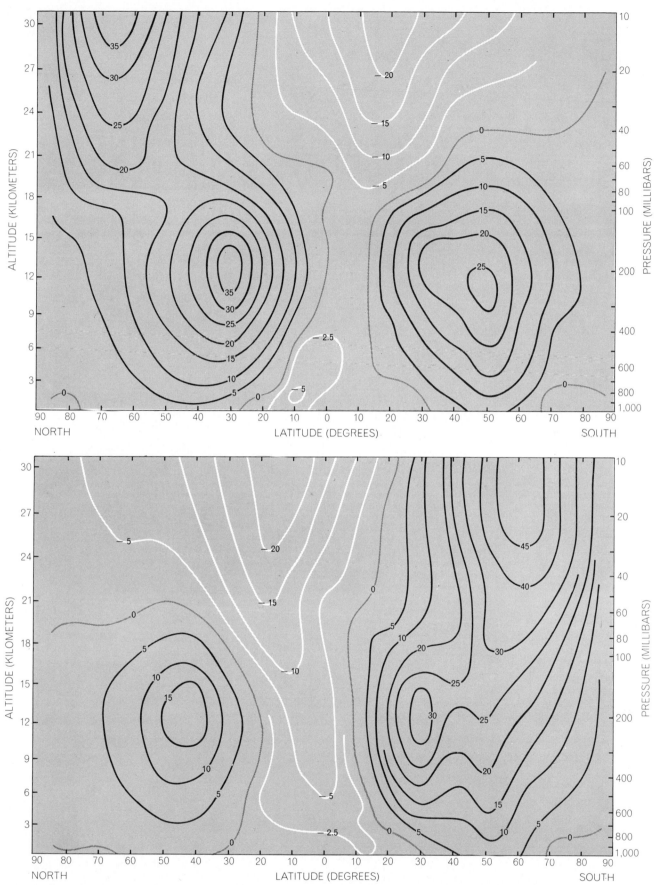

PREVAILING WINDS are revealed by averaging observations over a period of time. Here the mean east-west wind speed is shown as computed for two three-month periods: December–February (*top*) and June–August (*bottom*). The speed is given in meters per second, with positive numbers indicating winds blowing from west to east (*black contour lines*) and negative numbers for east-to-west winds (*white lines*). Note that seasonal changes in the lower atmosphere are larger in Northern Hemisphere than in Southern.

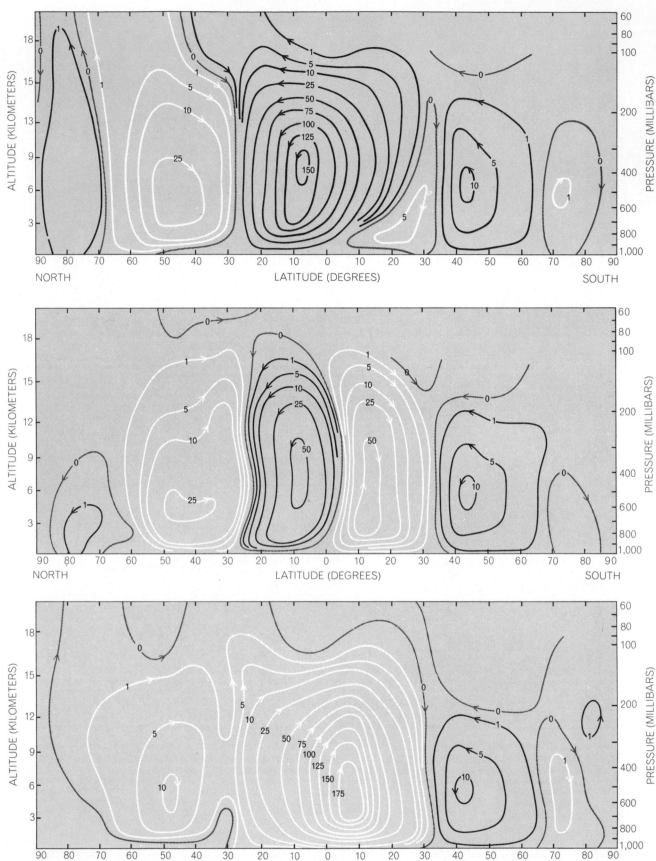

MASS FLUX IN THE TROPICS takes the form of Hadley-cell circulation, the result of a combination of north-south and up-and-down motions. It is shown here for three seasons; December–February (*top*), March–May (*middle*) and June–August (*bottom*). The difference between adjacent contour values gives the flux in millions of metric tons per second based on average values from all longitudes; actually the pattern varies with longitude. The charts are based on work of Dayton G. Vincent and John W. Kidson.

stratosphere and from two to three months in the troposphere.

Most of the steps of the cycle have been verified in detail by a "dynamical-numerical" model that was developed at the Environmental Science Services Administration's Geophysical Fluid Dynamics Laboratory at Princeton University and has been applied to the ozone problem by Syukuro Manabe and B. G. Hunt. In such a model equations governing temperature, wind speeds and ozone concentration are solved with a computer to trace the evolution of the fields over a period of time. The models have been developed to help in weather forecasting and in understanding the general circulation of the global atmosphere and theories of climatic change. Radiative effects, the earth's surface properties, clouds and even ocean temperatures all have to be included to achieve a good representation of the atmosphere, and the necessary degree of resolution in space and time can only be obtained with the largest computers. As far as can be seen, the influence of man-made ozone (in city smog) is small compared with the natural cycle of production, transport and loss, which totals about two billion tons per year. (This is on a global scale, of course, and is of small comfort to people whose smarting eyes make them quite aware of the ozone in the smog they live with.)

The water-vapor cycle in the troposphere depends on the difference between precipitation and evaporation in a given air column [see illustration on page 110]. In the subtropics there is an excess of evaporation over precipitation, so that these latitudes can be regarded as source regions; the opposite situation occurs over the middle and low latitudes. Water vapor is therefore transported from subtropical latitudes toward the Pole and toward the Equator. Large-scale eddy processes govern the movement toward the Pole and the mean Hadley-cell circulation governs the movement toward the Equator. The average rainfall at a point on the globe is about 100 centimeters per year and the average residence time of a water-vapor molecule in the troposphere is about 10 days.

The water vapor produced by man—by fuel combustion, for example—is not a significant fraction of the natural tropospheric cycle. (Again, people who live near power-plant cooling towers will think otherwise.) Nevertheless, more subtle effects could well be produced by interference with the evaporation-precipitation cycle. (Efforts have already been made to do this on a small scale by

	PARTIC-ULATES	SULFUR OXIDES	NITROGEN OXIDES	CARBON MONOXIDE	HYDRO-CARBONS
POWER AND HEATING	8.1	22.1	9.1	1.7	.6
VEHICLES	1.1	.7	7.3	57.9	15.1
REFUSE DISPOSAL	.9	.1	.5	7.1	1.5
INDUSTRY	6.8	6.6	.2	8.8	4.2
SOLVENT EVAPORATION					3.9
TOTAL	16.9	29.5	17.1	75.5	25.3

POLLUTANTS emitted in U.S. in 1968 are shown. Estimates, in millions of metric tons, are by George B. Morgan and colleagues of National Air Pollution Control Administration.

spreading a thin film on the water surface of some reservoirs in Australia to prevent excessive evaporation.) One would need a complete dynamical-numerical model to see what effect a given change would have if extensive regions were altered. Less water vapor evaporated would lead to less liberation of latent heat—but also to less cooling by the radiative effect; only a full model can give a proper idea of the interlocking feedbacks and their net effect.

The water-vapor balance of the stratosphere is much more delicate. Alan W. Brewer of the University of Toronto has suggested that most of the water vapor in the stratosphere enters through the region near the tropical tropopause in the rising branch of the Hadley-cell circulation. The temperature is close to −80 degrees Celsius in that region, and so the air can hold only minute amounts of moisture; in the process of passing through, most of the moisture from the troposphere is frozen out and precipitates, staying in the troposphere as cirrus clouds. (The fact that the frost point throughout the lower atmosphere even at middle latitudes is close to the temperature near the tropical tropopause forms the observational basis for Brewer's suggestion.) Henry Mastenbrook of the Naval Research Laboratory has been monitoring the water vapor in the stratosphere since 1964, flying a frost-point device (developed by Brewer) on high-altitude balloons. He finds that the content in the lower stratosphere varies sea-

sonally in phase with the temperature at the tropical tropopause, whereas the content at 30 kilometers varies only slightly, with average values throughout of only two to three micrograms of water vapor per gram of air.

From the mean rising motion and the water-vapor content near the tropical tropopause one can calculate the mass of water entering the stratosphere; it is about seven million grams per second. Now, 500 supersonic transport planes flying at 21 kilometers (70,000 feet) would inject about two million grams of water vapor per second directly into the stratosphere. Since this water is introduced above the cold trap, and since it is introduced at a rate that is of the same order of magnitude as the natural rate, it is clearly going to lead to a significant increase in the water-vapor content at high levels. Wherever rising motion and low temperatures exist together in the stratosphere, clouds may form as expansion and concomitant cooling allow the air to reach the local frost point. Such clouds are occasionally observed near 25 kilometers over Norway and Iceland and over the Antarctic in winter, and also near 80 kilometers at high latitudes in summer and near the tropical tropopause —regions in which temperatures become very low. Increased water-vapor content in the stratosphere, then, would be expected to produce increased cloudiness and therefore a change in the albedo, or reflectivity, of the earth. Again, however, a full dynamical-numerical model

	NATURAL	MAN-MADE
OZONE	2×10^9	SMALL
CARBON DIOXIDE	7×10^{10}	1.5×10^{10}
WATER	5×10^{14}	1×10^{10}
CARBON MONOXIDE	?	2×10^8
SULFUR	1.42×10^8	7.3×10^7
NITROGEN	1.4×10^9	1.5×10^7

NATURAL AND MAN-MADE trace-gas cycles are compared (in metric tons). Sulfur and nitrogen data are from E. Robinson and R. C. Robbins of the Stanford Research Institute. (The man-made carbon dioxide input is not really a cycle; some remains in the atmosphere.)

is required in order to study the possible changes.

The largest pollutant by mass is carbon dioxide, and here there is evidence that man's activities are indeed altering the concentrations formerly controlled by nature. Some 70,000 million tons per year are involved in the natural cycle, corresponding to a fluctuation of nine parts per million by volume in the carbon dioxide concentration of 320 parts per million. Plants and trees start taking up carbon dioxide in the spring and continue to do so until the fall; then the leaves drop, vegetable matter begins to decay and the direction of the net transfer is from matter to air. This seasonal cycle can be monitored on the ground or in the lower troposphere. Superimposed on the seasonal cycle is a long-term increase, produced by the burning of fossil fuels. The carbon dioxide increase expected from the fuel that has been burned is about 1.8 parts per million per year, yet the observations show an increase of only about .7 part per million per year [see "The Carbon Cycle," by Bert Bolin, beginning on page 81]. Where does the rest of the carbon dioxide go? Some of it may be incorporated in the biosphere, but it is thought that the largest fraction is dissolved in the oceans, which all together contain about 60 times as much carbon

dioxide as the atmosphere does. The solubility diminishes as the temperature of the water increases. There has been some concern that as carbon dioxide in the air increases, presumably raising the air temperature through the "greenhouse effect," the water temperature will rise also, releasing some of the carbon dioxide from the ocean back into the air. Proponents of this view argue that a runaway effect will occur, with the additional carbon dioxide producing still more atmospheric heating.

It is well to bear in mind, however, that the net radiative contribution of carbon dioxide is to cool the atmosphere. With computer programs developed by Thomas G. Dopplick, my colleagues at the Massachusetts Institute of Technology and I have rerun the radiative-heating computations, assuming a tripled carbon dioxide concentration of 1,000 parts per million and the same temperature and cloudiness distributions. We find that the cooling rate diminishes by only a small fraction of a degree per day in the lower levels and actually increases in the stratosphere. Other things being equal, one could interpret a smaller cooling rate as effective heating in the troposphere. The outgoing infrared flux emitted by carbon dioxide is smaller for higher concentrations; for the atmosphere to radiate the original amount of infrared radiation back to space it would have to

radiate at a higher effective temperature. Other things are not equal, however. If the temperature distribution changed, it is likely that the clouds, albedo and therefore the radiation-stream balance would change also. For example, a slightly higher temperature would mean more evaporation and hence more water-vapor radiative cooling. Although it is tempting to argue from our results that little net temperature change should be expected, it is again clear that the proper way to proceed is to run a complete dynamical-numerical model with higher carbon dioxide levels and see what happens.

Another substance that is copiously produced by combustion is carbon monoxide. Levels of 50 parts per million are not uncommon in city streets, with values up to several hundred parts per million in traffic tunnels and underground garages; weekly average levels of 20 parts per million are sometimes found in the Los Angeles basin. The toxic effect is proportional to the ambient-air concentration and the time of exposure. Carbon monoxide that enters the bloodstream combines with hemoglobin, forming carboxyhemoglobin, and thus reduces its capacity to carry oxygen. Impairment of mental function, as measured by visual performance and ability to discriminate time intervals, occurs when carboxyhemoglobin in the blood

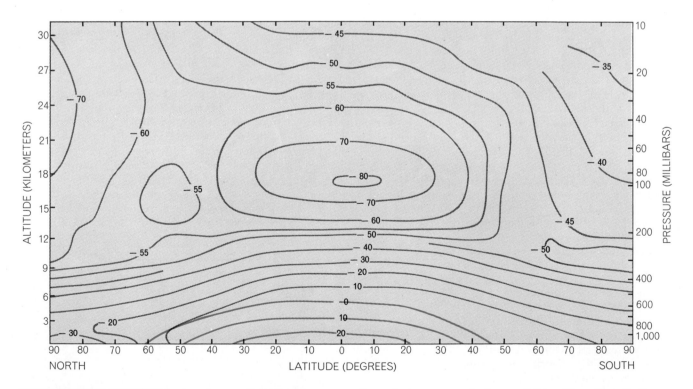

TEMPERATURE PATTERN is given for a north-south cross section of the atmosphere for the three-month period December–February. The isotherms, or contour lines of equal temperature, indicate the temperature in degrees Celsius. The coldest air is over the Equator and near the winter pole. In the latter region the temperature sometimes reaches −85 degrees C. at certain longitudes.

goes above about 2.5 percent (compared with a normal level of about .5 percent). Such values accompany exposure to 200 parts per million for about 15 minutes or 50 parts per million for about two hours.

The 200 million tons injected each year would correspond to an increase of .03 part per million per year. No such increase has been observed, and so a search for sinks of carbon monoxide is under way.

Christian E. Junge of the University of Mainz and his student Walter Seiler have reported that near sea level there is a boundary close to the Equator with high carbon monoxide values to the north and low values to the south; in the upper troposphere aircraft measurements show no such boundary [see bottom illustration on page 111]. Such a carbon monoxide discontinuity is compatible with the fact that there are more automobiles in the Northern Hemisphere and with what is known of the Hadley-cell circulation. At low levels the air just north of the discontinuity is streaming toward the Equator, whereas the air to the south has come from the Southern Hemisphere, with a much smaller population of automobiles. (A similar boundary was frequently revealed by surface-air measurements of strontium 90 when nuclear tests were being held in the Northern Hemisphere.) Higher up in the troposphere eddy mixing eliminates the interhemisphere gradients.

There is still much to be learned about the carbon monoxide sink. Junge's data show a decrease in the stratosphere, and it has been suggested that carbon monoxide combines above the tropopause with hydroxyl radicals (OH) to produce carbon dioxide and hydrogen. There are no direct measurements of the hydroxyl radical in the lower stratosphere and the computed amounts, together with the reaction rate, are just barely sufficient to account for the loss of 200 million tons per year. Furthermore, the rate at which air is transferred from the upper troposphere to the lower stratosphere also seems a shade too low to ascribe the entire loss to this path. The ocean, which has been found to contain carbon monoxide, has been suggested as both a sink and a source; so little is known that it is difficult to say which! Clearly many more observations are needed all over the atmosphere and in the ocean at different latitudes.

Nitrogen constitutes the largest fraction of the atmosphere and is involved in a variety of reactions, including plant growth. The total atmospheric turnover in all forms is about 10 billion tons per

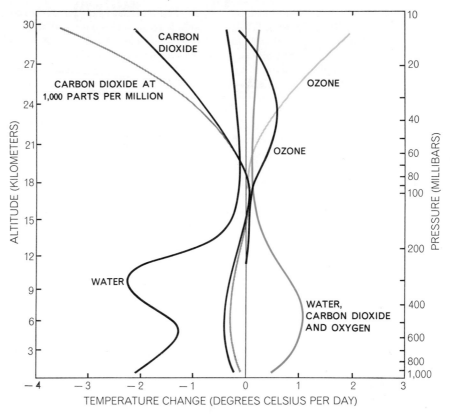

RADIATIVE TEMPERATURE CHANGE is brought about by the absorption of solar visible and ultraviolet (*light color*) and near-infrared (*dark color*) radiation and the absorption and reemission of thermal radiation from the earth (*black*) by various atmospheric gases. The direction and magnitude of each effect vary with altitude, as shown for at the Equator in January. If the carbon dioxide concentration were tripled to 1,000 parts per million, its effects would change as shown (*gray*), according to computations by author's group.

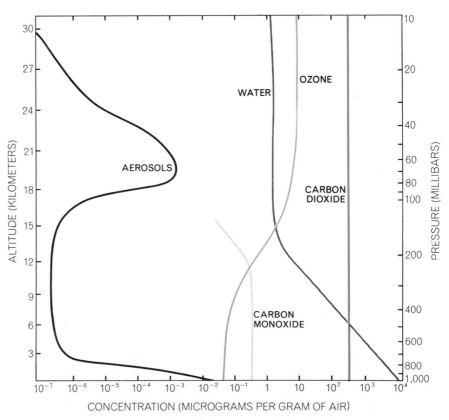

MIXING RATIO, or concentration, of the trace constituents of the atmosphere varies with altitude, decreasing with distance from their sources, except in the case of carbon dioxide.

year, whereas the amount introduced by man is only about 50 million tons. While this much pollution is very important in some regions (city smogs are produced by the action of sunlight on oxides of nitrogen from automobiles), on a global scale it seems that nature's contributions dominate.

There is no such clear distinction for the sulfur compounds. Sulfur is an abundant constituent of ocean water and is released to the atmosphere by sea spray and by biological decay; it is removed by precipitation, intake by vegetation and direct deposition. The total amount involved in the natural cycle is about 142 million tons per year; the 73 million tons injected by man as a pollutant is therefore a very significant amount. This additional sulfur is thought to end up in the ocean and, as Erik Eriksson of the University of Stockholm has stressed, increases the acidity of terrestrial waters. (Fish cannot live in waters of high acidity, and they already shun some inland waterways in Sweden; man has been given a warning.) Sulfur dioxide and particles together in the air seem to produce respiratory ailments.

A considerable amount—no one knows exactly how much—of sulfur dioxide emitted to the atmosphere ends up as sulfate ions or as ammonium sulfate par-

ticles. The time necessary for the sulfur dioxide to disappear as a gas and become incorporated into particles varies from about half an hour to a few days, depending on the air's moisture content and other factors. Measurements made by Junge show that the very small particles called Aitken nuclei (radius about .03 micron) decrease in concentration above the tropopause in a manner consistent with the view that their source is the troposphere. Their composition is unknown. In addition there is a layer of somewhat larger particles (mean radius about .3 micron, with some radii of a micron or more) in the lower stratosphere, composed of ammonium sulfate or sulfuric acid. James P. Friend of New York University and Richard D. Cadle of the National Center for Atmospheric Research have independently verified Junge's finding of this layer of larger particles. Junge has suggested that the large particles in the lower stratosphere may grow on the Aitken nuclei from the gases injected there, and gradually coagulate to form larger particles before they fall, or are transported by exchange, into the troposphere.

When the distribution of sizes of particles of various kinds in the lower troposphere is measured, it is found that most of the mass is accounted for by particles

whose radius is between .1 micron and 10 microns. Observations over land away from major cities have shown an increase in worldwide particle mass in the past 10 years, even though over some city areas there has been a decrease. There has been considerable speculation that an increase in the concentration of particles in the troposphere would cause more solar radiation to be scattered back to space and thus contribute to the lowering of terrestrial temperatures. There are no measurements to support this speculation. In fact, George Robinson of the British Meteorological Office has found that solar radiation is absorbed, rather than scattered, by tropospheric aerosols.

The efforts now being made to reduce particulate pollution, coupled with the relatively short washout time for aerosols, provide reason to hope the long-term effect of large man-made aerosols may be small. If particulate material is not removed but is simply more finely divided, however, more will find its way into the stratospheric regions where the residence times are long and the influence of aerosols generated at the surface could be appreciable.

In the case of hydrocarbons, very little is known about the natural cycle and consequently about the worldwide effect of man's interference. Methane (marsh gas) is abundantly produced by nature and finds its way to the stratosphere, but there are no global-scale measurements. Yet hydrocarbons do appear to play a role in smog formation and cannot be ignored at the local level.

Great volcanic eruptions, such as that of Krakatoa in 1883 or Mount Agung on Bali in 1963, increase the layer of stratospheric aerosols, so that colorful sunsets are produced all around the world. After Krakatoa, investigators suggested that the sunsets were due to the injection of sulfuric acid as well as volcanic dust, basing this suggestion on the fact that there is a strong odor of sulfur near active volcanoes. Samples taken in the lower stratosphere over Australia after the Bali eruption showed that this was indeed the case.

As the concentration of particles (and perhaps their size too) increases, solar radiation is intercepted and less arrives at the ground. Such blocking has often been proposed as a major cause of climatic change [see "Volcanoes and World Climate," by Harry Wexler; SCIENTIFIC AMERICAN Offprint 843]. We examined stratospheric temperature patterns after the Bali eruption in March, 1963, and found there was an immediate rise in stratospheric temperatures, with values

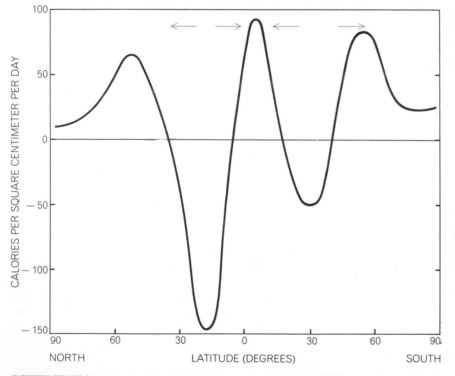

DIFFERENCE between precipitation and evaporation at each latitude is given in terms of the amount of heat lost through evaporation or gained through precipitation in each column of air. (About 600 calories are required to evaporate a column of water one centimeter high covering one square centimeter.) The curve, for December–February, is based on three sets of data. Imbalances of precipitation (*top*) and evaporation (*bottom*) are redressed by air motions that transport water toward Equator and toward poles (*arrows*).

as high as eight degrees C. above average being recorded; the rise was global in extent [see illustrations on next page]. It was mainly concentrated in the region above 15 kilometers and there were concomitant changes in the wind field. Presumably the small particles intercepted and absorbed solar radiation. (Their size is of the same order as the wavelength of light.)

The variation of the temperature change with latitude and height was very similar to the patterns that have been found for clouds of radioactive debris in the stratosphere. Such clouds persist longest over the Equator and slope downward toward the poles. The volcanic cloud from Bali did likewise—and the same slope was discussed in a report on Krakatoa published in 1888. The 1963 temperature increase is the largest climatic change ever observed by man. (There were no balloon observations of the stratosphere after Krakatoa.) It serves to warn us that we should watch the sulfur cycle carefully and try to learn soon whether the increased amounts injected into the troposphere by man can find their way into the stratosphere.

Investigators are able to base deductions about atmospheric motions on a network of 800 balloon sounding stations, at many of which temperature is also measured; global maps of temperature constructed from satellite observations are also becoming available. Yet the pollutants in the air we breathe are measured at very few places, and there are no systematic measurements above the surface layers. Moreover, most of the surface measurements are made near cities. It is perfectly well established that if a city has oil-fired or coal-fired power plants, there will be sulfur in the air; if it has automobiles on crowded streets, there will be carbon monoxide and oxides of nitrogen. The most intensive monitoring efforts near cities cannot provide information concerning the global buildup of pollutants.

Water vapor and ozone have been measured above the surface because of their interest as natural trace substances. Moisture sensors are included on the same balloon flights that collect wind and temperature information, but the sensors only operate up to about six to eight kilometers. Mastenbrook's special frost-point hygrometer, which records moisture content up to about 30 kilometers, makes only one flight a month from Washington, D.C.—pitifully inadequate for a study of the global distribution of stratospheric moisture. About six balloon stations measure ozone up to about

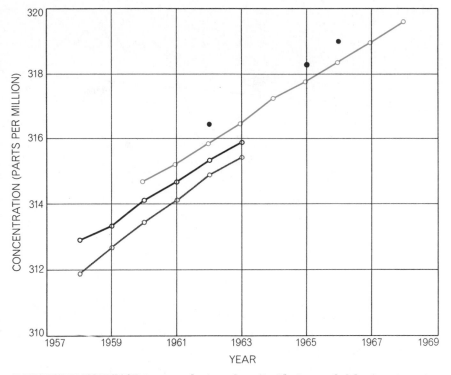

LONG-TERM INCREASE in atmospheric carbon dioxide is revealed by four investigations. The data are from measurements at Barrow, Alaska, by John Kelley of the University of Washington (black dots), aircraft observations by Bert Bolin and Walter Bischof of the University of Stockholm (color) and measurements at Mauna Loa in Hawaii (black curve) and in the Antarctic (gray) by Charles Keeling of the Scripps Institution of Oceanography.

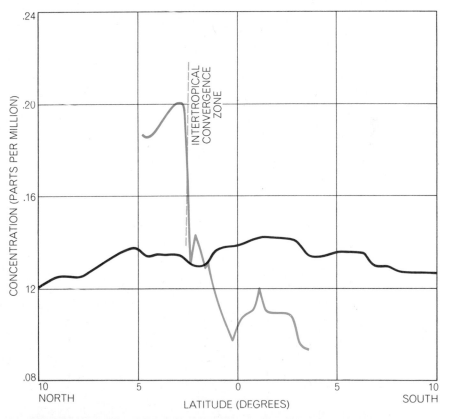

CARBON MONOXIDE measurements made from a ship at sea (color) and from an aircraft at 10 kilometers (black) by Christian E. Junge and Walter Seiler of the University of Mainz reveal a change in concentration at sea level but not in the upper troposphere. The boundary is explained by the larger number of automobiles in the Northern Hemisphere and the rising motion of the Hadley-cell circulation at the Intertropical Convergence Zone (broken line). Higher up eddy currents mix the air, eliminating the interhemisphere gradient.

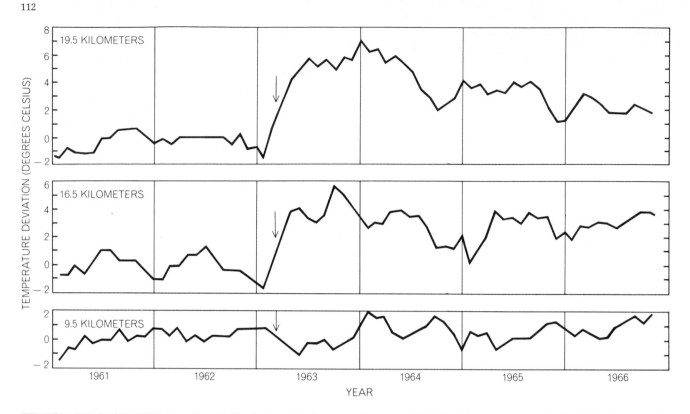

TEMPERATURE CHANGES above Port Hedland, Australia, show the heating effect (first noted by James G. Sparrow of the University of Adelaide) of particles from the 1963 eruption of Mount Agung on Bali. Monthly means were calculated for five years before the eruption (*arrows*); deviations from the means were computed and three-month running averages were plotted.

30 kilometers once a week. The Atomic Energy Commission launches balloons from four sites to obtain samples from up to 35 kilometers, mainly for analysis of radionuclides, although carbon dioxide has sometimes been included. High-altitude sampling aircraft such as the U-2 have also been used, but as nuclear tests in the atmosphere have decreased these sampling programs have naturally been pared. Practically all the published data on carbon monoxide measured away from the surface is represented in Junge's results. (The instrument he used was made portable and carried on a commercial jet flight; it occupied one seat, the observer another.) A few spot measurements of sulfur dioxide and sulfate above the surface layers have been made by Hans Georgii of the University of Frankfurt; again no extensive coverage is available.

There are many opportunities to collect data on atmospheric trace substances by taking advantage of commercial airline flights and regular ocean voyages as well as special oceanographic cruises, the unique AEC balloon network, the surface-air sampling networks established to monitor global radioactivity levels, mountaintop observatories and so on. Now that some important trace substances are being introduced into the atmosphere by man at rates comparable to those in the natural cycle, it seems appropriate to start making these measurements.

I must mention in closing that one cannot take the global view of pollutants without feeling some concern about the rate of use of natural resources and the rate of generation of pollutants. Both could be slowed considerably by a serious effort to use every pound of fuel in the most efficient manner possible.

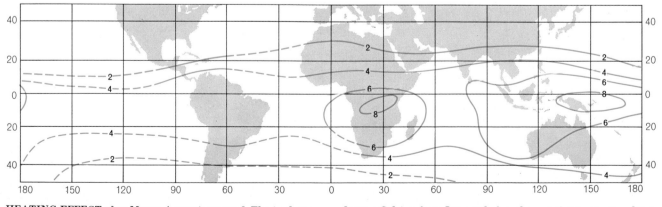

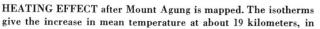

HEATING EFFECT after Mount Agung is mapped. The isotherms give the increase in mean temperature at about 19 kilometers, in degrees Celsius, from January before the eruption to one year later. (Broken lines indicate uncertainty due to scarcity of observations.)

10

Mercury in the Environment

INTRODUCTION

Never has there been a bigger outcry about the dangers of toxic metals in the environment than that about mercury in the late 1960s and early 1970s. The newspapers were filled with stories of high mercury concentrations in various foods, bans on the catching and sale of some kinds of fish, and the fearful consequences of mercury poisoning. In quick succession scientists of all kinds, toxicologists, public health physicians, chemists, geologists, geochemists, limnologists, and oceanographers, made pronouncements on the quantities of mercury in the environment, its concentration, and its chemical forms. Along with new and better data—for chemists quickly realized that accurate determinations of very small quantities of mercury in various chemical forms was no mean task—came the realization that we were dealing with another complex natural elemental cycle and that man's effect on it might not be as profound as once thought. It also brought us up sharply to the recognition that we had little or no historical baseline from which we could estimate how things had changed in the last half century. The best we could do was to analyze specimens of fish preserved in museums—and some of them seemed to be as high in mercury as fresh ones. But even here we are limited because the preservatives used may have had some contamination from mercury in their manufacture.

In this article, "Mercury in the Environment," Leonard Goldwater, a physician long involved in public health, brings clarity and reason to the discussion. He considers the nature of mercury toxicity in humans and observes that mercurials have been used for centuries as medicine. The mercury cycle in nature is well illustrated and serves as a background for a sober evaluation of how much man-made mercury is being added to the natural cycle and at what points. As with so many other environmental concerns, we are reminded that our precise quantitative knowledge about mercury and its effects is small compared with the amount of knowledge we need to predict confidently where and how mercury poisoning might be a serious public health matter, locally or globally. As we accumulate that information, we would be well advised to be prudent but not hysterical about the dangers of mercury.

Mercury in the Environment

by Leonard J. Goldwater
May 1971

The metal is widely distributed, mostly in forms and amounts that do no harm. The question is whether its concentration by industrial and biological processes now endangers animals and human beings

In the early 1950's fishermen and their families around Minamata Bay in Japan were stricken with a mysterious neurological illness. The Minamata disease, as it came to be called, produced progressive weakening of the muscles, loss of vision, impairment of other cerebral functions, eventual paralysis and in some cases coma and death. The victims had suffered structural injury to the brain. It was soon observed that Minamata seabirds and household cats, which like the fisherfolk subsist mainly on fish, showed signs of the same disease. This led to the discovery of high concentrations of mercury compounds in fish and shellfish taken from the bay, and the source of the mercury was traced to the effluent from a factory.

Since then there have been several other alarming incidents. In 1956 and 1960 outbreaks of mercurial poisoning involving hundreds of persons took place in Iraq, where farmers who had received grain seed treated with mercurial fungicides ate the seed instead of planting it. There were similar outbreaks later in Pakistan and in Guatemala. In Sweden, where poisoning of game birds and other wildlife, apparently by mercury-treated seeds, began to be noticed in 1960, the Swedish Medical Board in 1967 banned the sale of fish from about 40 lakes and rivers after it was found that fish caught in those waters contained high concentrations of methyl mercury. In 1970 alarm rose to a dramatic pitch in North America. Following the discovery of mercury concentrations in fish in Lake Saint Clair by a Norwegian investigator working in Canada, restrictions on fishing and on the sale of fish were imposed in many areas in the U.S. and Canada, and government agencies in both countries began to take action to control the discharge of mercury-containing wastes into lakes and streams.

Suddenly, almost overnight, mankind has become acutely fearful of mercury in the environment. The alarm is understandable. Quicksilver has always been regarded as being magical and somewhat sinister, in part because of its unique property as the only metal that is a liquid at ordinary temperatures. Mercury's peculiarities have been recognized since medieval times, when the alchemists took a keen interest in the element's fascinating properties. Its toxic properties became so well known that some mercury compounds came to be used as agents of suicide and murder. There are indications that Napoleon, Ivan the Terrible and Charles II of England may have died of mercurial poisoning, either accidental or deliberate. (Charles II experimented with mercury in his laboratory.) It has been suggested (incorrectly) that mercury is what made Lewis Carroll's Hatter mad (since it is used in the manufacture of felt hats). And it is authentically recorded that as early as 1700 a citizen of the town of Finale in Italy sought an injunction against a factory making mercuric chloride because its fumes were killing people in the town.

Nevertheless, although the recent incidents give us justifiable concern about the potential hazards of mercury in the environment, a panicky reaction would be quite inappropriate. Mercury, after all, is a rare element, ranking 16th from the bottom of the list of elements in abundance in the earth and comprising less than 30 billionths of the earth's crust. There are comparatively few places in the world where it occurs naturally in more than trace amounts, and the ore-bearing deposits of commercial value are so limited that a handful of mines scattered over the globe account for most of the world production. The uncompounded element in liquid form is not a poison; a person could swallow up to a pound or more of quicksilver with no significant adverse effects. Nor should it be forgotten that certain compounds of mercury have been used safely for thousands of years, and some are still prescribed, as effective medications for various infections and disorders. We need not shrink from mercury as an unmitigated threat. What is now required is detailed investigation of how mercury is being redistributed and concentrated in the environment by man's activities and in what forms and compounds it may be harmful to life. Extensive research on these questions is under way.

We consider first what might be called the normal distribution of mercury in nature. The element is found in trace amounts throughout the lithosphere (rocks and soil), the hydrosphere, the atmosphere and the biosphere (in tissues of plants and animals). In the rocks and soils (apart from ore concentrations) mercury is measured in fractions of one part per million, except in topsoils rich in humus, where the amount may run as high as two parts per million. In the hydrosphere (the seas and fresh waters) it generally occurs only in parts per billion. In the atmosphere mercury is present both as vapor and in the form of particles. Under natural conditions, however, the amount is so small that extremely sensitive methods are required to detect and measure it; the measurements that have been made at a few locations indicate that this atmospheric "background" amounts to less than one part per billion. The situation is somewhat different when we come to the biosphere. Plants and animals tend to concentrate mercury; it has been found, for example, that some marine algae contain a concentration more than 100 times higher than that in the seawater in which they live, and one study of fish in the sea showed mercury concentrations of up to 122 parts per billion. There is considerable variation, as we shall see, in the amounts of mercury found in plants and animals, depending on circumstances. Under natural conditions, however, the concentration in the earth's vegetation (aside from cultivated plants) averages no more than a fraction of one part per million.

Thus the natural cycle of circulation of mercury on the earth [see top illustration on page 116] disperses it widely through the habitable spheres in trace amounts that pose no hazard to life. How seriously has man altered its distribution?

The only ore containing mercury in sufficient concentration for commercial extraction is cinnabar, or mercuric sulfide (HgS). There are minable cinnabar deposits in many regions around the world, and man was attracted to its use as early as prehistoric times. There is evidence that cinnabar was mined in China, Asia Minor, the Cyclades and Peru at least two or three millenniums ago. Cinnabar, a brilliant red mineral, came into use at first as a pigment, but it was not until medieval times that physicians and other investigators became interested in extracting quicksilver from ore to produce medicines and other useful compounds. Hippocrates is believed to have prescribed mercury sulfide as a medication, and this was probably one of the first compounds of a metal to be employed therapeutically.

By the Middle Ages, when alchemists had synthesized chlorides, oxides and various other inorganic compounds and mixtures of mercury, its use in medications began to spread. Calomel (mer-

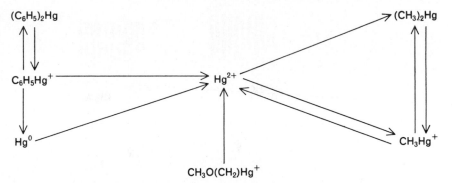

METHYL MERCURY COMPOUNDS are the most injurious ones. According to Arne Jernelöv of the Swedish Water and Air Pollution Research Laboratory, mercury discharged into water in various forms can be converted by bacteria in detritus and sediments into methyl and dimethyl mercury (*right*). Phenyl mercurials, metallic mercury and methoxyethyl mercury (*left and bottom*) are converted into methyls primarily through ionic mercury.

curous chloride, or HgCl) came into wide use as a cathartic, and in the 16th century mercury compounds were introduced as a treatment for syphilis. By the 19th century scores of mercurials were being employed in medicine. Many are still in the pharmacopoeias; the most useful ones today are the diuretics. It has been found that even among the organic compounds of mercury there are some that can be used safely. As much as 78.5 grams of mercury in the form of an organic compound has been given to a patient without harmful side effects!

The use of cinnabar as a coloring agent and of mercury compounds in

pharmaceuticals under careful control introduced no threat to the quality of the environment. With the development of other applications, however, particularly in industry and agriculture, came serious problems. The extraction of mercury from the ore by heating (that is, distillation) dangerously contaminates the air in localized areas with mercury vapor and dust (as the protesting citizen of Finale observed nearly three centuries ago). Mercury today is used on a substantial scale in chemical industries, in the manufacture of paints and paper and in pesticides and fungicides for agriculture. The world production of mercury

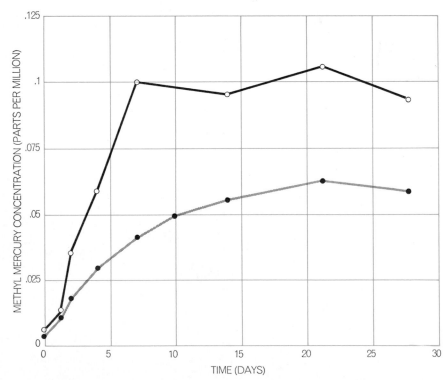

CONVERSION of inorganic into methyl mercury in sediment was measured by Sören Jensen and Jernelöv at intervals after the addition of 10 (*gray*) and 100 (*black*) parts per million of inorganic mercury. At lower mercury concentrations methylation may not occur.

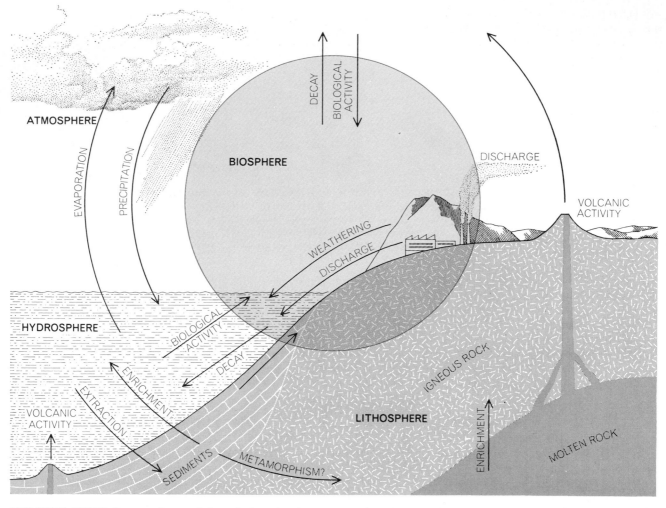

MERCURY CYCLE disperses the metal through the lithosphere, hydrosphere and atmosphere and through the biosphere, which interpenetrates all three. Mercury is present in all spheres in trace amounts, but it tends to be concentrated by biological processes. Man's activities, in particular certain industrial processes, may now present a threat by significantly redistributing the metal.

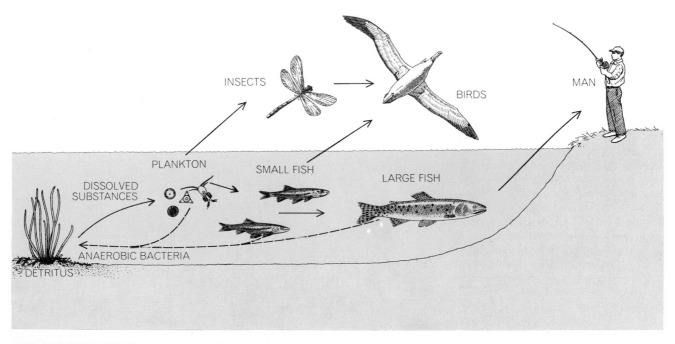

AQUATIC FOOD CHAIN is a primary mechanism by which mercury is concentrated. At each trophic level less mercury is excreted than ingested, so that there is proportionately more mercury in algae than in the water they live in, more still in fish that feed on the algae and so on. Bacteria and the decay chain (*broken arrows*) promote conversion of any mercury present into methyl mercury.

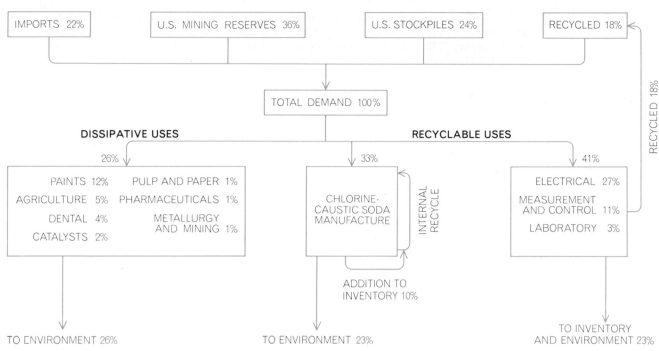

MERCURY FLOW through the U.S. is shown for 1968. The chart is based on one prepared by Robin A. Wallace, William Fulkerson, Wilbur D. Shults and William S. Lyon of the Oak Ridge National Laboratory. Major use of mercury has been as a cathode in the electrolytic preparation of chlorine and caustic soda. In this process a large inventory of mercury is continuously recycled, but in 1968 23 percent of total mercury demand still went to make up what was wasted. Another 10 percent went for start-up of new plants. Since then legislation and lawsuits have required manufacturers to increase recycling sharply, reducing emissions to the environment.

now amounts to about 10,000 tons per year, of which about 3,000 tons are used in the U.S. (The principal producers are Spain, whose mines at Almadén are the richest in the world, Italy, the U.S.S.R., China, Mexico and the U.S.) From the large-scale uses a considerable amount of mercury wastes is flowing into the air, the soil and streams, lakes and bays.

In agriculture, for example, corrosive sublimate (HgCl$_2$) is used to disinfect seeds and to control many diseases of tubers, corms and bulbs (including potatoes). The chlorides of mercury (both mercuric and mercurous) are also employed to protect a number of vegetable crops. In recent decades farmers in Europe and the U.S. have adopted the use of organic compounds of mercury, some of which are highly toxic, principally to prevent fungal diseases in seeds and in growing plants, fruits and vegetables. These chemicals may present a potential threat to health through the ingestion of treated seeds by birds (and people), through concentration in food plants and through percolation or runoff from the fields into surface waters. The U.S. Geological Survey, after analyzing the concentration of mercury in a number of U.S. rivers in 1970, reported that although the mercury content was only one part per 10 billion or less in most rivers and streams, "it may be several thousand times this concentration in some natural waters."

In order to evaluate the hazards of mercury in the environment we must examine the forms in which it occurs there and the relative toxicity of its various compounds. Liquid mercury itself, as we have already noted, is not ordinarily toxic to man. Inhalation of mercury vapor, however, can be injurious. In acute cases it causes irritation and destruction of the lung tissues, with symptoms including chills, fever, coughing and a tight feeling in the chest, and there have been reports of fatalities from such exposure. The acute exposures, however, usually come about not from the general environment but by accident, such as heating a household mercurial. More common is a chronic form of injury resulting from occupational exposure to mercury vapor, for example among mercury miners and workers in felt hat factories employing mercury nitrate for processing. These exposures, as we have found in examinations of miners, are not necessarily incapacitating; they produce tremors, inflammation of the gums and general irritability.

The soluble inorganic salts of mercury have long been known to be toxic. Mercury bichloride (corrosive sublimate), which has been used on occasion for suicide and homicide, produces corrosion of the intestinal tract (leading to bloody diarrhea), injury to the kidneys, suppression of urine and ultimately death from kidney failure when it is taken by mouth in a substantial dose. Its former use in moderate doses by mouth for treatment of syphilis did not, however, result in observable poisoning in most cases. Mercurous chloride is less soluble than the mercuric salt and therefore is less dangerous. It is still used medicinally, but some of its uses have been abandoned because it was found to cause painful itching of the hands and feet and other symptoms in children. Among other inorganic mercurials, some of the oxides, such as the red oxide used in antifouling paint for ship bottoms, may be potentially hazardous. In general, however, these inorganic mercurials are not important factors in contamination of the general environment.

What does cause us concern now with regard to the environment is the presence of some of the organic compounds of mercury, specifically the alkyls: the methyl and ethyl compounds. In Minamata Bay the substances that had poisoned the fish and people were identified as methyl mercurials. The grain that caused outbreaks of illness and death among the farmers of Iraq had been treated with ethyl mercury p-toluene sulfonanilide. And alkyls of mercury were similarly incriminated in Sweden and other places.

It has been known for some time that alkyl mercury can cause congenital mental retardation, and recent laboratory studies have shown that it can produce abnormalities of the chromosomes and, through "intoxication" of the fetus in the uterus, can bring about cerebral palsy. The alkyl mercurials attack the brain cells, which are particularly susceptible to injury by this form of mercury. The chemical basis of this effect seems to be mercury's strong affinity for sulfur, particularly for the sulfhydryl groups (S-H)

in proteins (for which arsenic and lead have a similar affinity). Bound to proteins in a cell membrane, the mercury may alter the distribution of ions, change electric potentials and thus interfere with the movement of fluid across the membrane. There are also indications that the binding of mercury to protein disturbs the normal operation of structures such as mitochondria and lysosomes within the cell. Alkyl mercury appears to be especially dangerous because the mercury is firmly bonded to a

carbon atom, so that the molecule is not broken down and may maintain its destructive action for weeks or months. In this respect it differs from the inorganic and phenyl (aryl) mercurials, and that may explain why it produces permanent injury to brain tissue, whereas the injury caused by inorganic and aryl mercurials is almost invariably reversible.

At one time it was thought that aryl mercurials (compounds based on the phenyl group) might act like the alkyls; however, in an extensive series of studies I initiated in 1961 at the Columbia University School of Public Health we found that chemical workers who continually handled phenyl mercurials and experienced exposures far above the supposedly safe limit did not show any evidence of toxic effects.

Having examined the nature of the mercury "threat" and its quantitative presence in the environment, we should now look at the other side of the equation: the extent of man's exposure and his response to this factor up to now.

Without question the major source of man's intake of mercury is his food. Alfred E. Stock in Germany initiated analyses of the mercury concentration in foods in the 1930's, and there have been several follow-up studies since then, including one by our group at Columbia in 1964. (Numerous further investigations are now in progress.) The measured concentrations in samples from various sources are in the range of fractions of one part per million [see illustration on facing page], but the concentrations in fish in contaminated waters may run several hundred times higher than that. A joint commission of the Food and Agriculture Organization and the World Health Organization proposed in 1963 that the permissible upper limit for mercury in foods (except fish and shellfish) should be .05 part per million; there is as yet no firm basis, however, for determining what the safe standard ought to be. Perhaps the most significant conclusion that can be drawn from these sets of samplings is that in general the concentration of mercury in foods does not appear to have changed substantially over the past 30 years. The comparisons may not be entirely valid, however, because of differences in analytical methods.

In addition to food, there are other possible everyday sources of exposure to mercury. It is used fairly commonly in antiseptics, paint preservatives, floor waxes, furniture polishes, fabric softeners, air-conditioner filters and laundry preparations for suppression of mildew; no doubt there are other such exposures

ATMOSPHERIC MERCURY LEVELS were measured by S. H. Williston at a station south of San Francisco. The concentration averaged about .0002 microgram of mercury per cubic meter of air when the wind blew from the Pacific (a) and was somewhat higher when the wind was from the generally nonindustrial southeast (b). The average was .008 microgram, with many peaks that went off the record at .02 microgram, when the wind was from the industrial area to the northeast (c). The mercury was often associated with dust particles.

of which we are not aware. In view of all these factors, it is not surprising to find that 20 to 25 percent of the "normal" population—persons who have apparently had no medicinal or occupational exposure to mercury—show easily measurable amounts of mercury in their body fluids. Several studies of this matter have been made, including a fairly extensive international investigation we carried out at Columbia in 1961–1963 as a joint project with the WHO. Analyzing 1,107 specimens of urine collected from "normal" subjects in 15 countries, we found that except in rare instances the mercury content in the urine ran no higher than about 20 to 25 parts per billion. A similar examination of blood samples showed that the highest mercury concentration in the blood among "normal" subjects was 30 to 50 parts per billion. And analyses of human tissues made at autopsy have indicated that similar traces of mercury are present in the body organs.

It is important to consider these findings in the context of the evolutionary relationship of life on our planet to the presence of mercury. Unquestionably the element has been omnipresent in the sea, where life originated, from the beginning, and presumably all plants and animals carry traces of mercury as a heritage from their primordial ancestry. Man as the top of a food chain must have added to that heritage by eating fish and other mercury-concentrating forms of food. Over the millions of years he presumably has built up an increased tolerance for mercury. (The development of tolerance for chemicals is of course well recognized today. It was put to use more

than 2,000 years ago by Mithradates the Great, king of Pontus, who armed himself against poisoners by taking small and increasing doses of toxic agents.) Tolerance for a potent substance not infrequently grows into dependence on it, and it is reasonable to suppose that man, as well as other forms of life, may now be dependent on mercury as a useful trace element. Whether its effects are beneficial or harmful may be influenced decisively by the form in which it is incorporated in tissues, by the dose and probably by other factors. It has been found, for instance, that the highly toxic element arsenic is sometimes present in healthy shrimp in concentrations of close to 200 parts per million (dry weight) in the form of trimethylarsine. Methylation in this case apparently suppresses the element's toxicity. The biochemical behavior of mercury has much in common with that of arsenic, which suggests that there may be harmless forms of methyl mercury as well as toxic ones in fish.

Our concern, then, must be with any disturbance of the environment that alters the natural balance of mercury in relation to other substances or that generates virulent forms of mercurials. In the case of the fish in Minamata Bay apparently both factors were at work. The polluting effluent from the chemical plant itself contained methyl mercury, and elemental mercury in the effluent was methylated by microorganisms in the mud on the bottom of the bay. This conversion was fostered by the enrichment of the water with a high concentration of mercury and organic pollutants that promoted the growth of the methylating bacteria. The result was the accu-

mulation in the fish of concentrations of methyl mercury as high as 50 parts per million (wet weight), which is 100 times the *total* mercury concentration currently accepted as "safe" in the U.S. and Canada. The effect on the fishermen and their families was compounded by the fact that their diet consisted largely of the bay fish and may have been deficient in some essential nutrients; dietary deficiencies are known to enhance the adverse effects of toxic agents.

The current journalistic outcry on the "mercury problem" has produced a state of public alarm approaching hysteria. "Protective" measures are being proposed and applied without basis in established knowledge. Research on mercury poisoning in the past has focused primarily on occupational hazards involving prolonged exposure, principally by way of inhalation. Mercury in the general environment, however, presents an almost entirely different problem. Discharge of mercury into the atmosphere, either as vapor or as particulate matter, is not likely to become a serious general hazard. The main threats to which we shall have to give attention are solid and liquid wastes that may ultimately enter bodies of water, thus threatening fish and eaters of fish, and agricultural uses of mercury that may dangerously contaminate food. We do not yet have enough information to estimate the magnitude of these threats or to establish realistic standards of control.

To begin with, we need a better understanding of what should be considered a toxic level of mercury in the human body. Analysis of the mercury con-

	A	B	C	D	E
MEATS	.001 — .067	.005 — .02	.0008 — .044	.31 — .36	.001 — .15
FISH	.02 — .18	.025 — .18	.0016 — .014	.035 — .54	0 — .06
VEGETABLES (FRESH)	.002 — .044	.005 — .035	0	.03 — .06	0 — .02
VEGETABLES (CANNED)			.005 — .025		.002 — .007
MILK (FRESH)	.0006 — .004	.0006 — .004	.003 — .007	.003 — .007	.008
BUTTER	.002 (FATS)	.07 — .28			.14
CHEESE	.009 — .01				.08
GRAINS	.02 — .036	.025 — .035	.002 — .006	.012 — .048	.002 — .025
FRUITS (FRESH)	.004 — .01	.005 — .035		.018	.004 — .03
EGG WHITE				.08 — .125	.01
EGG YOLK				.33 — .67	.062
EGG (WHOLE)	.002	.002	0		
BEER	.00007 — .0014	.001 — .015			.004

CONCENTRATION IN FOODS was reported by Alfred E. Stock in Germany in 1934 (*A*) and 1938 (*B*), O. S. Gibbs in the U.S. in 1940 (*C*), Y. Fujimura in Japan in 1964 (*D*) and the author's group in the U.S. in 1964 (*E*). A listing of "0" means simply a concentration too low to be detected by the method used. The World Health Organization proposed a permissible upper limit of .05 part per million for foods other than fish; the U.S. Food and Drug Administration has set an upper limit of .5 part per million for fish.

centration in the urine or in the blood has not given much enlightenment on this point. Among workers exposed to mercury in their occupations it has been found that the mercury content in the urine varies greatly from day to day and from one individual to another. As a general rule the body's excretion of mercury does tend to reflect the amount of exposure, but recent studies of workers in the chemical industries have disclosed that individuals who have been exposed to mercury in high concentrations or for long periods often show no sign of adverse effects. Furthermore, high levels in the urine or the blood do not necessarily indicate poisoning; many cases have been observed in which the individual had a mercury concentration in the blood amounting to 10 to 20 times the "normal" upper limit and yet showed no indications of illness or toxic symptoms! All in all there is substantial evidence that host factors may be more important than the amount of exposure, up to a point, in determining the individual's response to mercury in the environment. In any case, urine or blood analysis is of no value for early diagnosis of mercury poisoning. Other possible indicators, such as disturbances of the blood enzyme system, are being investigated, but no reliable diagnostic test has yet been produced. Nor can we define a precise threshold for the toxic level, either for exposure to or for absorption of mercury.

A good deal of useful information is available, on the other hand, about the sources and avenues of possible danger. Mercury in one form or another can invade the human system by way of the lungs, the skin or the ingestion of food. (Incidentally, recent studies have shown that the mercury in dental fillings is not a hazard; most people with amalgam fillings have negligible amounts of mercury in their urine or blood.) Mercury in the air, as we have noted, is a local problem confined to certain industries. Attack by way of the skin is also a problem that does not apply to the general environment, although direct contact with organic mercurials can cause severe (second degree) local burns of the skin and can result in absorption of measurable quantities of mercury into the body from underclothing and bed linen.

With regard to food, we can identify the most important possible sources of trouble. There is abundant evidence that the inorganic compounds of mercury and the phenyl mercurials are relatively nontoxic compared with the alkyl mercurials. We need to be concerned, however, about the potential conversion of the inorganic forms and the phenyls to methyl

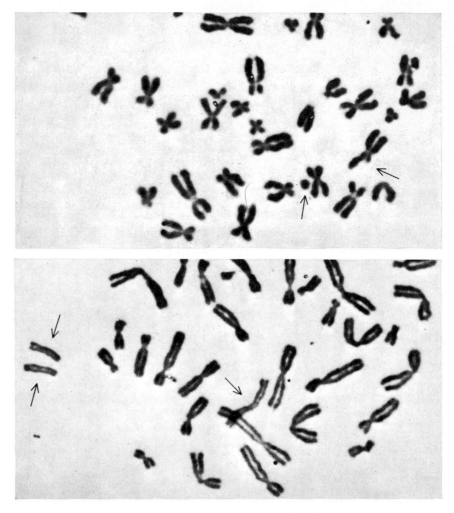

CHROMOSOME DAMAGE is found in persons with high blood levels of mercury after exposure to methyl mercury in fish. Photographs of lymphocytes made by Staffan Skerfving of the Swedish National Institute of Public Health show a broken chromosome and extra fragment (top) and three sister-chromatid fragments that lack centromeres (bottom).

mercury, which on the basis of the experience so far must be rated as the prime hazard in the environment. A number of alkyl mercury fungicides have already been eliminated, quite properly, from use on food crops. Government agencies are now beginning to move to ban other organic mercurials as well from any use that might contaminate our food or water.

A calm view of the present state of affairs regarding mercury in the environment suggests that the best way to deal with the problem is to apply the techniques of epidemiology, preventive medicine, public health and industrial hygiene that have been effective in meeting hazards in the past. A system should be set up for frequent monitoring of the environment for the detection of significant increases in mercury contamination. Research should be carried forward to establish measures for the levels and forms of mercurial pollution that signal a

threat to health. Techniques for mass screening of the population to detect mercurial poisoning should be developed. Controls should be applied to stop the discharge of potentially harmful mercury wastes at the point of origin. Toxic mercurials in industry and agriculture should be replaced by less toxic substitutes. To implement such a program we shall need, of course, realistic education of the public and legislative action with adequate enforcement. And these measures should be applied to all contaminants that threaten man's environment, not only mercurials.

It would be foolish to declare an all-out war against mercury. The evolutionary evidence suggests that too little mercury in the environment might be as disastrous as too much. In the case of mercury, as in all other aspects of our environment, our wisest course is to try to understand and to maintain the balance of nature in which life on our planet has thrived.

III

PLATE TECTONICS, SEA-FLOOR SPREADING, AND CONTINENTAL DRIFT

The ascendancy of the concept that the earth's outermost shell is broken into about a dozen rigid plates is one of the highlights of science of the past decade. The relative motions of the plates lead to a concentration of deformational forces at plate boundaries. Plate tectonics deals with the geological manifestations—the mountain making, earthquake activity, volcanism, and so forth—that occur when plates collide, drift apart, or slide by one another.

The plates are ephemeral on the geological time scale; they are mostly created and destroyed in a time scale of some 200 million years. The plates evolve along mid-ocean ridges in a process of sea floor spreading, where new sea floor is created. Old sea floor is destroyed elsewhere to accommodate the new material.

The old idea of continental drift—that continents can split and drift apart—has been revitalized as part of the theories of plate tectonics and sea-floor spreading. The continents are raft-like inclusions carried passively by the plates. Sea-floor spreading can begin with the splitting of a continent and new plate material will fill the widening gap.

We include five recent articles to review these concepts and show something of the impact they have had on the geosciences. The first article sets the stage with a description of the geometry of plate motions, a summary of the supporting evidence, and a discussion of continental drift and geological phenomena at plate boundaries.

The second provides us with an example of mountain making on a grand scale, the uplifting of the Andes at the boundary between the colliding South American and Nazca plates. As an example of sea-floor spreading we use a newly developed history of the Indian Ocean. While not as simple a story as that of the Atlantic, the evolution of the Indian Ocean is an excellent example because it shows how sea-floor spreading can become complicated when plate boundaries shift and the direction of spreading changes. Paleontology is one of the many disciplines that will benefit from these new concepts. The diversity of species is affected by the degree of isolation, a factor strongly influenced by continental drift and the article by Hallam shows how the plate tectonics concept can aid in the study of evolution and how the fossil record can contribute to our ability to reassign continents to their former positions.

The final article raises the interesting possibility that the distribution of mineral deposits may be associated with present and former plate boundaries. Should this prove true, a valuable new tool would become available to exploration geologists and an untouched area, the deep-sea floor, would become a candidate for major new mineral discoveries.

11

Plate Tectonics

INTRODUCTION

Geoscientists will long remember the decade of the sixties, which saw the development and growing acceptance of the concept of plate tectonics. The theory holds that the outermost spherical shell of the earth, to a depth of about 100 kilometers, is broken into about a dozen rigid plates whose shifting leads to a changing pattern of earth's surface features. Plates are created from magma rising from the interior along mid-ocean ridges. Plates spread from the region of their creation, cool off, and become rigid and strong, only to be reabsorbed into the interior by subduction under deep-sea trenches. Ocean basins have been created and destroyed by this process of sea-floor spreading for the last 200 million years. The continents embedded in the plates drift like rafts as the plates move. Perhaps it is the fact that they are made of relatively light rock that enables continents to survive for as long as 4 billion years, even though they have been split, re-assembled, crumpled, uplifted and down-warped while floating on the moving plates. To the geologist trying to fit together continental history, it is like working on a jigsaw puzzle whose pieces continually change shape.

Many large-scale geologic features are associated with plate boundaries. Earthquake belts, chains of volcanoes, young mountain systems, deep-sea trenches occur where plates separate or collide. The San Andreas Fault in California, along which catastrophic earthquakes have occurred in the past and will occur in the future, marks a boundary along which the Pacific Plate is sliding northward alongside the North American Plate.

John F. Dewey's article, "Plate Tectonics" reviews the geometry of plate motions and describes the diverse geological phenomena that are associated with plate boundaries of various types. He shows how the old theory of continental drift can be integrated into modern concepts of sea-floor spreading. Although the evidence for both theories is overwhelming in its ability to explain earth's history since Jurassic time, that is, for the last 200 million years, history earlier than that is less complete and more difficult to decipher. Dewey speculates on the possibility of piecing together pre-Jurassic plate tectonics so that even earlier episodes of sea-floor spreading and continental drift could be recognized.

Plate Tectonics

by John F. Dewey
May 1972

The earth's surface is divided into a mosaic of rigid, shifting plates. As they move apart, slide past each other and converge, new crust is generated, continents drift and mountains are formed

It has long been observed that mountains, volcanoes and earthquakes are not randomly distributed over the earth's surface but are found in distinct and usually narrow zones. To account for these evidences of instability in the earth's crust many hypotheses have been put forward. They have included such diverse notions as global expansion, global contraction, the effect of lunar tidal forces and wholesale uplift or foundering of large segments of the earth's crust. One other explanation—continental drift—was advanced from time to time but was unpalatable to most geophysicists because it seemed to violate what was known about the mechanical properties of the earth's crust. Nevertheless, continental drift seemed to explain geological similarities between continents thousands of miles apart. It also explained why some continental margins, for example those of South America and Africa, match each other so precisely.

Within the past 10 years continental drift has been placed on a firm foundation by the development of the concept of sea-floor spreading, originally proposed by the late Harry H. Hess of Princeton University. Sea-floor spreading involves the notion that the floor of the ocean is continuously being pulled apart along a narrow crack that is centered on a ridge that can be traced through the major ocean basins. Volcanic material (liquid basalt) rises from the earth's mantle to fill the crack and continuously create new oceanic crust.

The concept might have been difficult to confirm except for the fortunate fact that the polarity of the earth's magnetic field periodically reverses. It had been observed from magnetometer surveys that rocks of the ocean floor exhibit a zebra-stripe pattern in which the intensity of magnetization changes abruptly in linear ribbons parallel to the nearest oceanic ridge. In 1963 F. J. Vine and D. H. Matthews of the University of Cambridge proposed that the magnetic pattern was evidence of both sea-floor spreading and reversals of the earth's magnetic field. To many geologists the one seemed almost as improbable as the other. Vine and Matthews argued that as the basaltic liquid rose into the axial crack of the oceanic ridges and solidified it would become magnetized in the then prevailing direction of the earth's magnetic field. If new oceanic crust was continuously generated, as Vine and Matthews believed, one should find that each ridge axis is bordered symmetrically by pairs of parallel strips whose direction of magnetization is the same, that is, both normal or both antinormal. The hypothesis was strikingly confirmed in many traverses across oceanic ridges. Furthermore, a time scale of magnetic reversals has been developed showing that the rate of sea-floor spreading is between two and 18 centimeters per year.

It is now clear that virtually all of the present area of the oceans has been created by sea-floor spreading during the past 200 million years, or during the last 5 percent or so of the recorded geologic history of the earth. The creation of new surface area means either that the earth has expanded dramatically or that surface area is somewhere being destroyed at the same rate at which it is being created. There is good evidence that the earth has not expanded more than about 2 percent in the past 200 million years. Thus there must be, in general terms, a global conveyor-belt system or surface motion that links zones of surface creation and surface destruction.

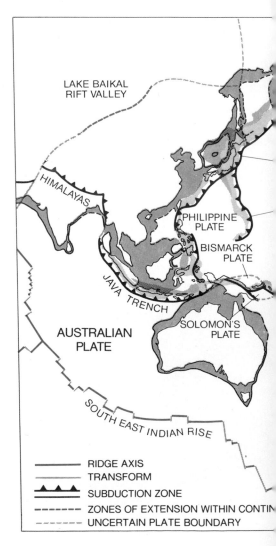

MOSAIC OF PLATES forms the earth's lithosphere, or outer shell. According to the recently developed theory of plate tectonics, the plates are not only rigid but also in constant relative motion. The boundaries of plates are of three types: ridge axes, where plates are diverging and new oceanic floor is generated; transforms, where plates slide past each other, and subduction zones, where plates converge and one plate dives under the leading edge of its neighbor. Triangles indicate the leading edge of a plate.

LAKE BAIKAL RIFT VALLEY

HIMALAYAS

PHILIPPINE PLATE

BISMARCK PLATE

JAVA TRENCH

AUSTRALIAN PLATE

SOLOMON'S PLATE

SOUTH EAST INDIAN RISE

—————— RIDGE AXIS
············· TRANSFORM
▲▲▲▲ SUBDUCTION ZONE
- - - - - ZONES OF EXTENSION WITHIN CONTIN
– · – · – UNCERTAIN PLATE BOUNDARY

The concept of sea-floor spreading has now been joined to the earlier idea of continental drift in a single unifying theme called the theory of plate tectonics. The geometric part of the theory visualizes the lithosphere, or outer shell, of the earth as consisting of a number of rigid plates. The kinematic part of the theory holds that the plates are in constant relative motion: they can slide past each other, they can move apart on opposite sides of an oceanic ridge or they can converge, in which case one of the plates must be consumed. Let us now see how various instabilities in the earth's crust can be visualized in terms of plate tectonics.

Earthquakes

Most earthquakes occur in narrow zones that join to form a continuous network bounding regions that are seismically less active. The seismic network is associated with a variety of characteristic features such as rift valleys, oceanic ridges, mountain belts, volcanic chains and deep oceanic trenches [see illustra-

tion below]. The seismic areas mark the boundaries between plates, which are largely free of earthquakes. There appear to be four types of seismic zone, which can be distinguished by their characteristic morphology and geology.

The first type is represented by narrow zones of high surface heat flow and basaltic volcanic activity along the axes of mid-oceanic ridges where earthquakes are shallow (less than 70 kilometers deep). The axes of the ridges, of course, are the active sites of sea-floor spreading. In Iceland, where the Mid-Atlantic Ridge rises above sea level, the spreading rate has been measured at about two centimeters per year.

The second type of seismic zone is marked by shallow earthquakes in the absence of volcanoes. A good illustration is the region around the San Andreas fault in California and around the Anatolian fault in northern Turkey, along both of which large surface displacements parallel to the fault have been measured [see "The San Andreas Fault," by Don L. Anderson; SCIENTIFIC AMERICAN Offprint 896].

The third type of seismic zone is intimately related to deep oceanic trenches associated with volcanic island-arc systems, such as those around the Pacific Ocean. Earthquakes there can be shallow, intermediate (70 to 300 kilometers) or deep (300 to 700 kilometers), according to where they take place in the steeply plunging lithospheric plate that borders the trench. Thus the earthquake epicenters (the points on the surface above the initial break) define a geologic structure dipping down into the earth away from the trench. These inclined earthquake zones, called Benioff zones, underlie active volcanic chains and have a variety of complex shapes.

The fourth type of seismic zone is typified by the earthquake belt that extends from Burma to the Mediterranean Sea. It consists of a wide, diffuse continental zone within which generally shallow earthquakes are associated with high mountain ranges that clearly owe their existence to large compressive forces. Earthquakes of intermediate depth occur in some areas such as the Hindu Kush and Romania. Although deep-focus

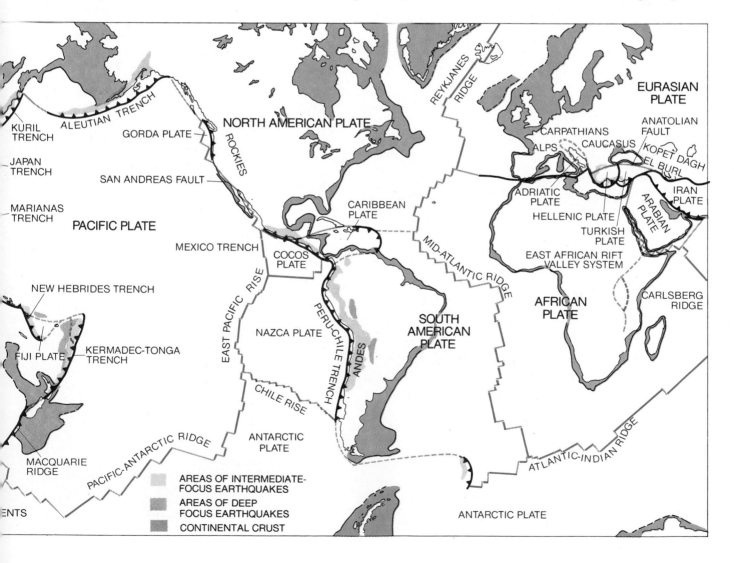

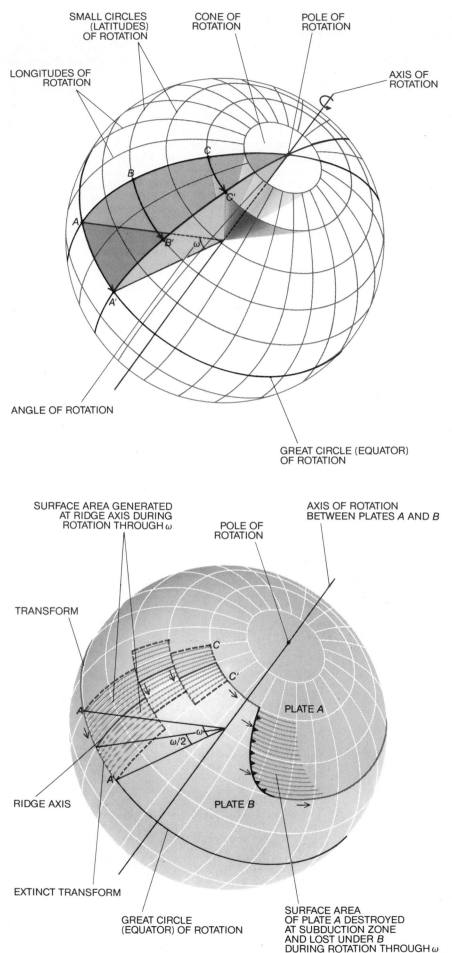

SMALL CIRCLES
(LATITUDES)
OF ROTATION

CONE OF
ROTATION

POLE OF
ROTATION

LONGITUDES OF
ROTATION

AXIS OF
ROTATION

ANGLE OF ROTATION

GREAT CIRCLE (EQUATOR)
OF ROTATION

SURFACE AREA GENERATED
AT RIDGE AXIS DURING
ROTATION THROUGH ω

POLE OF
ROTATION

AXIS OF ROTATION
BETWEEN PLATES A AND B

TRANSFORM

PLATE A

RIDGE AXIS

PLATE B

EXTINCT TRANSFORM

GREAT CIRCLE
(EQUATOR) OF ROTATION

SURFACE AREA
OF PLATE A DESTROYED
AT SUBDUCTION ZONE
AND LOST UNDER B
DURING ROTATION THROUGH ω

earthquakes are rare, they have been recorded in a few places, for example just
north of Sicily under the Tyrrhenian volcanoes.

An earthquake results when stresses
accumulate to the point that rocks in the
earth's crust break. The breakage is the
consequence of brittle failure of the rock
(in contrast to plastic deformation, which
can relieve stresses slowly). The first seismic waves to leave the region of the
break (the hypocenter) are waves of
alternate compression and rarefaction
generated by the sudden release of elastic energy. After an earthquake one finds
that the seismological stations that have
received the first waves can be assigned
to one of four geographic quadrants.
In two of the quadrants, lying opposite
each other, the first waves are compressional; in the other two quadrants the
first waves are rarefactional.

The quadrants define the orientation
of two nodal planes on one of which a
sudden slip has presumably produced
the earthquake. The intersection of the
nodal planes is the null direction, or intermediate stress axis, parallel to which
effectively no strain occurs. The line bi-
secting the quadrant in which the first
motion is compressional defines the direction of least principal stress, parallel
to which there is extensional strain. The
bisector of the rarefaction quadrant defines the direction of the maximum principal stress along which there is compressional strain.

Lynn R. Sykes of the Lamont-Doherty
Geological Observatory of Columbia
University has applied this analysis to
the seismic belts of the world and has
shown systematically that the ridge axes
are in tension, that there is lateral movement along the second type of seismic
zone and that compression dominates
the third and fourth types. Thus seis-

AXIS OF ROTATION can be selected in
such a way (*top illustration at left*) that a
set of two or more points lying on the surface of a sphere (A, B, C) can be moved by
a rigid rotation around that axis to new positions (A', B', C'), preserving the original
geometry of the set. A unique axis can be
found only if the initial and final positions
of two or more points are known. Similarly,
the relative motion of two rigid plates can
be described as a rigid rotation around a
suitably selected axis of rotation (*bottom illustration at left*). Plate A is designated
as fixed while plate B is rotated anticlockwise,
as viewed down the axis of rotation. As plate
B rotates through angle omega (ω), new
surface area is added symmetrically to both
plates at the ridge axis, which itself travels
through an angle equal to one-half omega.

mology emphasizes that there are three kinds of plate boundary: boundaries across which plates are pulled apart, boundaries along which plates slide past each other and boundaries across which plates converge. Since rock material does not pile up indefinitely in the compressional zones, it follows that somewhere there must be zones in which plates are consumed.

The Plate Mosaic

One can therefore construct a model of global surface displacement involving a mosaic of plates each of which exhibits one or more of the three types of boundary. At ridge axes plates separate and new surface area is generated by the continuous accretion of new oceanic crust at the trailing edges of the plates. At transform faults plates slide past each other and surface area is neither created nor destroyed. At subduction zones one plate is consumed and slides down into the mantle under the leading edge of another plate.

Plates vary greatly in size from the six major plates, such as the plate that carries virtually the entire Pacific Ocean, down to very small plates, such as the plate that is essentially coextensive with Turkey. Moreover, plate boundaries do not always coincide with the margins of continents; many continental margins are peaceful earthquake-free nonvolcanic regions. Hence plates can be partly oceanic and partly continental or they can be entirely one or the other. This fact overcomes one of the traditional objections to continental drift, namely the mechanical difficulty of having a geologically weak continent plow its way across a strong ocean floor. According to the plate-tectonic view, continents and oceans are rafted along by the same crustal conveyor belt.

A look at the boundary around the African plate reveals two important consequences of plate motion. The greater part of the boundary is a ridge axis extending from the North Atlantic into the Indian Ocean and the Red Sea; thus the entire African plate must be growing in area. This behavior in turn means that plates elsewhere on the globe must be getting smaller. The second consequence of the growth of the African plate is that the Carlsberg Ridge in the Indian Ocean is moving away from the Mid-Atlantic Ridge, illustrating one of the essential corollaries of plate kinematics: plate motion is relative. There is no coordinate system within which absolute plate motion can be defined except a system defined in relation to a particular plate or

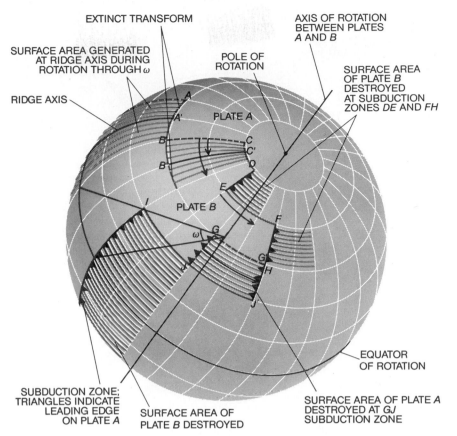

THREE KINDS OF TRANSFORM can exist as segments of a single plate boundary: ridge axis to ridge axis (*AB*), ridge axis to subduction zone (*CD*) and subduction zone to subduction zone (*EF, GH, IJ*). Plate *A* is again assumed to be fixed while plate *B* rotates anticlockwise. Ridge-axis-to-ridge-axis transforms (*AB, A'B'*) maintain a constant length because new surface area is generated symmetrically at ridge axes. Transforms joining ridge axes to subduction zones decrease or increase in length at half the transform slip rate. In the example depicted here *CD* shortens to *C'D*, but if the leading edge at subduction zone *DE* were on plate *A* (as in the case of *GJ*), *CD* would have lengthened. Transform *EF* maintains a constant length, whereas *GH* shortens to zero length and *IJ* lengthens to *IJ'*.

plate boundary that is arbitrarily chosen as being "fixed."

The basic assumption that plates are rigid is essential to plate tectonics and appears to be justified by the fact that excellent restorative fits can be made between many pairs of continental margins. (In making such a fit the margin is typically defined as the 1,000-fathom isobath on the continental shelf adjacent to the continent.) Similar fits can be made with even greater precision between pairs of magnetic anomalies symmetrically disposed on each side of a ridge axis. If plates had been distorted as they evolved, these fits could not be made. As further confirmation of rigidity, profiles produced by seismic reflection have shown that sediments laid down on the oceanic crust as it moves away from a ridge axis form undistorted flat layers.

The fact that rigid plates are in relative motion on a spherical earth means that a displacement between any two plates can be described by a rotation around an axis passing through the center of the earth. The intersection of this

axis with the earth's surface is termed the pole of rotation [*see illustration on opposite page*]. This concept was first applied by Sir Edward Bullard, J. E. Everett and A. G. Smith of the University of Cambridge to demonstrate the fit of the continental margins around the Atlantic Ocean. Relative surface motion between two plates proceeds along circles of rotation around the axis of rotation. The circles can be considered as latitudes of rotation from zero radius at the pole of rotation to a maximum at the equator of rotation. Relative plate motion is best described, however, as an angular velocity, since the velocity along rotation circles increases from zero at the pole of rotation to a maximum at the equator of rotation. The nature of displacement across a plate boundary is therefore entirely dependent on its orientation with respect to the circles of rotation.

Of particular interest are boundaries parallel to rotation circles. At these boundaries are faults where surface area is conserved; such faults are called trans-

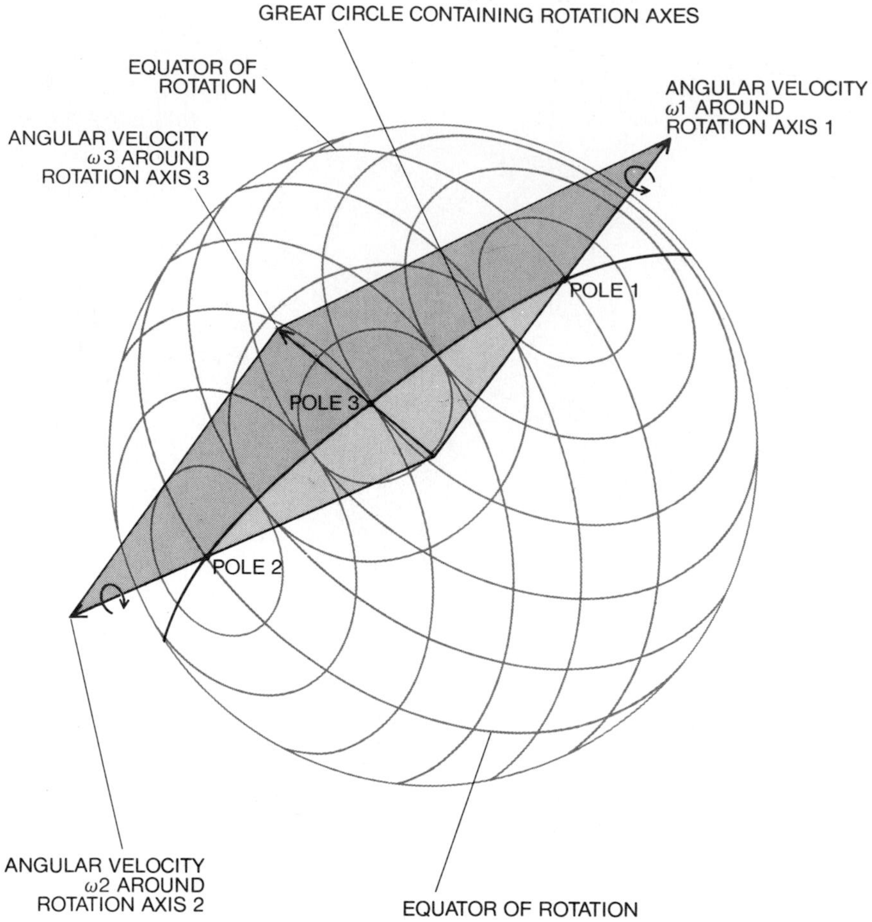

GREAT CIRCLE CONTAINING ROTATION AXES

EQUATOR OF ROTATION

ANGULAR VELOCITY ω3 AROUND ROTATION AXIS 3

ANGULAR VELOCITY ω1 AROUND ROTATION AXIS 1

POLE 1

POLE 3

POLE 2

ANGULAR VELOCITY ω2 AROUND ROTATION AXIS 2

EQUATOR OF ROTATION

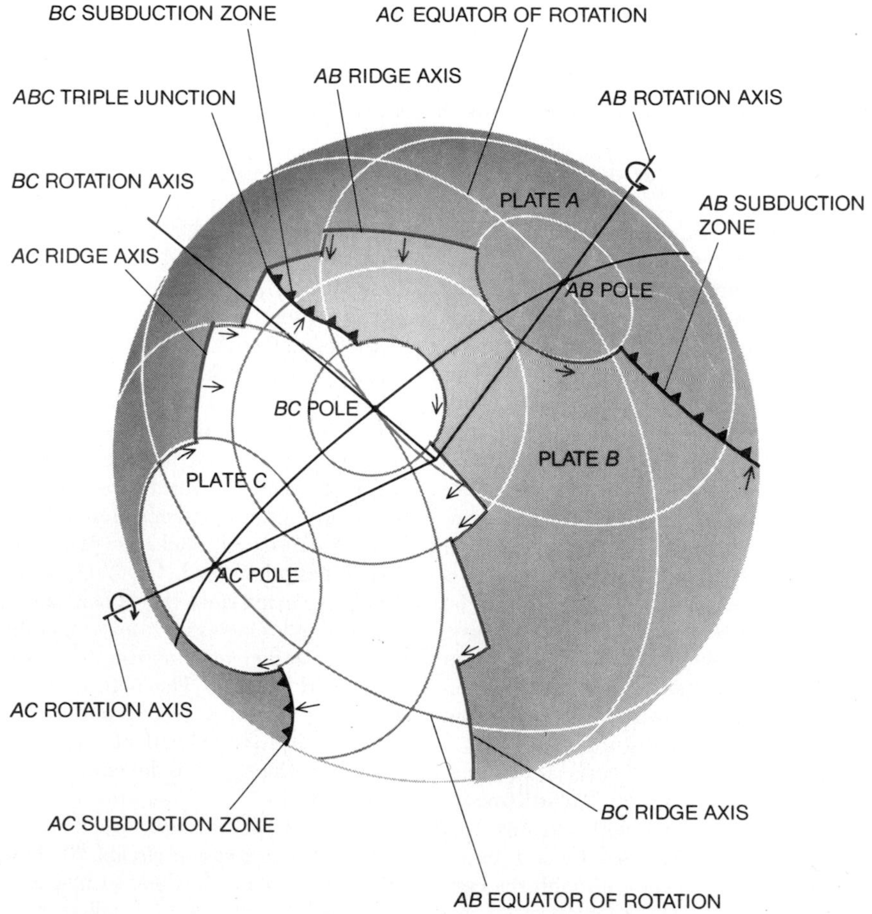

BC SUBDUCTION ZONE

AC EQUATOR OF ROTATION

ABC TRIPLE JUNCTION

AB RIDGE AXIS

AB ROTATION AXIS

BC ROTATION AXIS

AB SUBDUCTION ZONE

AC RIDGE AXIS

PLATE A

AB POLE

BC POLE

PLATE B

PLATE C

AC POLE

AC ROTATION AXIS

AC SUBDUCTION ZONE

BC RIDGE AXIS

AB EQUATOR OF ROTATION

form faults. Great circles drawn perpendicularly to transform faults that are segments of one plate boundary will intersect to define the rotation pole. Plate boundaries oblique to circles of rotation are either ridges or subduction zones, depending on whether plates are separating or converging across them. The increasing rate of plate separation across ridge axes with increasing distance from the pole of rotation is reflected by a progressively increasing distance between particular magnetic anomalies and the ridge axis. Similarly, the rate of plate convergence at subduction zones increases away from the pole of rotation. A particularly good illustration is afforded by the New Zealand–Tonga seismic zone. That part of the zone south of New Zealand has only shallow earthquakes; intermediate-focus earthquakes start in New Zealand and deep-focus earthquakes north of New Zealand [*see illustration on pages 124 and 125*]. This suggests a progessive northward increase in convergence rate across the subduction zone, so that the downgoing plate reaches progressively deeper levels.

Separation rates across ridges where plates are moving apart can be directly deduced from magnetic-anomaly patterns, but there is no direct method for deducing convergence rates across subduction zones where one plate is diving under another and creating a trench. Close attention has therefore been given to plate boundaries where individual segments are a ridge, a transform fault and a subduction zone, because the angular velocity of relative motion deduced for a ridge segment also applies to a trench segment. The angular velocity can be directly translated into a circle-of-rotation velocity for any circle of rotation that crosses the subduction zone,

THIRD AXIS and angular velocity of rotation are defined (*top illustration at left*) as the vector sum of two others. If one knows the angular velocities of two rigid rotations around two axes (*1 and 2*) passing through the center of a sphere, one obtains their vector sum to determine the angular velocity around a third axis (*3*) lying in the same plane as the first two. In the example illustrated the poles of rotation axes *1* and *2* are 90 degrees apart and the angular velocities around them (ω1 *and* ω2) are equal, so that the pole of the third axis lies midway along a great circle between poles *1* and *2*. Similarly (*bottom illustration at left*), if one knows the axes of rotation and the angular velocities that describe the relative motion of plates *A* and *B* and plates *B* and *C*, one can ascertain the rotation axis (*BC*) that describes the relative motion of plates *A* and *C*.

depending on its "rotational latitude." Although there is no apparent geometric reason why ridges should lie parallel to longitudes of rotation, they appear in most cases to approximate this relation. Furthermore, the symmetrical distribution of paired matching magnetic anomalies with respect to the ridge axis where they were generated indicates that new crustal material is added symmetrically to the trailing edges of the diverging plates.

For some reason, perhaps one related to the driving mechanism for plate motion, straight ridge-transform boundaries are a mechanically stable configuration. Subduction zones are generally curved, perhaps also for mechanical reasons, and convergence directions can be right-angled or oblique to them, according to whether they are right-angled or oblique to rotation circles. One may pass in gradations from pure subduction to pure transform motion along a single plate boundary; therefore the orientation of a subduction zone, unlike the orientation of a ridge or a transform fault, is a poor guide to the relative direction of displacement.

The axial spreading zone of oceanic ridges is not a continuous feature. It is interrupted and offset by transform faults that in some places create high submarine cliffs. Transform faults were formerly thought to be lines along which the ridge axis had been displaced from a once continuous zone, since they continue as bathymetric features beyond the offset ends of the ridge axis. J. Tuzo Wilson of the University of Toronto argued, however, that they are simply offsets of the spreading axis and form an integral part of a single boundary. He coined the term transform fault to describe them because they merely transform relative motion between the two ridge segments. Seismic first-motion studies confirmed Wilson's prediction of transform motion; further support is provided by the observation that earthquakes are restricted to that portion between the offset ends of the ridge axis.

The active portion of a transform fault defines part of a circle of rotation. Similarly, the inactive continuation of a transform fault beyond the offset ridge axis defines circles of rotation for the previous history of plate divergence across the ridge axis; it represents earlier circles of rotation "frozen" into adjacent oceanic crust generated earlier. This is of fundamental importance for two reasons. First, the excellent circle-of-rotation lines described by inactive transform faults justify the assumption that relative motion between two plates can be described in

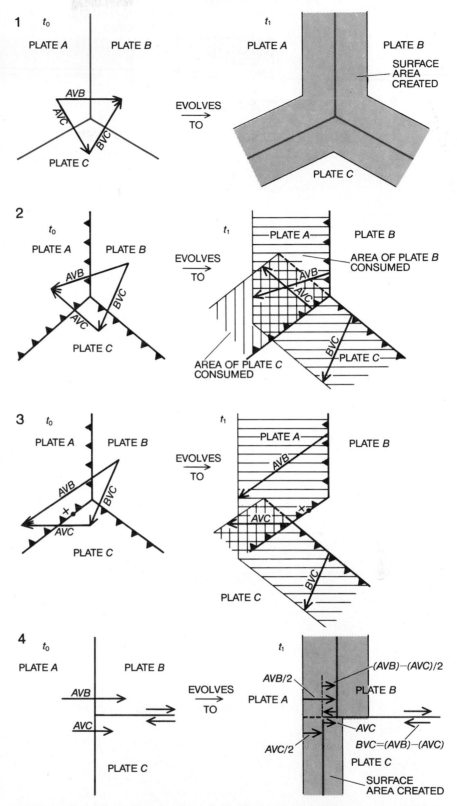

FOUR TRIPLE JUNCTIONS are depicted with their velocity triangles at time t_0 and their configuration at time t_1. Ridge axes are solid color; subduction zones are solid black; transforms are gray. A triple junction involving three ridges (1) is always stable. When three subduction zones meet (2 and 3) and two leading edges lying on plate A do not form a straight line, the triple junction is stable only when the A v. C velocity vector is parallel to the leading edge of plate C (2). Otherwise (3) the triple junction moves. Thus at some time between t_0 and t_1 the triple junction moves past point X. Before the time of movement the relative motion at X is A v. C and thereafter A v. B. The last diagram (4) depicts how it comes to pass that a triple junction involving two ridges and a transform can have the configuration at t_0 only instantaneously since the triple junction evolves immediately to t_1, in which plates B and C are separated by the transforms and the ridge is thus offset.

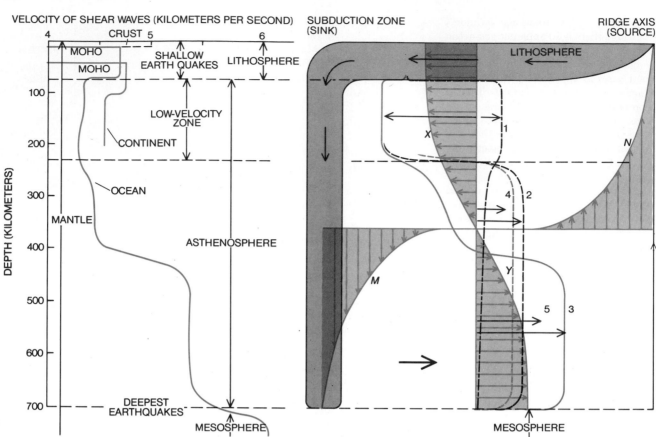

VARYING VELOCITY OF SHEAR WAVES in lithosphere and asthenosphere (*left*) suggests models showing how mass-transfer circuits (*right*) introduce new plate material at ridge axes and consume it at subduction zones. The velocity of shear waves decreases at 70 and 150 kilometers under oceans and continents respectively. Evidently mass-transfer circuits extend to a depth of at least 700 kilometers. Material descending at sinks (under curve *M*) must be balanced by material rising at sources (under *N*), and the upper lateral transfer (under *X*) must be balanced by return flow (under *Y*). Five shapes are suggested for lateral-transfer gradients.

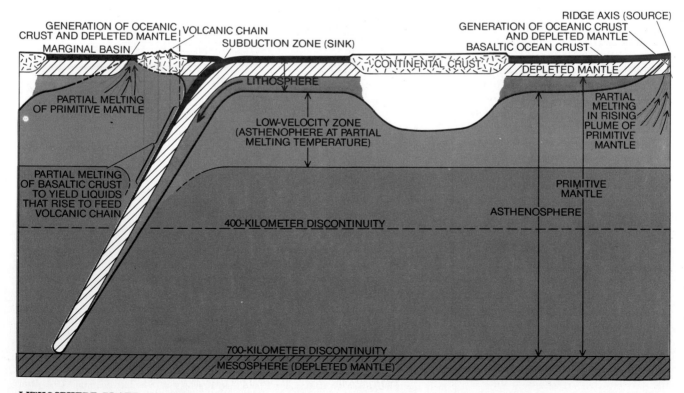

LITHOSPHERE PLATE of solidified rock serves as a thermal boundary conduction layer "floating" on molten or semimolten rock of the asthenosphere. In this schematic view the lithosphere is thicker under the continent it is rafting toward a subduction zone, where the plate descends into the mantle under leading edge of another plate. New oceanic crust is steadily being added at ridge axis.

terms of a rigid rotation around a fixed pole. Second, it provides the key to working out past plate motions. The inactive transform faults give the direction but not the rate of old displacements. Rates, however, can be deduced from the spacing of magnetic anomalies of known ages.

Walter C. Pitman III and Manik Talwani of the Lamont-Doherty Observatory have devised a simple but elegant technique using well-defined pairs of magnetic anomalies of known ages and fracture-zone orientations for computing plate displacements in the central Atlantic Ocean for the past 180 million years. First they assumed that successive pairs of magnetic anomalies were generated at the ridge axis and moved apart on rigid plates. They then found a series of rotation poles around which they could fit the pairs of progressively older magnetic anomalies; the series ended with a rotation that fitted the continental margins of Africa and North America. Reversing this sequence gives the kinematics of plate divergence that records the history of the opening of the central Atlantic Ocean.

A consequence of the symmetrical generation and asymmetric destruction of surface area, at ridge areas and subduction zones respectively, is that transform faults can retain a constant length or can change their length [see illustration on page 127]. Transform faults that join ridge axes, or subduction zones with leading edges on the same plate, stay the same length. Where transform faults join subduction zones with leading edges on different plates, they lengthen or shorten, depending on whether the leading edges face away from or toward one another. A transform fault that joins a ridge axis to a subduction zone increases or decreases in length depending on which plate the leading edge lies on.

If the axes and angular velocities of rotation between two pairs of plates (A and B and A and C) are known, the axis and angular velocity of rotation between the third pair (B and C) can be calculated [see illustration on page 128]. This means that if ridge-axis segments fall along the boundaries between A and B and A and C, the relative motion of B and C can be found. Xavier le Pichon of the Center for the Study of Oceanography and Marine Biology in Brittany developed this technique for computing the relative motions between the six largest plates and was thus able to work out convergence directions and rates across all the major subduction zones.

With the same technique Pitman has computed the relative motion of Africa and Europe for the past 80 million years. During this time North America and Africa have been parts of separate plates moving apart around a sequence of rotation axes as the central Atlantic has widened. North America and Europe have been similarly moving apart but around a different sequence of rotation axes. Therefore there has been relative motion between Africa and Europe. This motion has been complex but the net effect has been to nearly eliminate a formerly wide oceanic region between the two continents.

Since relative displacements are along circles of rotation, the relative motion between three plates cannot be described by the customary velocity vector triangle except instantaneously at a point. If, however, we are interested in relative motions in an area of the earth's surface so small that it may be regarded as a flat surface, with the result that circle-of-rotation segments are virtually straight lines, the velocity vector triangle is a convenient device for illustrating relative motion. A small area of great interest is one where three plate boundaries join to form a triple junction. Triple junctions are demanded by plate rigidity; this is the only way a boundary between two rigid plates can end. D. P. McKenzie of the University of Cambridge and W. Jason Morgan of Princeton University have analyzed all possible forms of triple junctions with velocity vector triangles and have shown that they can be stable or unstable depending on whether they are able or unable to maintain their geometry as they evolve [see illustration on page 129].

Plate Thickness and Composition

So far we have considered those essential aspects of plate geometry and evolution that can be treated from a surface point of view. We have not yet inquired into the thickness and composition of lithosphere plates. It has been known for many years from gravity and seismic-refraction measurements and from general considerations of mass balance that the continents are underlain by a relatively light "granitic" crust about 40 kilometers thick and the oceans by a denser "basaltic" crust only about seven kilometers thick. Both continental and oceanic crust are underlain by a mantle of denser material. The junction between crust and mantle is the "Moho," or Mohorovicic discontinuity. It was the goal of the now abandoned Mohole project to drill through the relatively thin oceanic crust into the mantle.

Plates must be at least as thick as the oceanic and continental crust because some plates have oceanic and continental portions between which there is no differential motion. It was thought for many years that the Moho might be an important physical discontinuity of mechanical decoupling on which large crustal displacements proceed. It is now clear that if there is a zone of decoupling between an outer rigid shell and a less viscous layer below, it is considerably deeper than the Moho.

The best evidence for the thickness of plates comes from seismology. The velocity of seismic waves is dependent on the density and flow properties of the rock through which they pass; it is high in rigid, dense rocks and low in less rigid, lighter rocks. Moreover, an increase in confining pressure increases the velocity of waves and an increase in temperature decreases it. Although confining pressure must increase with depth, recent studies indicate that the velocities of shear waves suddenly decrease below a surface about 70 kilometers under the oceans and about 150 kilometers under the continents [see top illustration on opposite page]. Shear-wave velocities then increase with depth, with marked increases between 350 and 450 kilometers and just above 700 kilometers.

These data suggest that an outer rigid layer 70 to 150 kilometers thick (the lithosphere) lies above a weaker and hotter layer (the asthenosphere) that becomes increasingly viscous with depth. The thickness of the lithosphere therefore probably constitutes the thickness of the rigid plates, and the lithosphere is discontinuous at plate boundaries. Earthquakes present a test of this hypothesis, since the cold, rigid lithosphere is probably their source. The distribution of earthquakes should thus provide a guide to the thickness of the lithosphere and to its distribution where it descends in subduction zones into the interior of the earth.

Ridges and transform faults are characterized by earthquakes whose depth extends to about 70 kilometers. The inclined zone of intermediate and deep-focus earthquakes indicates the descent of the lithosphere into the asthenosphere, where it is consumed at subduction zones.

Bryan L. Isacks and Peter Molnar of the Lamont-Doherty Observatory, working with seismic first-motion records, have analyzed the stresses in descending lithospheric plates. They find that the stresses are consistent with those that would be expected if a cold slab of bending lithosphere were to meet increasing resistance as it descended into

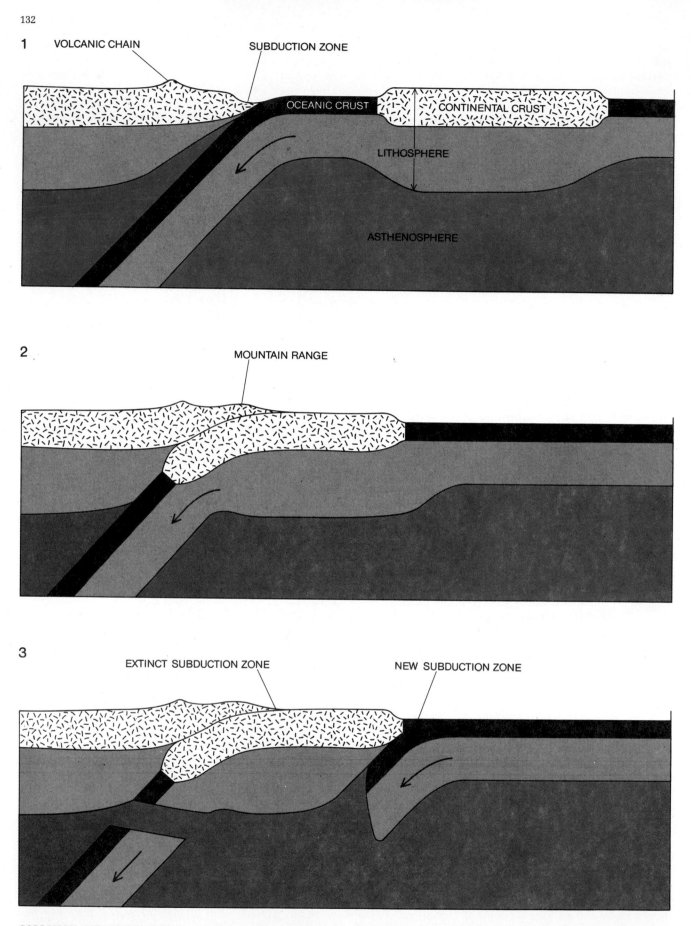

1 VOLCANIC CHAIN   SUBDUCTION ZONE

OCEANIC CRUST   CONTINENTAL CRUST

LITHOSPHERE

ASTHENOSPHERE

2 MOUNTAIN RANGE

3 EXTINCT SUBDUCTION ZONE   NEW SUBDUCTION ZONE

COLLISION OF CONTINENTS occurs when a plate carrying a continent is subducted at the leading edge of a plate carrying another continent (1). Since the continental crust is too buoyant to be carried down into the asthenosphere, the collision produces mountain ranges (2). The Himalayas evidently formed when a plate carrying India collided with the ancient Asian plate some 40 million years ago. The descending plate may break off, sink into asthenosphere and a new subduction zone may be started elsewhere (3).

an increasingly viscous asthenosphere. Where the cross section of the lithosphere is curved (at subduction zones) its upper part is under tension, suggesting elastic bending. Where the lithosphere has descended only a short distance into the asthenosphere it is under tension along its length, indicating low resistance to its descent. Continuously inclined seismic zones, representing slabs of lithosphere that have descended to lower parts of the asthenosphere, are characterized by compression along the slab. This fact suggests that the downgoing lithosphere is put into compression as it meets increasing resistance to its descent. An instructive case is where there is a gap in the inclined seismic zone, suggesting a discontinuity in the downgoing slab. Earthquakes above the gap indicate tension; earthquakes below it indicate compression. Evidently part of the slab has broken off and is descending at a faster rate than the remainder.

The general kinematics of plates—their growth and consumption—requires some form of mass convection, or mass-transfer circuits, in the mantle. Heat flow to the earth's surface is highest along ridge axes; it declines rapidly to a low plateau value across the plates and falls to a minimum at subduction zones. Therefore the lithosphere may represent a cold, rigid boundary conduction layer that is created at the hot ridge "sources" and destroyed at the cold subduction-zone "sinks." Any acceptable model for the geometry of the mass circuits in the earth's mantle must satisfy a number of conditions.

Conditions to Be Met

First, there must be a gross balance between vertical mass transfer at sources and sinks and lateral mass transfer by plate motion and motions in the asthenosphere. Second, the 700-kilometer limit on the depth of earthquakes and the abrupt increase in shear-wave velocity that marks the bottom of the asthenosphere imply that the mass-transfer circuits involve only the lithosphere and the asthenosphere. Third, the boundaries along which new crust is being generated on the earth exceed in length the boundaries where crust is being destroyed, so that plates are generally consumed at individual sinks faster than they are created at individual sources. Fourth, a simple geometric consequence of the fact that plates can change their surface area is that plate boundaries can move in relation to other plate boundaries. This fact implies that mass-transfer

circuits must also change their geometry as the plates evolve. Fifth, mass-transfer circuits cannot take the form of simple, regular convection cells linking sources and sinks with an upper lateral transfer and a lower return flow because there is no simple one-to-one relation between sources and sinks.

There are several great circles around the earth along which one can cross two ridge axes without an intervening subduction zone or can cross two subduction zones without an intervening ridge axis. Circuits involving mass-transfer rates of up to 10 centimeters per year must be accompanied by convective heat transfer, because thermal inertia prevents the elimination by conduction of temperature differences between different parts of the circuit. This condition is reflected in the persistence of earthquakes in a plunging slab down to depths approaching 700 kilometers. It is not clear, however, whether convective mass transfer and heat transfer is the cause or a consequence of plate motion. The models for relative lithosphere-asthenosphere motion at the top of page 130 illustrate this difficulty. They are all certainly far too simplistic. Indeed, the relative surface motions of plates may not be a guide to motions in the asthenosphere.

Let us consider a model in which the crust, lithosphere and asthenosphere are linked in a simple source-to-sink system [see bottom illustration on page 130]. The lithosphere in this model acts as a cold boundary conduction layer to a hotter asthenosphere, the upper part of which (the low-velocity zone) is probably near its melting temperature. Tension induced by plate separation at the ridge axis reduces the confining pressure in the low-velocity zone under the ridges. Reduction of confining pressure causes partial melting of primitive mantle in the low-velocity zone and the rising of a mush of crystals and liquid, with the result that the ridges are broadly uplifted. As the column of partially molten material rises it undergoes further partial melting; eventually the basaltic liquid rises to fill the crack continuously generated by plate separation. The liquid cools and crystallizes to form the basaltic oceanic crust and leaves under it a depleted layer of mantle.

Where plates descend into the asthenosphere, their leading edge carries a chain of volcanoes; thus one infers that the volcanic rocks are linked in some way with the descending plate. Because the volcanic rocks are less dense than the basalts of the oceanic crust, it is likely that they are formed by partial fusion of

the oceanic basalts with other material as the basalts are carried down into the hot asthenosphere. The depleted mantle in the plunging plate is denser than the primitive asthenosphere through which it sinks, both because it has had a lighter basaltic fraction removed from it under the ridge axis and because it is cooler. Therefore once a plate has begun to descend in a particular subduction zone it is likely to continue until the plunging plate meets increasing resistance deep in the asthenosphere.

Since continental crust is only about 40 kilometers thick, whereas plates are 70 kilometers or more thick, the continents ride as passengers on the plates. In the framework of plate tectonics continental drift is no more significant than "ocean-floor drift." Nevertheless, continents, unlike oceans, impose certain important restraints on plate motion. The narrow, sharply defined trenches and the regularly inclined earthquake zones sloping away from trenches indicate that oceanic lithosphere is easily consumed by subduction, probably because it has a thin, dense crust. Intracontinental seismic zones associated with mountain ranges exhibit compressional deformation over a wide area, which implies that continental lithosphere is hard to consume because it has a thick, relatively buoyant crust.

Within the Alpine-Himalayan mountain belt are narrow zones characterized by a distinctive assemblage of rocks, known as the ophiolite suite, whose composition and structure suggest that they are slices of oceanic crust and mantle. If they are, ophiolite zones mark the lines along which continents collided following the contraction of an ocean by plate consumption [see illustration on opposite page]. The small oceanic areas within the Alpine belt, such as the Mediterranean Sea and the Black Sea, may be remnants of larger oceans that once lay between Africa and Europe. Evidently lithosphere carrying light continental crust is difficult to consume, as is indicated by the marked scarcity of intermediate-focus and deep-focus earthquakes in zones where continents have collided. Thus it seems that continental collision terminates subduction along the collision zone. This implies that mass-transfer circuits must be drastically rearranged after the collision of continents, since a major sink is eliminated. As a result new sinks may form in oceans elsewhere.

As we have seen, any hypothetical driving mechanism for plate motion must meet a number of conditions. At present some form of thermal convection in the upper mantle seems to hold the most promise, but other mechanisms have been suggested that may play some role in plate dynamics. These mechanisms include the retarding effect of earth tides raised by the gravitational attraction of the moon, the possible pull exerted by a plate "dangling" in the asthenosphere and forces created by plates sliding down the slight grade between sources and sinks. It is also possible that some small plates are mechanically driven by the effects of relative motion between adjacent larger plates. For example, the westward motion of the wedge-shaped Turkish plate with re-

spect to the Eurasian plate may be caused by its being squeezed like an orange seed between the Arabian and the Eurasian plate.

Extinct Plates

It is now certain that plate tectonics has operated for at least the past 200 million years of earth history. During this time virtually all the present oceans were created and others were destroyed. Two hundred million years ago the major continental masses were assembled into the single supercontinent Pangaea [see illustration below]. It is therefore

legitimate to ask if the breakup of Pangaea some 180 million years ago marked the beginning of plate tectonics. Geologic studies of mountain belts older than 200 million years strongly indicate that they owe their origin to processes operating at plate boundaries that are now extinct. The Ural and Appalachian-Caledonian mountain belts, which lie within ancient Pangaea, have narrow zones where ophiolites are found. These old ophiolite zones, like those in the Alpine-Himalayan mountain belt, thus mark the sites of vanished oceans. This implies that the Urals, for example, were created by the collision of two conti-

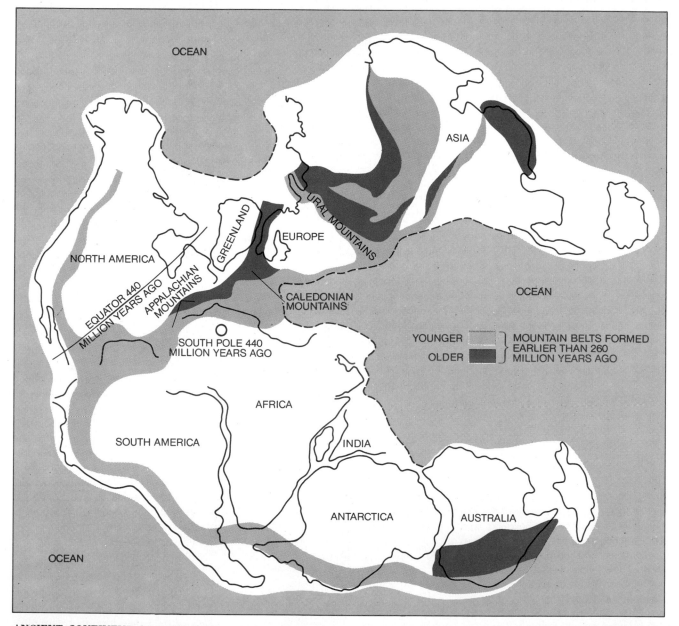

ANCIENT CONTINENT OF PANGAEA is reconstructed by fitting together the major continental land masses. Pangaea started to break up about 200 million years ago with a rift between Africa and Antarctica. Other rifts allowed South America, Australia and India to drift into their present positions. Mountain belts formed more than 260 million years ago are shown in shadings that distin-

guish younger and older belts. These mountain belts indicate lines of collision between continental fragments antedating Pangaea. Thus a prior collision of North America and Africa formed the younger part of the Appalachians 260 million years ago. Such a collision would explain how Equator and South Pole of 440 million years ago were brought close together after formation of Pangaea.

nental masses and that the ophiolites were generated by sea-floor spreading at a ridge axis before the continents were brought together.

Large-scale horizontal motions of continents before 200 millon years ago are supported by other lines of reasoning. Glacial deposits and other data indicate that about 400 million years ago a south polar ice cap covered the Sahara. At the same time eastern North America lay near the Equator. On the Pangaea reconstruction these close polar and equatorial positions are incompatible: they indicate that Africa and North America were separated by an ocean some 10,000 kilometers wide. The contraction of this ocean and the resulting collision of North America and Africa were probably largely responsible for the growth of the Appalachian mountain belt [see "Geosynclines, Mountains and Continent-Building," by Robert S. Dietz; SCIENTIFIC AMERICAN Offprint 899]. It seems a reasonable assumption that long, narrow, well-defined zones of mountain-building were established along zones of plate convergence. If this is the case, plate tectonics has been operating for the past two billion years.

The absence of well-defined zones of mountain-building older than two billion years suggests, however, that some mechanism other than plate tectonics, at least as we know it at present, was responsible for the evolution of the earth's crust in earlier epochs. The ancient "shield" regions of the continents, which contain rocks older than 2.4 billion years, are characterized by rocks distributed in swirling patterns over areas so wide they can hardly be explained by processes arising at the boundaries of rigid plates. Evidently the shield areas were stabi-

lized about 2.4 billion years ago, and by some 400 million years later a lithosphere with sufficient rigidity to crack into a plate mosaic had developed.

This does not necessarily mean that plate tectonics as we know it today began two billion years ago. Mountain belts older than 600 million years do not have ophiolite complexes like those of the younger mountain belts, indicating that sea-floor spreading before 600 million years ago generated a different type of oceanic crust and mantle. Geological data suggest that plates may have been getting thicker and that plate boundaries may have become more narrowly localized with time.

An exciting corollary of plate tectonics is that it provides a means whereby the total volume of continental crust can increase with time. We have seen that the primitive mantle of the asthenosphere undergoes partial melting to liberate a basaltic liquid that rises and cools to form the oceanic crust on the ridge axis, and that partial melting of the oceanic crust on a descending plate may yield the liquids that erupt to build the volcanic chains on the leading edge of plates. Volcanic rocks, with their deep-level intrusions of liquids that crystallized before they reached the surface, have the same bulk composition as that of the continental crust. The volcanic chains may therefore be sites where strips of embryonic continental crust are generated. Since they lie on the leading edge of plates, their destiny is to collide with other volcanic chains or with various kinds of continental margin. In this way new strips of light continental crust will be added to continental margins.

As we have seen, the arrival of a continental margin at a subduction zone

blocks further plate consumption at that site. Thus the oceanic ridge provides an effective means of growing continental crust, but there is no means of destroying such crust. This implies that the total volume of continental crust has been increasing for the past two billion years. One should not conclude, however, that strips of new crust have been added to the continents as a succession of regular concentric rings. Rather, discontinuous strips have been added at different times, reflecting the complex interaction of continental margins with the plate-boundary mosaic.

Although there are geologic phenomena that plate tectonics does not yet obviously explain, and although the driving mechanism is obscure, these deficiencies do not constitute rational objections to the theory of plate tectonics. One of the serious mistakes that many earth scientists made in the past was to reject continental drift because it was not clear how or why it occurred. The remarkable success of plate tectonics has been not only that it provides a consistent logical framework that draws together such diverse phenomena as sea-floor spreading, continental drift, earthquakes, volcanoes and the evolution of mountain chains but also that it has been successfully quantified and tested to the point where its essential core can no longer be questioned.

The essential core of plate tectonics is the geometric evolution of plates and the kinematics of their relative motion. It is of paramount importance to fully explore all the geometric and kinematic aspects of plate evolution if we are ever to understand the dynamics of plate motion and the geologic corollaries of plate tectonics.

12

The Evolution of the Andes

INTRODUCTION

Mountain belts are not scattered randomly over the earth's surface. They occur in patterns that had been recognized long before the concept of plate tectonics had been advanced. A topographical or geological map of North America, for example, shows a stable interior that has been free from major orogenic (mountain-making) episodes for over a billion years. Fringing the interior shield are the cordilleran and Appalachian systems —orogenic belts that testify to the crumpling of the ocean margins of the continent during the last 500 million years. Geologists could tell from reading the rocks that the deformation forces that made these mountain systems progressed almost as a series of waves from the ocean toward the interior. Beyond this, the textbooks on the past could only speculate in vague terms about the reasons for the pattern and the nature of the deformational forces.

David E. James's article on "The Evolution of the Andes" shows how the theory of plate tectonics provides a natural explanation of the entire sequence of events of an orogenic episode. The story begins in Paleozoic time with the deposition of a thick wedge of sediment on the slowly subsiding but otherwise inactive western margin of South America. From the end of the Paleozoic to modern times, this margin has been the site of collisions between the Nazca Plate under the Pacific Ocean and the South American Plate to the east. (The convergence of these two plates is but one part of the great episode of seafloor spreading, still underway, that broke the ancient supercontinent of Pangaea into the Americas, Africa, and Eurasia, separated by the widening Atlantic Ocean.)

With the activation of the margin, the gently dipping sedimentary formations were crumpled, uplifted, and invaded by magma. The geologist can trace the sequence of deformation, volcanism, metamorphism, and uplift from the beginnings of plate convergence to modern times. The geophysicist can peer more deeply, using echosounders to explore the offshore deep-sea trench, recording of earthquakes and seismic waves to delineate the subducted Nazca Plate beneath the Andes, and records of gravity changes to trace the thickening of the continental crust in response to compression and magmatism.

James documents the story of the Andes, but a similar tale of plate convergence could be told of the Appalachians, the Alps, the Himalayas, the Sierra Nevada and the Coast Ranges of the western United States, and possibly other great mountain belts.

The Evolution of the Andes

by David E. James
August 1973

*The geology of the central Andes indicates that the
history of the range can be understood in terms of the
consumption of a plate of the earth's crust plunging
under South America*

A great series of mountain belts, the
Andean cordillera, sweeps down
the west coast of South America
from Venezuela and Central America
nearly to the southern tip of Chile. A
long arc-shaped deep in the ocean floor,
the Peru-Chile trench, is clearly allied
with the mountain chain and runs roughly parallel to it from about four degrees
north latitude to 40 degrees south. Some
15,000 meters (more than nine miles) of
vertical relief separates the deepest part
of the trench from the highest Andean
peak. The Andean arc, comprised of
both mountains and trench, is a living,
evolving system. It is a part of the cir-
cumpacific "ring of fire," and the live
volcanoes that dot the length of the cordillera and the devastating earthquakes
that punctuate the flow of South American life are ever present reminders that
the mountain-building processes that
raised the Andean chain are still very
active today.

Many earth scientists, of whom I am
one, believe that the Andean arc is a
modern analogue of many older mountain belts: that if we had lived in the region of the Sierra Nevada 100 million
years ago or the northern Appalachians
450 million years ago, we would have
witnessed mountain-building activity of
a similar kind. Until recently this view
was not widely accepted. It was generally supposed that the processes through
which older mountain belts evolved are
not being duplicated today. That supposition is rapidly giving way to the concept that mountain-building activity has
proceeded in accordance with consistent
geologic patterns that have been repeated over and over again during at
least the past two billion years of earth
history, and that it continues to do so today. In this view the Andean chain is
the foremost modern example of those

mountain-building processes at work,
and so, by understanding the Andes, we
seek to learn how ancient mountain belts
were born, matured and in some cases
died long before man arrived to record
their origin or their passage.

Concepts of orogeny, or mountain
building, have recently been revolu-
tionized by the theory of plate tectonics
[see "Plate Tectonics," by John F.
Dewey, beginning on page 124]. This
theory, which has been developed
with startling rapidity over the past
decade, holds that the outer rind of
the earth, the lithosphere, consists of a
mosaic of rigid plates that are in motion
with respect to one another. These
plates, some 100 kilometers (60 miles)
thick, include not only the earth's solid
crust but also part of the denser upper
mantle. Plate boundaries, or junctions,
rarely coincide with continental margins,
and so the relative movements of parts
of the earth's surface are now viewed in
terms of plate motions rather than "continental drift." The proud continents,
which were once thought to plow
through oceanic crust like great ships,
are now reduced to the status of passive
passengers on the lithospheric plates.
Volcanic magma welling up from deep
within the earth's mantle creates the
lithospheric plates along oceanic ridges.
Newly generated lithosphere moves
away from the ridges to yield to constantly replenished volumes of magma
injected along the axes of the ridge.
These spreading plates are consumed at
trenches, where they bend down and
plunge into the earth's mantle.

Most of the world's tectonic activity—
earthquakes, volcanoes and mountain
building—is concentrated along plate
junctions. The west coast of South
America is such a junction [see *illustra-*

tion on next page]. Here the oceanic
Nazca plate, generated along the East
Pacific Ridge, is consumed in the Peru-
Chile trench, where it bends down and
slides under the South American plate at
a rate of about six centimeters per year.
Examination of the structure and geology of the central Andes reveals that most
of the present-day orogenic activity and
the geologic evolution of the past 200
million years can be understood in terms
of the subduction, or consumption, of the
oceanic plate under South America. This
interaction of the two plates accounts for
the crumpling of the stable continental
margin to form belts of fold mountains
that now constitute the eastern ranges of
the Andes, for the birth of the great Andean volcanic cordillera to the west and
for the continental growth of western
South America.

Lured by the sheer scale of the Andes
mountains, the Carnegie Institution of
Washington, in concert with a number of
South American institutions, set out
more than a decade ago to study the
nature and evolution of the central
Andes of southern Peru, Bolivia and
northern Chile. Many years of work have
led progressively to a rather complete
understanding of the physical properties
of the earth under the central Andes, but
it is only recently that the conceptual
tools have been available with which to
understand the forces and sequences of
orogenic events that produced the Andes
and that continue to generate volcanism
and seismicity along the Andean chain
even today.

To understand the evolution of the
central Andes it is necessary to draw on
two distinct kinds of evidence, geophysical and geological. Important geophysical evidence includes the distribution of
earthquakes and the distribution within
the crust and upper mantle of certain

138

GEOPHYSICAL SETTING of the Andes is portrayed. According to the theory of plate tectonics, the lithosphere, or outer shell of the earth, consists of several rigid plates that are moving with respect to one another. The juncture between the Nazca plate and the South American plate represents the most important modern example of interaction of an oceanic and a continental plate. The Nazca plate is generated along the East Pacific Rise and consumed in the Peru-Chile trench. The interaction of the two plates at the trench

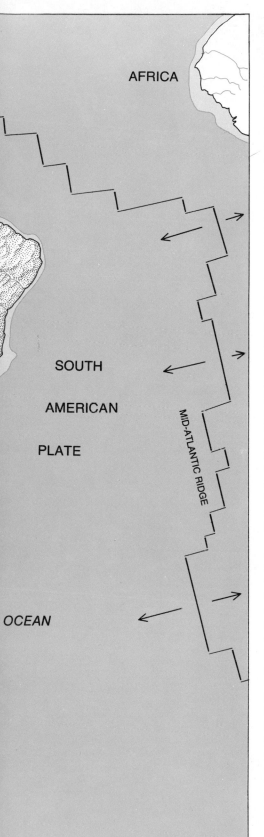

physical properties: the velocity of seismic waves, the absorption of their energy (their attenuation) and the density of rock. From these properties one can deduce information on the kinds of rock in the earth's interior and whether they are rigid or mobile. Geological evidence derives from the study of rock types and their structures as observed on the earth's surface; it is only from geological evidence that one can read the history of past events. I shall consider the geophysical evidence first, since it provides the observational basis for most of our knowledge concerning the plate-tectonic framework of the juncture between the converging South American and Nazca plates.

Earthquake distribution provides key evidence regarding plate interactions. It is a general tenet of plate theory that the inclined earthquake zones (the Benioff zones, named for Hugo Benioff) that bend down under trenches and volcanic arcs define the upper part of the descending oceanic plates. Descending lithospheric plates are cooler and hence more rigid than the asthenosphere (the hot mantle) through which they sink, and it is widely supposed that only within the lithosphere are rocks rigid enough to support brittle fracture, or earthquakes. This supposition is supported by the observation that in most island arcs earthquakes appear to be confined almost entirely to the descending plate.

Under the central Andes, however, earthquakes occur not only in the descending plate but also in a continuous wedge between the earth's surface and the top of the sinking oceanic plate to a depth of from 200 to 300 kilometers [*see top illustration on page 145*]. This observation leads one to suspect that the leading edge of the overriding South American lithosphere may be an abnormal 200 to 300 kilometers thick—much thicker than the usual 100 kilometers observed for most plates and in startling contrast to the 50-kilometer thickness measured seismically for the undersliding Nazca plate. More direct evidence that the Andean lithosphere is 200 to 300 kilometers thick derives from seismic results, which show that there is no low-velocity zone or zone of high seismic-wave attenuation in that region.

Both low-velocity zones and zones of high attenuation in the mantle are commonly interpreted as being regions of softer rock and are associated with the asthenosphere; conversely, the absence of these zones implies that the rock is comparatively rigid. An important consequence of the thick lithosphere under the Andes is that it precludes convective overturn in the mantle above the Benioff zone. It is this overturn that is believed to produce the secondary centers of seafloor spreading that develop behind island arcs but that are conspicuously absent behind the continental Andean arc.

A second important set of geophysical observations, on crustal thickness, provides an additional linkage between the present plate-tectonic regime of the Andes and the orogenic history of the Andes as it is read in the geologic record. Seismic data show that the crust of the central Andes is extremely variable in thickness. Over a distance of little more than 500 kilometers it varies from about 11 kilometers (including water) in the Pacific basin to 30 or 35 kilometers along the coast (a "normal" continental crust), more than 70 kilometers under the western-range volcanic crest, 50 to 55 kilometers under the eastern fold ranges and finally 35 kilometers under the Brazilian shield. Only in the Himalayas have crustal thicknesses of as much as 70 kilometers been observed elsewhere.

Yet one could have predicted these results obtained from seismic measurements simply on the basis of topography and the principle of isostasy. This principle states that mass excess above sea level—a mountain chain, for example—must be compensated by an equal mass deficiency at depth: the mountain chain's crustal root, which displaces mantle rock, must be less dense than the rock it displaces. It is evident, of course, that the reverse expression of the principle of isostasy is equally valid: Any mass deficiency at depth must be compensated by a mass excess at the surface; any growth or thickening of the crust, such as by the injection of magma into the crust from below, will result in surface uplift. From seismic studies of the crust it is known that areas that have long remained at sea level (and are thus presumed to be in isostatic balance) are underlain by "normal" continental crust 30 to 40 kilo-

has caused deformation and growth of the continental margin of South America, thereby forming the Andean mountain system. Arrows indicate flow away from ridges.

REGION OF THE ANDES near Lake Titicaca, on the boundary between Peru and Bolivia, is shown on the following page in a photograph made by the Earth Resources Technology Satellite. The lake is on the altiplano, or high plateau, nearly four kilometers above sea level. Rising above the altiplano to the west (*bottom*) are the volcanic mountains of the western cordillera; to the east (*top*) are the fold mountain belts of the eastern cordillera.

meters thick. Areas below sea level are underlain by thinner crust, those above by thicker crust. This fact becomes central when considering crustal evolution, since geology indicates that many of the highest parts of the Andes were once near sea level.

One additional line of geophysical evidence, seismic velocity, can be used for guessing the composition of the rocks within the Andean crust and the adjacent oceanic crust. In laboratory experiments the velocity of seismic waves in common rock types has been measured over a large range of temperature and pressure. Seismic velocities of rock in the lower oceanic crust are appropriate to rock of basaltic composition at that temperature and pressure. Interestingly, even though the pressure and the temperature in the crustal root under the Andes are much greater than in the lower oceanic crust, the seismic velocities of the rocks are the same. Yet it is known that if basaltic rock were subjected to the pressures and temperatures found in the deep continental crust, it would undergo metamorphic transformation to a rock type with higher seismic velocities and densities than are observed.

When different kinds of rock are examined in detail, only lighter crustal rocks, similar in composition to those observed at the surface, exhibit the appropriate seismic velocities at the temperature and pressure of the lower crust. This leads to the inference that the Andean crust is rather homogeneous vertically, the rocks in the Andean root being similar in composition to the volcanic and plutonic rocks that form the upper levels of the crust. Exceptions to vertical crustal homogeneity occur in the altiplano (the high plateau between the two great volcanic cordilleras) and the eastern cordilleras. Variations in seismic velocity show that the altiplano could be underlain by up to 30 kilometers of sedimentary rock, perhaps mixed with volcanic deposits, and that the eastern cordillera is underlain by at least five to 10 kilometers of sedimentary rock.

In order to reconstruct the progressive development of the Andean system one must turn to the information that can be read from the geologic record. Paleozoic sedimentary rocks laid down some 250 to 450 million years ago are among the oldest rocks of the central Andes. Time has not treated them kindly, and we find them collapsed and crumpled to form the fold mountains of the eastern ranges. These rocks, some 10 kilometers of monotonously repetitive muddy, sandy sedimentary beds, are called geo-synclinal rocks [see "Geosynclines, Mountains and Continent-Building," by Robert S. Dietz; SCIENTIFIC AMERICAN Offprint 899]; once they formed the quiet Paleozoic continental margin of western South America. (A modern analogue to the western seaboard of Paleozoic South America is the inactive continental margin of the east coast of North America, where a great sedimentary wedge some 250 kilometers wide and up to about 10 kilometers thick has formed between the continental shelf and the abyssal plain of the ocean.)

During Permian and Triassic time, between about 200 and 250 million years ago, the quiet of the western coast of South America gradually gave way to rumblings brought on by the incipient breakup of the supercontinent Pangaea and the onset of the plate-tectonic cycle that is still under way. The continental edge became unstable and the geosynclinal strata were warped and buckled upward. Magma, possibly derived from partial melting of deeply buried geosynclinal strata, invaded and poked through the sedimentary pile to form batholiths (bodies of intrusive igneous rock) at depth and volcanoes at the surface. The old volcanoes have long since disappeared, but volcanic rocks from this period are found interlayered with sedimentary beds. The intrusive rocks, probably representing the deeper feeder levels of the volcanoes, are commonly found today exposed in the cores of the eastern-fold ranges, nestled among crumpled geosynclinal rocks.

About 190 million years ago, in earliest Jurassic time, the major axis of tectonic activity shifted several hundred kilometers oceanward, to the west. The lithosphere broke along the junction between the South American continent and the Pacific Ocean basin and the oceanic plate began its descent into the mantle under western South America. One can only speculate now on the reasons for the catastrophic rupture of the lithosphere, although there is some evidence that it may have been in response to the onset of sea-floor spreading along an ancestral East Pacific Rise. We do know that the South Atlantic Ocean had not yet begun to open and that South America and Africa were still one.

As the Pacific plate plunged under South America, andesitic magma welled to the surface, sweated out of the descending basaltic oceanic crust. Andesite is the characteristic volcanic rock of the Andes. It is richer in silica, and hence less dense, than basalt; its composition is what is termed intermediate, that is, between basalt and granite in chemical composition. These andesitic rocks of Jurassic age are well preserved all along the coastal regions of southern Peru and northern Chile. It is still somewhat problematical, however, whether the andesites were extruded on sialic (continental, shallow-water) crust or simatic (oceanic, deep-water) crust. The lavas themselves appear to have been extruded below water, since many have been highly altered by seawater.

One might infer from this that the volcanic arc of the central Andes began as an island arc in the ocean off the coast of ancestral southern Peru and northern Chile [see illustration on next two pages]. Yet the picture of the Jurassic arc is not simple, because Jurassic volcanic rocks in southern Peru are wedged in among crystalline metamorphic rocks at least 400 million years old. Just what these remnants of ancient sialic crust are doing some 300 kilometers west of the currently exposed geosynclinal rocks of the Paleozoic continental margin is not known. The presence of sialic rocks does not necessarily imply, however, that the Jurassic arc formed on continental crust. The rocks could be part of a Paleozoic microcontinent or peninsula that lay to the west of the South American coastline. Or they could be sialic flotsam swept into and plastered to the edge of South America, buoyant debris scraped from the top of the oceanic plate as it dived down at the trench.

These difficulties aside, formation of the Andean volcanic arc was well in progress during Jurassic time. The fact that Jurassic volcanic rocks are still widely preserved indicates that the Jurassic arc never stood much above sea level and consequently suffered little erosional destruction. Furthermore, it appears that the entire region to the east of the Jurassic arc, extending as far east as and encompassing the area of the present-day Andes, must have been near sea level, since marine sedimentary deposits are mapped throughout the area. This observation is important because it shows that the crust of the Andes was still thin, probably no more than 35 kilometers thick.

I have noted that the Jurassic arc formed during a period when South America and Africa were still a single continent. Dating of the magnetic pattern of the South Atlantic by Walter C. Pitman III and others at the Lamont-Doherty Geological Observatory of Columbia University indicates that it was only about 135 million years ago that the South Atlantic began to open by spreading along the Mid-Atlantic Ridge. Formation of the Jurassic arc inaugurated the impressive series of mountain-building episodes that produced the Andean belt, but it was not until sometime after the break-off of South America from Africa that Andean orogeny began in earnest.

About 100 million years ago, during Cretaceous time, a second major volcanic arc began to form parallel to and continentward of the Jurassic arc. (My treatment of Andean development as a series of a few distinct episodes is an oversimplification, but it serves to trace the growth of the mountain system; igneous activity has actually been rather continuous from about 200 million years ago to the present day.) The lavas of this volcanic chain were extruded above sea level—clear evidence that the volcanoes developed astride continental crust. Activity along the Cretaceous arc reached maximum intensity about 50 to 60 million years ago with the invasion of the crust by massive amounts of magma. The invading magmas crystallized to form

enormous batholiths that now lie exposed all along the western flanks of the western ranges.

The contemporaneous volcanoes, which must have towered far above sea level, have long since been stripped off by erosion, exposing the intrusive underpinnings of the volcanic arc: the Andean batholith. In some areas, such as southern Peru, erosion has not eaten as deeply, so that small batholiths are mantled by volcanic debris of similar age and composition. These small intrusive bodies are probably parts of the feeder pipes that supplied the volcanoes. In time the volcanic roofs of the batholiths will be further stripped away, leaving only the deep-seated intrusive foundation of the vanished Cretaceous volcanic arc exposed to view.

The crustal swelling that accompanied the invasion of magma produced major traumas in the easily deformed geosynclinal rocks of the eastern ranges. Upwelling magmas dilated the crust along the line of the Cretaceous arc. The geo-

synclinal rocks were alternately sloughed off and pushed aside by the growing orogenic welt and were crumpled and thrown up into the fold mountains of the eastern ranges. Enormous volumes of sediments were eroded from the flanking mountain ranges and dumped into the intervening altiplano basin; some places in the altiplano received up to seven kilometers of sediment in Cretaceous time alone.

It is significant that the Cretaceous arc lies on the continent side of the older Jurassic arc because it has usually been supposed that, as debris from the oceanic crust is stuffed under volcanic arcs, the trench and arc will migrate oceanward. Quite the contrary is true in the Andean region: during the course of Andean evolution the axis of the volcanic arc has marched ever inland, away from the sea. Even on a fine scale one finds successive intrusions of the Andean batholith shifted continentward. One can only speculate on the reasons for the eastward mi-

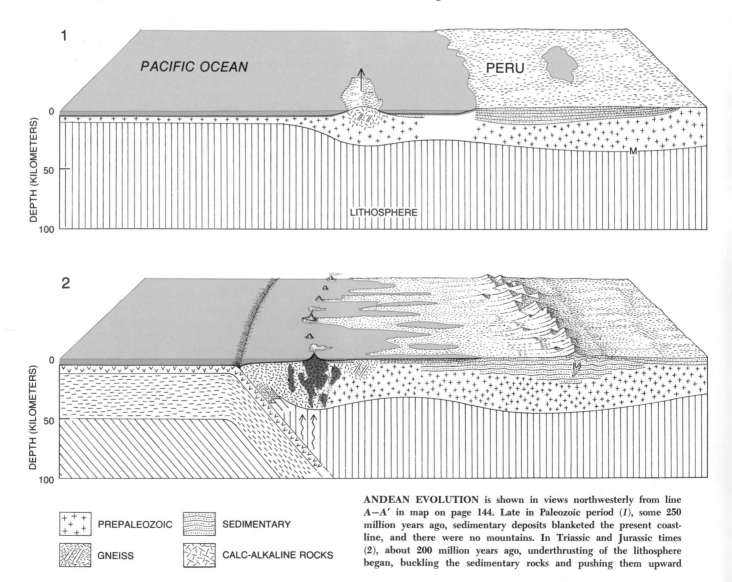

ANDEAN EVOLUTION is shown in views northwesterly from line A–A' in map on page 144. Late in Paleozoic period (*1*), some **250 million years ago**, sedimentary deposits blanketed the present coastline, and there were no mountains. In Triassic and Jurassic times (*2*), about **200 million years ago**, underthrusting of the lithosphere began, buckling the sedimentary rocks and pushing them upward

PREPALEOZOIC

GNEISS

SEDIMENTARY

CALC-ALKALINE ROCKS

gration, but I shall return to this question when I consider the origins of the volumes of magma that have built the Andean crust.

The present-day Andean volcanic edifice began to emerge about 15 million years ago. In northern Chile and southern Peru huge volumes of silicic volcanic ash exploded out of fissures and flowed out from eruptive centers, eventually to blanket hundreds of thousands of square kilometers. Even now, when most of these ash deposits have long since been eroded away, it has been estimated that their remnants cover some 100,000 to 150,000 square kilometers to an average depth of 500 meters. The ash eruptions continued steadily until about four million years ago, when they ended abruptly and were followed closely by the outpouring of andesitic lavas from volcanic vents. These later lavas form the great stratovolcanoes, some still active, that dominate the Andean chain and rise to well over six kilometers in the central Andes.

The massive influx of magma into the crust under the modern Andean volcanic crest swelled the crust and produced extensive folding and thrust-faulting in the altiplano and the eastern ranges. The geosynclinal sediments were further pushed aside by the expanding magmatic welt and the eastern ranges were jammed up to form narrow, high mountain chains. Sedimentary debris flooding in from both the eastern and the western range during Tertiary time piled up in the altiplano to thicknesses as great as 15 kilometers. We are now witnessing at least a temporary waning of activity, and through the relaxation of compressive stresses some extensional features have formed in the altiplano. Lake Titicaca is a notable site of one such feature: a graben, a long depression between fault lines.

So far I have considered Andean evolution from the point of view of surface geology, describing the inland march of the volcanic arc, the resulting progressive crumpling of the Paleozoic geosyn-

cline and the gradual uplift of the Andean chain. I have not yet inquired in detail into the growth of the crust and the origin of the crustal rocks below the surface.

Classical concepts of mountain building, formulated long before the advent of plate tectonics, held that geosynclinal rocks formed in elongated basins of deep subsidence and that these basins were necessary precursors of later mountain belts, which were presumed to form through the deformation and melting of the thick geosynclinal sedimentary strata. The tectonic mechanics whereby these sedimentary rocks were melted and deformed was always a mystery, as was the origin of the subsiding basin itself. A number of problems are posed by classical concepts of geosynclines when they are applied to the central Andes. The crust under the volcanic crest is more than 70 kilometers thick, and yet no sedimentary basin can ever sit for long on crust more than about 35 or 40 kilo-

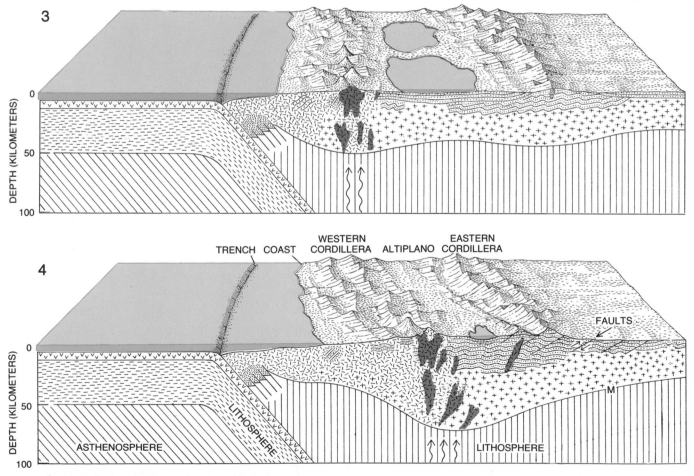

and eastward. **Rising magma from the descending oceanic plate formed an arc of volcanoes in the coastal waters of western South America. Some batholiths, or bodies of intrusive igneous rock, formed in the sedimentary layers of the eastern ranges. In Cretaceous and early Cenozoic times (3), 100 to 60 million years ago, a second volcanic arc began to form eastward of the Jurassic arc. Up-** **welling magma swelled the crust, pushing aside the ancient sedimentary rocks, which crumpled to form the fold mountains of the eastern cordillera. Material eroded from these mountains poured into altiplano region. Formation of the present volcanic range began 10 to 15 million years ago, reaching by Pliocene or Pleistocene time (4), one or two million years ago, the present structure.**

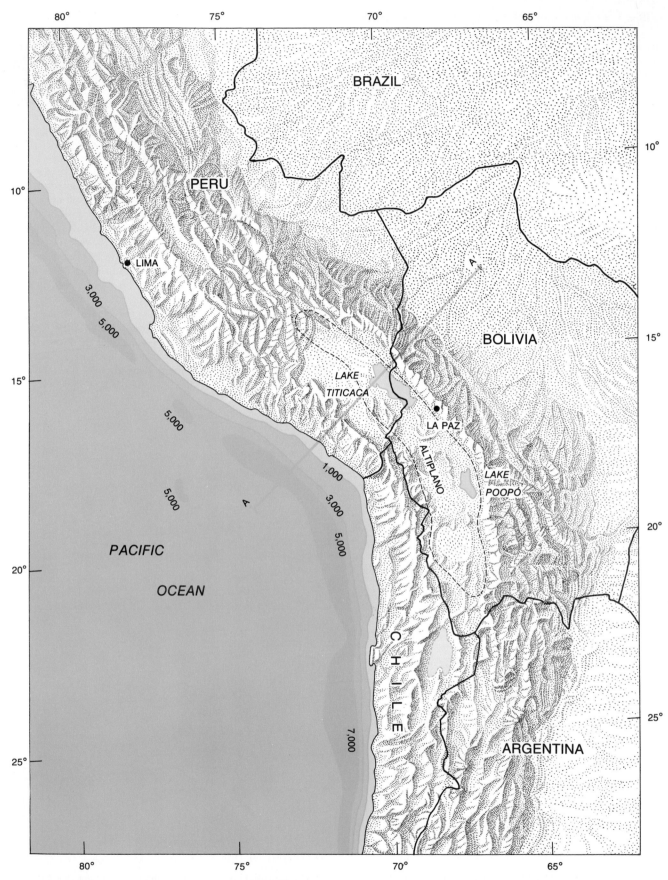

CENTRAL ANDES is a region where mountain-building forces appear to be still at work. North of Lima the Andes form a single belt of closely spaced mountain chains running parallel to the coast. South of Lima, however, the mountains branch, with the easterly fold belt running hundreds of kilometers inland and the westerly volcanic chain continuing parallel to the coast. Between these ranges lies the altiplano, a broad flat plain underlain by an enormous wedge of sedimentary debris eroded from the adjacent cordilleras. The two ranges merge again in northern Chile. Line A–A' shows region portrayed in several succeeding illustrations.

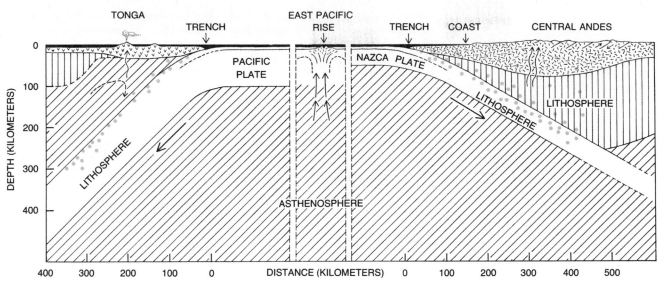

STRUCTURAL CONTRASTS between the Andean volcanic arc and the Tonga-Fiji island arc account for differences in physiographic settings and earthquake distributions. In the Tonga arc most earthquakes occur within the descending slab, which has mobile asthenosphere material above it and below it. There is also evidence of minor sea-floor spreading west of the arc. These observations suggest that drag from the sinking plate causes convective overturn (*arrows*) in the overlying asthenosphere, which in turn causes the spreading behind the arc. Earthquakes in the Andes are not confined to the descending plate, nor is there any evidence of crustal spreading east of the volcanic arc. Evidently the thick lithosphere under the leading edge of South America prevents convection and secondary spreading, and the forces generated by the descending slab produce earthquakes rather than convection above the slab. The intensity of earthquake activity under the Andes is indicated by the density of dots. Few dots show lighter activity.

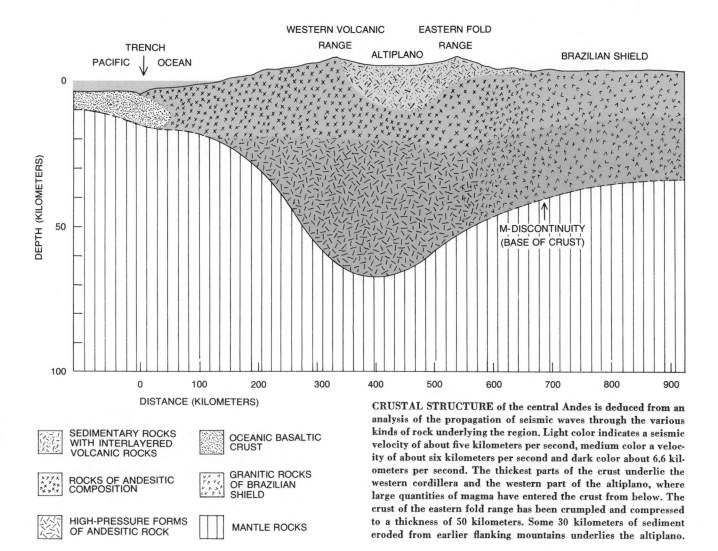

SEDIMENTARY ROCKS WITH INTERLAYERED VOLCANIC ROCKS

OCEANIC BASALTIC CRUST

ROCKS OF ANDESITIC COMPOSITION

GRANITIC ROCKS OF BRAZILIAN SHIELD

HIGH-PRESSURE FORMS OF ANDESITIC ROCK

MANTLE ROCKS

CRUSTAL STRUCTURE of the central Andes is deduced from an analysis of the propagation of seismic waves through the various kinds of rock underlying the region. Light color indicates a seismic velocity of about five kilometers per second, medium color a velocity of about six kilometers per second and dark color about 6.6 kilometers per second. The thickest parts of the crust underlie the western cordillera and the western part of the altiplano, where large quantities of magma have entered the crust from below. The crust of the eastern fold range has been crumpled and compressed to a thickness of 50 kilometers. Some 30 kilometers of sediment eroded from earlier flanking mountains underlies the altiplano.

meters thick without bobbing above sea level and thus ending sedimentation.

If remobilized geosynclinal rocks are to make up the crust of the western cordillera, they must therefore have undergone twofold foreshortening to produce a doubling in crustal thickness. Yet in the volcanic chain we find few tectonic patterns such as thrust faults and tightly compressed folds that would suggest crustal shortening. On the contrary, studies of seismic earthquake mechanisms and modes of faulting indicate that the deformational style is extensional: the earth under the volcanic arc is swelling rather than contracting. Clearly knowledge of the Andean crust could not be easily mated with classical concepts of geosynclinal mountain building.

Only the advent of plate-tectonic theory has adjusted this unhappy mismatch between mountains and mountain-building theory, and it has done so by providing mountain-building mechanisms and processes that bear a clear relation to the Andean orogenic belt. It now appears to be evident (although some rearguard scientific battles still rage) that the geosynclinal sedimentary rocks are simply victims trapped at the continental margins and caught up in the tectonic processes that accompany subduction. The geosynclinal wedge must participate in, but can never be a prime mover of, the mountain-building process itself. If this analysis is correct, it follows that the magmas that built the central Andean volcanic arc were derived from below the crust and that the great volumes of rock that solidified from these magmas have produced crustal thickening and massive uplift of the arc.

We are left, then, with the question of the sources of magma below the crust. There are two principal possibilities: partial melting of the dense mantle rocks in the wedge between the crust and the top of the underthrust plate, or partial melting of the basalts and sediments of the oceanic crust riding down into the mantle on the lithospheric conveyor belt. Most observations argue against a mantle source for the andesitic rocks of the Andes. If the thickness of the crust in the volcanic arc had been doubled by the addition of magma from below, then some 20 percent partial melting of mantle rock between the crust and the subduction zone would be required to provide the necessary volume of additions to the crust. Experimental studies of the partial melting of probable mantle materials do not readily support the extraction of this quantity of andesitic magma from dense mantle rocks. A

source in the mantle for the Andean rocks seems possible only if mantle flow provides a constant supply of rock undepleted in its low-melting-point fraction. Yet we have already seen that the upper mantle under the Andes is rigid lithosphere to a depth of from 200 to 300 kilometers, and therefore probably quite immobile.

An alternate proposition, that the magmas are derived from the descending oceanic crust, is currently most in favor and is a theory that I believe provides a viable explanation of Andean crustal genesis. Rocks of the oceanic crust have melting temperatures several hundred degrees Celsius below those of mantle rocks. As the cold oceanic crust descends into the mantle it is heated by the hot mantle in which it is enveloped, and at depths of about 100 to 150 kilometers the basaltic crust may well reach its melting point. The partly melted fraction of oceanic crust would be less dense than either the parent rock or the surrounding mantle rocks and would migrate upward into the crust. Most of the magma would be trapped and would solidify at the lower and intermediate levels of the crust; only a small fraction would ever make its way to the surface to come out through volcanoes, and so the volcanic pile on the earth's surface represents only a small part of the total increase in crustal volume.

A phenomenon related to the origins of the crustal rocks is the eastward migration of igneous activity. Assuming that the rocks originate through partial melting of the descending plate, there are a number of ways to explain the migration. If, for example, the depth in the earth at which melting occurs remains constant, then either a progressive decrease in the dip of the descending plate or a progressive continentward migration of the trench could cause an eastward migration of igneous activity. Or, if the position of the descending plate remains constant, the depth at which melting occurs on the slab may increase with time, thus pushing the source of magma deeper and eastward. A critical consideration here is the position and dip of the descending plate since the onset of subduction. Has the dip of the Benioff zone changed or has the trench migrated over the past 200 million years?

Examination of the chemistry of the rocks themselves provides a partial answer to this question. William R. Dickinson of Stanford University and Trevor Hatherton of the New Zealand Department of Scientific and Industrial Research have demonstrated a clear posi-

tive correlation between the amount of potash (K_2O) in modern volcanic rocks and the depth to the underlying Benioff zone. For particular rock types such as andesite collected from island arcs all over the world, the potash content is a more or less linear function of the depth to the underlying Benioff zone. By the inverse corollary of this relation one should be able to estimate the depth to defunct Benioff zones by measuring the potash content in older rocks.

We have just begun to apply this technique in the Andes. The data are by no means complete, but as we proceed from rocks of the Jurassic arc to those of the Cretaceous arc and on to those of the modern arc we find a progressive increase in potash content. This progressive increase, of course, is also correlated with increasing depth to the present Benioff zone. We find similar variations for strontium content, another chemical variable that has been shown to exhibit a positive correlation with the depth to the Benioff zone.

It is therefore possible to speculate with some basis in fact that the position of the Benioff zone has not changed greatly with time. If that is so, the eastward migration of the arc over the past 200 million years implies that the position in the mantle at which the oceanic crust melts has been pushed to progressively greater depths. It may well be that in the Andes the continued underthrusting of the cold oceanic slab into the mantle under the continent has gradually cooled the surrounding mantle rocks, so that the temperature at which melting occurs in the descending plate is attained only at greater and greater depths. If such a process accounts for the migration of igneous activity ever inland from the trench, then as long as subduction continues under the central Andes the volcanic chains of the future will impinge ever more on the old geosynclinal rocks, driving them even farther inland.

One day, however, Andean orogeny will end. The volcanic chain will stop growing and the inevitable destruction by erosion will set in. To reduce the Andes to sea level, some 35 kilometers of crust will have to be stripped away, exposing the center of the earth's crust. The sedimentary debris removed from the mountain edifice will be dumped on the continental shelf, the continental slope and the oceanic abyssal plain of the Pacific. There the sediments will pile up to await the next tectonic cycle, in which they will be crumpled, buckled upward and swept aside to give way to the next great volcanic mountain belt.

13

The Evolution
of the Indian Ocean

INTRODUCTION

I f the Atlantic Ocean were as complicated as the Indian Ocean, the verification of the concepts of sea-floor spreading and plate tectonics might have been delayed for years. The South Atlantic is marked by one major ridge, segmented by transform faults. This mid-Atlantic ridge is the site of the original cleavage that split South America from Africa about 135 million years ago. Since then, sea-floor has been created continually along the growing American and African plates, and the continents have drifted apart. Theoretically, the magnetic lineaments of the sea floor are parallel to the still-active spreading ridge. Earthquakes occur along the ridge and its associated transform faults as they have throughout the history of the Atlantic's spreading.

Things are not as simple in the Indian Ocean. There are a number of ridges and plateaus, some active, others quiescent. Only the youngest magnetic lineaments are parallel to present-day spreading ridges; the older lineaments show large offsets. Ancient continental fragments are found in the Indian Ocean (for example, Madagascar and the Seychelles); but not in the Atlantic.

It is a testimony to the power of the ideas of plate tectonics and sea-floor spreading that D. P. McKenzie and J. G. Sclater, in this article, "The Evolution of the Indian Ocean," could invoke the simple principles worked out in the Atlantic to unravel the more complex history of the Indian Ocean. They were able to identify ancient transform faults, which, though long inactive, have left their topographic scars. Other transform faults have been altered to still-active spreading centers. The authors' reconstruction includes the northward motion of India until it collided with the land mass of Asia and the separation of Australia and Antarctica. A story this complicated may not be as convincing as the simple history of the South Atlantic. Nevertheless one cannot help but be impressed by the plausible fitting together of all of the pieces of evidence of this most difficult Indian Ocean puzzle.

The Evolution
of the Indian Ocean

by D. P. McKenzie and J. G. Sclater
May 1973

*A 75-million-year magnetic record in the floor of the
Indian Ocean reveals how India was rafted northward
for 5,000 kilometers before it collided with Asia. The
encounter gave rise to the Himalayas*

The creation of the Himalayas, the highest and most dramatic mountain range in the world, long presented a puzzle to geologists. Like the Alps and the heavily eroded Appalachians, it was clear that the Himalayas consisted of sedimentary rocks, laid down over many millions of years in shallow seas, then uplifted and heavily deformed by mighty tectonic forces. But in what sea were the Himalayan rocks deposited and how did they get sandwiched between the subcontinent of India and the great Asian landmass to the north? The geology textbooks of a decade ago could provide no satisfactory answer.

With the rapid development and acceptance of the concept of sea-floor spreading, continental drift and plate tectonics the origin of the Himalayas is no longer a mystery. The vast Himalayan range was created when a plate of the earth's crust carrying the landmass of India collided with the plate carrying Asia some 45 million years ago, having traveled 5,000 kilometers nearly due north across the expanse now occupied by the Indian Ocean. Thus the Indian Ocean and the Himalayas have a common origin.

The principal evidence supporting this theory has come from oceanography rather than from classical geology. A geologist's life is too short for him to study a large part of a major mountain system in any detail. Furthermore, the rocks that form the Himalayas have been

so strongly deformed and eroded that it is not possible to discover the extent of the relative motion of India and Asia from a study of the rocks themselves.

That left the ocean as the only other place to look for evidence of a collision. It seemed unlikely that a large continent could cross an entire ocean basin and leave no trace of its passage. Yet as recently as a decade ago oceanographers who had studied the Indian Ocean saw no evidence indicating the passage of a continent and therefore were mostly opposed to the idea of continental drift. The solution to this difficulty and the beginning of our present understanding was provided about 1960 by the late Harry H. Hess of Princeton University. Hess suggested that the continents do not plow their way through the ocean floor but move with it much like coal on a conveyor belt. He proposed that the sea floor is being continuously generated along linear fissures running for thousands of kilometers through the major ocean basins; molten rock wells up into the fissures from the mantle below and quickly hardens [*see top illustration on next page*]. Elsewhere on the earth the ocean floor plunges into a trench and then sinks to great depths in the mantle.

The difference between the concepts of sea-floor spreading and plate tectonics is mostly one of emphasis. Sea-floor spreading is primarily concerned with the processes by which the ocean

floor is created and destroyed, and much less with the continents and the passive parts of the ocean floor. The inactive areas, on the other hand, are the principal subject matter of plate tectonics, which starts from the idea that all areas of the earth's surface free from earthquakes are not at present being deformed. The concept is useful because most earthquakes occur within narrow belts that delineate the regions where large plates of the earth's crust are being subjected to irresistible tectonic forces. It turns out that most of the earth's surface can be divided up into only a few large, rigid plates.

Since the plates are rigid, one can work out the relative motions everywhere of any two plates simply by knowing how they move at any two points along a common boundary. One must emphasize "relative motion" because there is as yet no method for defining absolute motion. The simplest type of motion on a plane is when two plates slide past each other with no relative rotation. A more general type of motion also involves rotation, and it must then be everywhere parallel to circles centered on the center of rotation. Wherever plates are moving apart, the separating edges must be growing by the upwelling of material from below. Such accretions almost always add to each retreating plate at the same rate, and commonly they form boundaries known as ridges. Because fresh material usually adds to the two separating plates at the same rate, ridges normally spread symmetrically. The reason for this symmetric behavior is not fully understood. Furthermore, there is no geometric requirement that the ridge should form at right angles to the direction of motion of the plates, but that is usually the case.

When two plates simply slide past

HIMALAYAN RANGE looking to the west was photographed from *Apollo* 7 in October, 1968. The green area to the left of the range is Nepal; the area to the right is mostly Tibet. Mount Everest is among the peaks in the middle of the picture two and a half inches from the bottom. At the lower right edge is Kanchenjunga, the third highest peak in the world.

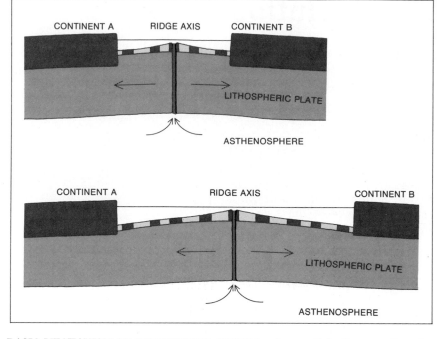

BASIC MECHANISM OF CONTINENTAL DRIFT involves a rift in the ocean floor that allows molten rock to well up from the asthenosphere and form a spreading ridge. The continents are rafted apart at rates up to 20 centimeters per year. As the molten rock solidifies at the ridge axis it becomes magnetized in stripes of alternating polarity (*light and dark bands*) as a consequence of intermittent reversals in the earth's magnetic field.

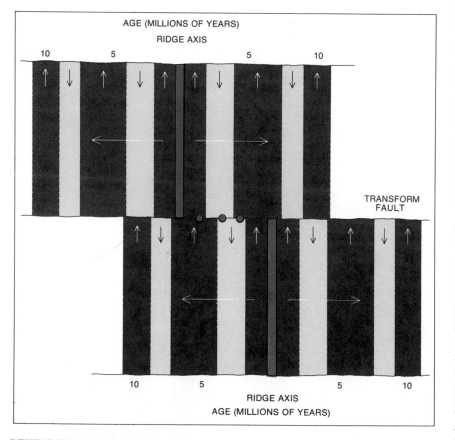

REVERSALS OF EARTH'S MAGNETIC FIELD produce a zebra-striped pattern of magnetization in the crust of the ocean floor. The reversals (*small arrows*) have been dated back about 80 million years by combining radioactivity measurements of rocks with identification of organisms in deep-sea cores. Transform faults often offset the pattern of magnetic anomalies. Friction between the sliding blocks causes shallow earthquakes (*colored dots*).

each other, nothing is added to or subtracted from either plate. Plate boundaries of this type, known as transform faults, are always parallel to the direction of relative motion between the two plates. Such faults must therefore be segments of circles around the center of rotation. When two plates are moving toward each other, one eventually underrides the other and is consumed by plunging into a trench. Simultaneously the edge of the overriding plate is typically uplifted, creating a belt of crumpled and folded rocks parallel to the trench. There is, however, no simple relation between the downward direction of a trench and the relative direction of motion between the plates.

The motion of plates on the earth is slightly more complicated because the earth is spherical. Their relative motion, however, can still be described by a rotation about an axis. Since the plates must remain on the earth's surface, this axis must pass through the earth's center. The motion must again be parallel to circles about the rotation axis. The axis for the opening of the South Atlantic is shown in the illustrations on the opposite page, together with circles drawn with the axis as the center. The earthquake epicenters mark the plate boundary, which is in places a ridge and in places a transform fault parallel to the direction of relative motion marked by the small circles. The original relative position of Africa and South America can be obtained by rotating the two continents toward each other until they meet. The earthquakes have not been moved, but each continent has been rotated through an equal angle toward the ridge axis.

The reason the earthquakes then mark the original break is that the ridge in the South Atlantic has added to the African and South American plates at the same rate. South America and Africa can thus be fitted together rather accurately by a simple rigid rotation. Neither has been deformed during the formation of the South Atlantic, and since both continents consist largely of very ancient rocks, neither has been appreciably deformed in the past 300 million years. Together they form a spherical cap whose diameter is about equal to the radius of the earth. If the earth had expanded appreciably in the past 600 million years, both continents would have been broken into many fragments, since the spherical cap they originally formed would no longer fit on an expanded earth. Plate tectonics therefore shows

AFRICA AND SOUTH AMERICA were locked together as part of the primitive continent of Gondwanaland until about 180 million years ago. The black dots mark the positions of earthquakes that have occurred on the Atlantic mid-ocean ridge in the past 10 years. Their location coincides with the location of the primitive cleavage line between South America and Africa when the two continents are rotated toward each other by an equal amount along the "latitude" lines centered on the axis of rotation (*arrow*). India, Antarctica and Australia were also part of Gondwanaland, but there is still no agreement on how several pieces fitted together.

PRESENT LOCATION of Africa and South America shows the two continents to be equidistant from the still-active spreading ridge (*color*) in the mid-Atlantic. Evidently molten rock flowing into the ridge has increased the size of each plate by an equal amount for the past 180 million years. The two continents are still being pushed apart at the rate of about three centimeters a year. The transform faults developed because the initial break was not everywhere at right angles to the circles of rotation. As required by the theory of plate tectonics, the transform faults lie parallel to the circles.

that geologically important expansion has not occurred, although it does not rule out changes in radius of a few tens of kilometers.

The South Atlantic also shows how transform faults can originate. The original break between South America and Africa was a jagged line that followed old lines of weakness in the Precambrian shield and did not form a plate boundary at right angles to the direction of motion between the plates. After separation had started, the plate boundary changed from a jagged ridge to a series of straight ridge sections joined by transform faults. This change maintained the general shape of the ridge but allowed the ridges to be at right angles to the direction of relative motion.

To a continental geologist it is quite surprising that the simple ideas of plate

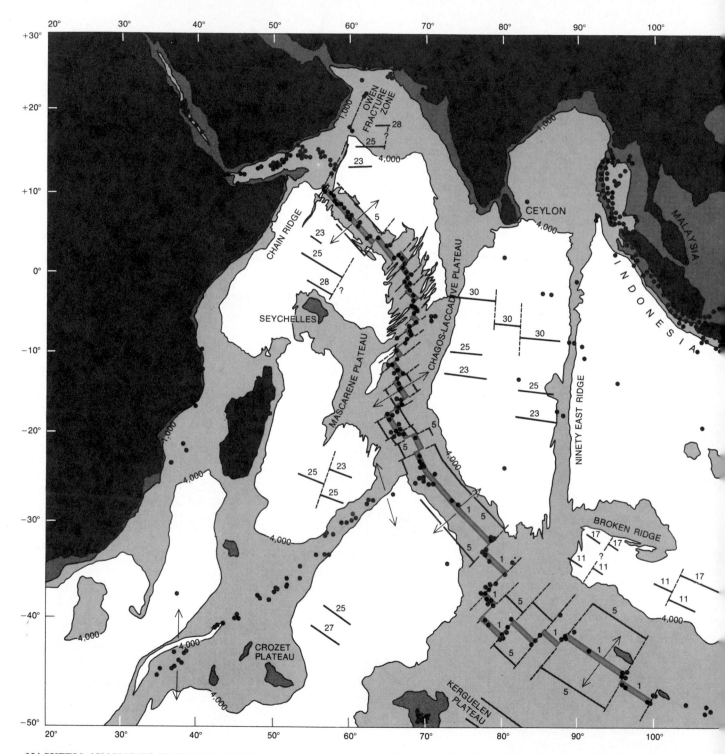

MAGNETIC ANOMALIES IN INDIAN OCEAN are represented by straight lines. The numbers on the lines start with No. 1 on the present ridge axis (*two closely spaced lines*) and extend to No. 33, which identifies molten rock that solidified about 80 million years ago. Earthquakes recorded over a recent seven-year period are indicated by black dots. They can be seen to cluster along the present ridge axis and on lines where the axis is offset by transform faults. The greatest number of earthquakes, however, take place around the island arc between Southeast Asia and Australia where the Indian Ocean plate and the Pacific Ocean plate plunge into deep trenches and are consumed. The major features on the floor of the Indian Ocean are the three large ridges that run north and

tectonics work so well. The rocks he studies were generally severely deformed by plastic flow when they were near a plate boundary. Thus there was opposition to the new ideas from investigators who had worked only on land. That opposition has now largely ceased.

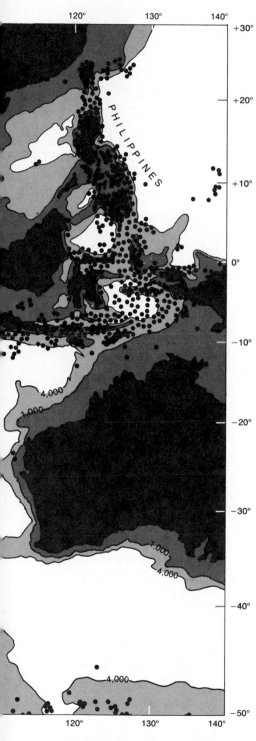

south: Ninety East Ridge, the Chagos-Laccadive Plateau and the Mascarene Plateau. Two parallel ridges, Broken Ridge and the Kerguelen Plateau, run approximately east-west near Australia. Such features aided in reconstructing evolution of Indian Ocean.

The early tests of plate tectonics all involved the present motions of plates measured at plate boundaries as defined by the location of earthquakes. This approach is obviously powerless to determine how plates may have moved in the geologic past. Providentially the rocks of the ocean floor carry a permanent record of their own history. It was noted about 1960 that the oceanic crust is not uniformly magnetized. Instead magnetometer surveys revealed a remarkably regular striped pattern of magnetic variation: stripes of normal polarization (with north pointing to the magnetic north pole) alternate with stripes of reverse polarization.

In 1963 F. J. Vine and D. H. Matthews of the University of Cambridge proposed that when lava wells up in the oceanic ridges, it is magnetized as it cools by the magnetic field of the earth. The alternations of polarity indicate that the magnetic field of the earth reverses at infrequent and irregular intervals. As a result the entire ocean floor is covered with long linear blocks of alternating magnetic polarity [see bottom illustration on page 150]. The shape of the blocks is exactly the shape of the ridge that formed them. We are doubly fortunate: the period between reversals of the magnetic field is generally sufficient to produce stripes at least 10 kilometers wide and, equally important, the reversals come at irregular intervals, so that the stripes vary in width. Hence the magnetic anomalies can be matched where they have been shifted by transform faults.

If the present ridge axis is designated No. 1, the magnetic stripes that flank it on each side can be numbered sequentially up to No. 33, formed about 80 million years ago. Older anomalies are known, but their age and relation to the most recent 33 are not yet established. Magnetic surveys can be made with a magnetometer towed behind either a ship or an airplane. The mapping and identification of magnetic anomalies now provides a fast and accurate method of studying the evolution of an ocean.

In regions where there are no magnetic data or where the anomalies are not clear the depth of the sea can often provide a rough estimate of the age of the sea floor. The youngest portion of a plate forms the peak of a ridge system that rises about 3,000 meters above the flatter and older portions of the sea floor. The reason is that the recently intruded rock is hotter and therefore less dense than the older material that has cooled and contracted as it moved away from the plate boundary. Sea-floor profiles

calculated on the basis of rock-cooling models agree well with profiles established by accurate depth surveys.

The separation rate recorded by the anomalies indicates the relative motion between two plates. It is also possible to measure the motion between one plate and the earth's magnetic pole. If the magnetic pole is assumed to coincide approximately with the rotational pole, the measurement of plate motion can be used to find the latitude and orientation of the plate on which the observations were made. If the rocks are magnetized near the Equator, the magnetization is horizontal; the closer the rocks are to the poles when they are magnetized, the more vertical is the pitch, or dip, of the magnetization. With many paleomagnetic measurements made over the past two decades one can plot the apparent migration of the pole going back tens of millions of years. One cannot decide, of course, whether the pole itself has moved or whether the pole has remained fixed while the plates have moved. For convenience we take the poles as being fixed and move the plates to a latitude corresponding to the pitch of their magnetization. Naturally any reconstruction of the positions of the plates must be consistent with respect both to latitude and to north-south direction of polarization.

When we started our work on the Indian Ocean, the theory of plate tectonics had been proposed and tested but its success was not yet generally recognized. Like many scientific projects, ours started by chance. When we began in 1968, the Indian Ocean was the only major ocean in which no one was trying very hard to use marine geological and geophysical data to construct a tectonic history. It happened, however, that Robert L. Fisher of the Scripps Institution of Oceanography was doing detailed work on the Central Indian Ridge and had organized an expedition that would spend six months collecting marine-geological and geophysical information in the Indian Ocean. One of us (Sclater) was chief scientist on that expedition for two months and proposed that the other (McKenzie) join the ship for a month, more to show a theoretical geophysicist how cruises are run than with the intention of putting him to work.

The magnetic anomalies observed on that cruise left little doubt in our minds that the plate carrying India had crossed the Indian Ocean. To see if we could strengthen the hypothesis we collected all the available magnetic profiles from ship and airplane traverses of the ocean.

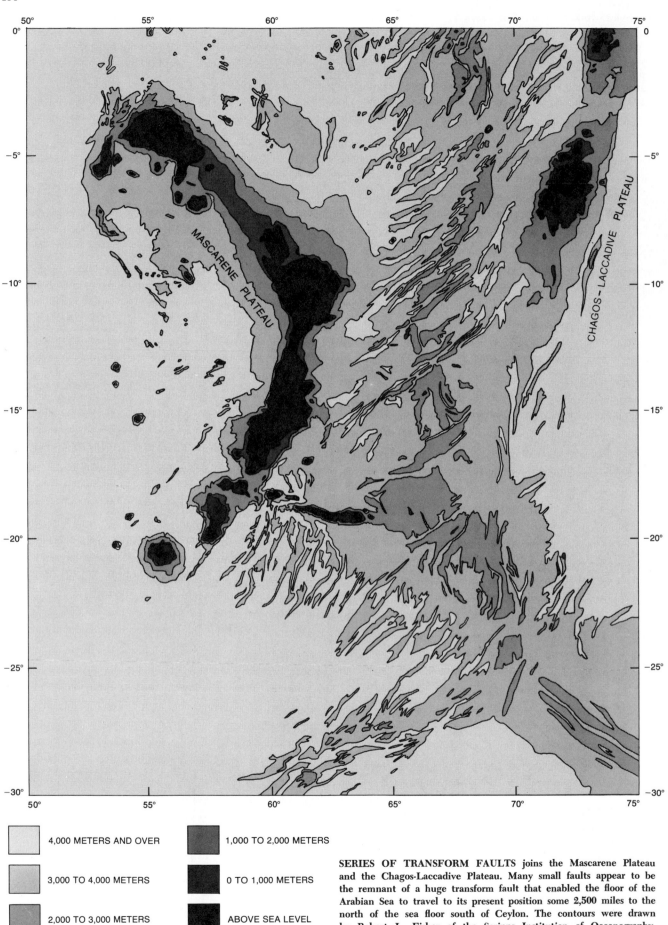

SERIES OF TRANSFORM FAULTS joins the Mascarene Plateau and the Chagos-Laccadive Plateau. Many small faults appear to be the remnant of a huge transform fault that enabled the floor of the Arabian Sea to travel to its present position some 2,500 miles to the north of the sea floor south of Ceylon. The contours were drawn by Robert L. Fisher of the Scripps Institution of Oceanography.

4,000 METERS AND OVER

3,000 TO 4,000 METERS

2,000 TO 3,000 METERS

1,000 TO 2,000 METERS

0 TO 1,000 METERS

ABOVE SEA LEVEL

Fortunately the theory of plate tectonics changed very little in the three years that the two of us and three co-workers spent interpreting the observations. When we were done, the evolution of the Indian Ocean emerged clearly.

The shapes of the magnetic anomalies show that the evolution of the Indian Ocean was much more complicated than the evolution of the South Atlantic. The topography of the floor of the Indian Ocean is correspondingly more complex. The floor shows four large north-south ridges, two of which are still actively spreading [see illustration on pages 152 and 153]. It is hard to see any relation between the 'two huge inactive ridges (Ninety East Ridge and the Chagos-Laccadive Plateau) and the ridges that have remained active. It is also difficult to understand how the various ancient continental fragments, such as the island of Madagascar, the Seychelles islands and perhaps also Broken Ridge west of Australia, became isolated from the surrounding landmasses. Islands and ridges of this type are not found in the Atlantic and the Pacific. Their existence in the Indian Ocean is baffling.

A feature of particular interest is the plate boundary that runs north and south between the Chagos-Laccadive Plateau and the Mascarene Plateau in the northwestern part of the Indian Ocean. In this area Fisher's work has shown that the short spreading ridge segments are offset by many transform faults, none of which crosses the two inactive ridges on each side [see illustration on opposite page]. These faults therefore cannot have been produced by the shape of the original break, as the faults in the South Atlantic were. Their origin becomes clear only when the anomalies between anomaly No. 23 and anomaly No. 29 are used to map the shape of the ridge axis that produced them.

The anomalies in the Arabian Sea south of Iran run east and west but are about 2,500 kilometers north of the similarly numbered anomalies south of Ceylon. Hence there must at one time have been a huge transform fault joining the ridge in the Arabian Sea to the one south of Ceylon [see illustration on this page]. Some 55 million years ago relative motion south of Ceylon slowed from 16 centimeters per year to about six centimeters and probably stopped altogether in the Arabian Sea. Then about 35 million years ago movement resumed in the Arabian Sea, but its direction was no longer parallel to the ancient transform fault. New sea floor was again generated, and the boundary changed from a

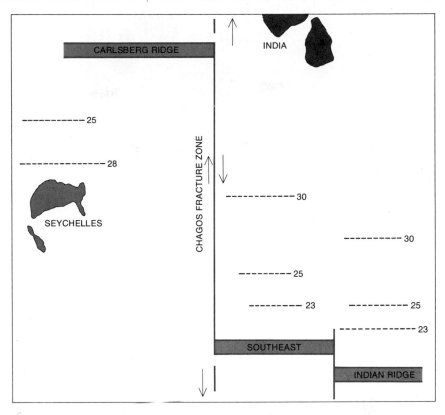

FIFTY-FIVE MILLION YEARS AGO a huge transform fault, the Chagos fracture zone, divided the floor of the Indian Ocean into two large segments. Between 80 million and 55 million years ago the segment carrying India traveled about 16 centimeters a year.

ridge oblique to the spreading direction to a series of ridge segments that are perpendicular to the direction of motion and are joined by transform faults [see illustrations on next page].

As in the Atlantic, the shape of the present plate boundary preserves the shape of the ancient boundary and the north-south section is the image of the old transform fault preserved by the new structures. This region and the central Atlantic thus show how the complex geometry of plate boundaries often results from the simple behavior of ridges and transform faults. To be sure, not all ridges behave so simply. In some cases a plate breaks and forms a ridge in a new place. Such "jumping" ridges are fortunately rare. Since they produce very muddled patterns of anomalies, they are difficult to identify.

A second area of much interest is the region south of Ceylon. Here we found anomalies from anomaly No. 32 to anomaly No. 21 that were formed between 75 million and 55 million years ago. These anomalies are now just south of the Equator, but the matching anomalies on the other side of the ridge are now in the far southwest Indian Ocean, northeast of the Crozet Islands. The two segments of anomaly No. 32 are now

separated by about 5,000 kilometers. India must therefore have traveled this distance at the rate of 7.5 centimeters per year after breaking away from Antarctica 75 million years ago.

The separation of the anomalies between anomaly No. 30 and anomaly No. 22 shows that the separation rate during this period was more than twice as rapid: 16 centimeters per year. At that rate a plate would be carried completely around the world in only 250 million years, which is a very short span of time by geological standards. Even faster rates, however, have been measured on the present spreading ridge in the southeast Pacific, where the plates are moving apart at nearly 20 centimeters per year.

We can now use the magnetic lineations to reconstruct the relative positions of the different plates underlying the Indian Ocean. We can also employ the paleomagnetic observations from the neighboring continent of Australia to place all the plates in their correct position with respect to the magnetic poles, which we believe coincide approximately with the earth's rotational poles. The motion of each plate can be described by a rotation, so that the complete reconstruction process consists of a series

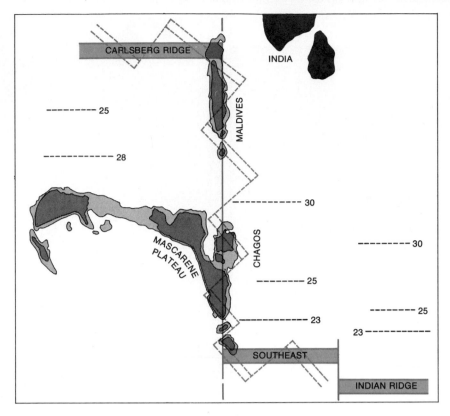

THIRTY-FIVE MILLION YEARS AGO, after a 20-million-year period during which India's northward drift slowed significantly, the single north-south fault of the Chagos fracture zone evidently broke up into a number of ridge segments connected by short faults.

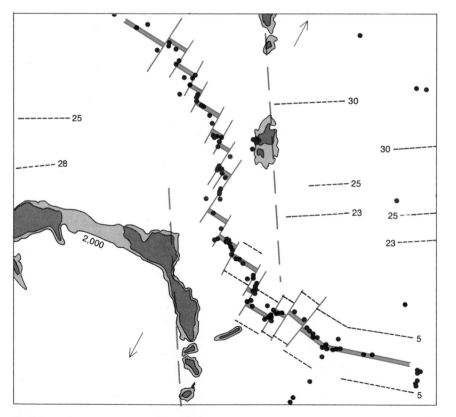

WHEN NORTHWARD MOTION RESUMED about 35 million years ago, new sea floor was extruded along ridges that cut obliquely across the line of the ancient Chagos fault. Thus the direction of sea-floor spreading was rotated about 45 degrees from the original direction. This map repeats a central portion of the ridge-and-fault structure illustrated on page 152.

of rotations. Unlike the single rotation that suffices to close the South Atlantic, however, the closing of the Indian Ocean requires the rotation of several plates that have moved at varying rates and in different directions since India started its rapid movement northward.

The earliest reconstruction we can obtain is governed by the oldest anomaly we can identify: anomaly No. 32, formed about 75 million years ago. At that time the ridge south of Ceylon was spreading in such a way as to propel India generally a little east of north, and Ninety East Ridge was an active transform plate, facilitating the northward slippage of the Indian plate [*see illustration on opposite page*]. Since Asia was evidently more or less stationary, or at least not moving north nearly as fast as India was approaching, the northern edge of the Indian plate must have been consumed in a trench lying somewhere to the south of Asia, but little is known about the position or history of that trench. Australia was then still joined to Antarctica, forming a single continent; Africa and South America had already separated.

During the next 20 million years India moved rapidly northward between two great transform faults: Ninety East on the east and a fault now called the Owen fracture zone on the west. Some 55 million years ago the relative movement between plates slowed down or stopped. Only the ridge between India and Antarctica continued to produce new plate, and the rate of separation dropped from 16 centimeters per year to six centimeters or less. Forty-five million years ago there was a rupture between Australia and Antarctica, setting Australia free for its northward migration. Soon afterward all the plate boundaries in the Indian Ocean became active again, but they were now moving in new directions [*see illustration on page 158*].

The result was the complicated pattern of ridge segments and transform faults that is still active in the central part of the western Indian Ocean. Motion on the Ninety East plate boundary ceased about 45 million years ago. The most recent events in the history of the Indian Ocean have been the formation of the Red Sea and the Gulf of Aden when Arabia separated from Africa, probably about 20 million years ago.

Beginning about 45 million years ago the northern edge of the Indian plate must have begun crumpling up the many layers of shallow-water sediments, known as a geosyncline, laid down over millions of years on the continental shelf that bordered the southern edge of Asia.

The result was the upthrusting of the Himalayas. The application of plate-tectonic concepts to both the evolution of the sea floor and the building of mountains is encouraging. It should eventually be possible to tie the two phenomena together by relating the composition and historical movement of rocks in the mountain ranges to records left in the sea floor.

Many attempts have been made to guess precisely how South America, Africa, India, Antarctica and Australia were once joined to form the primitive continent known as Gondwanaland. There is as yet no general agreement as to how this should be done. The fit between South America and Africa, as is well known, is excellent. The fit between Australia and Antarctica is good. The

arrangement of all five major units, however, is controversial, and the original position of Madagascar is unknown.

The principal difficulty is that no magnetic lineations have yet been discovered on the older parts of the floor of the Indian Ocean between the continents. We therefore cannot continue to reassemble the continents by the same methods we have used to trace the movement of

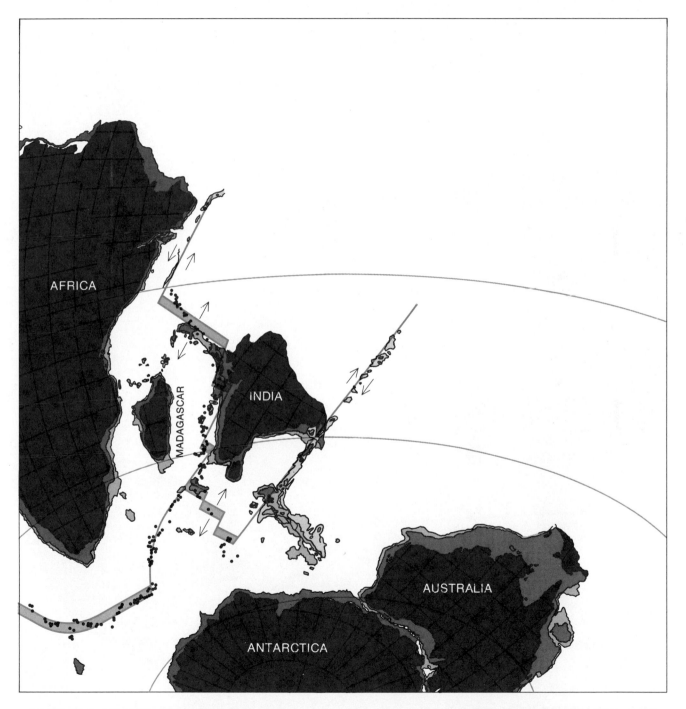

SEVENTY-FIVE MILLION YEARS AGO, after breaking away from the ancient primitive continent named Gondwanaland, the plate carrying India started to move rapidly toward the northeast. The motion was accompanied by the generation of new plate along the ridges indicated by pairs of parallel lines. Large transform faults east and west of the connected series of ridges enabled the Indian plate to move unimpeded. The original fit of India, Africa and Madagascar with Antarctica-Australia is not yet established.

India during the past 75 million years. There are also no other structures like Ninety East Ridge, which was recognized as a transform fault even before the magnetic lineations were mapped. Fortunately the area of sea floor in which the record presumably lies hidden is not great. Last year a series of deep holes were drilled in the floor of the Indian Ocean by the drilling vessel *Glomar Challenger*. The data from these holes have confirmed and amplified our reconstruction of the history of the ocean.

They have also added to the evidence needed to reconstruct Gondwanaland.

Meanwhile one can speculate about the original juxtapositions of India, Antarctica and Australia. One guess is that existing reconstructions are wrong because they have attempted to remove practically every piece of sea floor between the continents. That approach has been favored because all the continents believed to have formed Gondwanaland show evidence of having been covered

by a huge ice cap some 270 million years ago. We know from the recent glaciation in the Northern Hemisphere, however, that continental ice caps can simultaneously cover landmasses that are separated by oceans. It may be that a small ocean basin, comparable perhaps to the Arctic Ocean, was nestled somewhere among the southern landmasses 270 million years ago. It may be our ignorance of its existence and shape that is preventing the successful reconstruction of Gondwanaland.

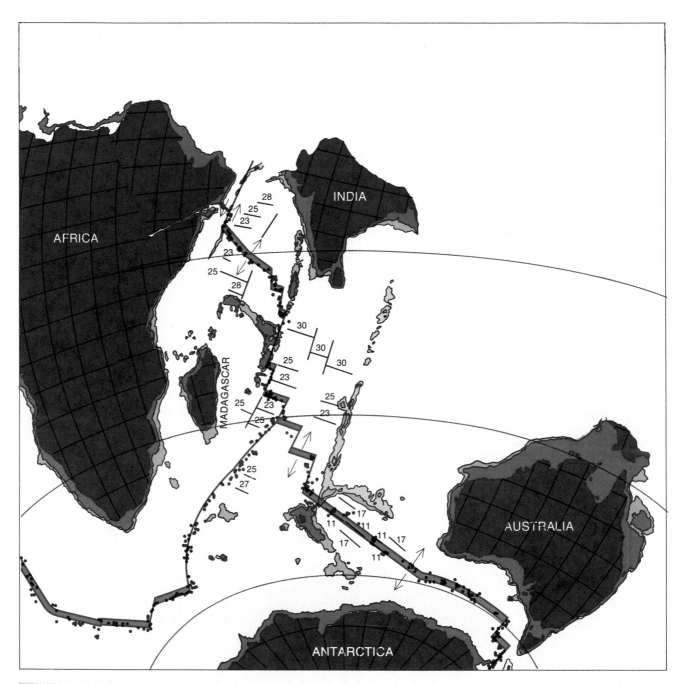

THIRTY-FIVE MILLION YEARS AGO India had been carried some 4,000 kilometers to the north at rates that evidently varied from 16 centimeters per year to seven centimeters or less. It was about this time that the direction of relative motion among several of the major plates changed quite drastically to produce the complicated sea-floor geometry we see today. Australia has been set adrift from Antarctica and has started on its way to its present position. At this point Arabia has still not split off from Africa.

14

Continental Drift and the Fossil Record

INTRODUCTION

Several decades before the theory of continental drift became generally accepted, the fossil record was used by proponents to buttress their case. However, for every example that "drifters" cited of similar fossils found in Africa and in South America to prove an ancient contact between the continents, their opponents would find a contradictory example.

In this article, "Continental Drift and the Fossil Record," A. Hallam discusses their role in evolution. The basic concepts of plate tectonics and sea-floor spreading have been shown in previous chapters to have a secure foundation based on oceanic magnetic data, in the results of seismology, and in deep-sea drilling. Now that we know the larger details of the positions of the continents during the past 200 million years, we are in a better position to understand such problems of evolution as reasons for similarity and differences of fauna of different regions and the roles of migration and isolation in producing such homogeneity and diversity.

This is not to say that paleontology cannot contribute to the unfolding story of ancient continental breakup and reassembly. Higher resolution in paleogeographic maps, more accurate dating of important events, such as the initial contact between India and Asia, or between Africa and Europe, may come from the study of the fossil record.

The nature of sea-floor spreading and continental drift in Pre-Jurassic time, more than 200 million years ago, has yet to be unraveled. The fossil record may have much to contribute to this problem now that we are learning to interpret the interplay between evolution and continental drift.

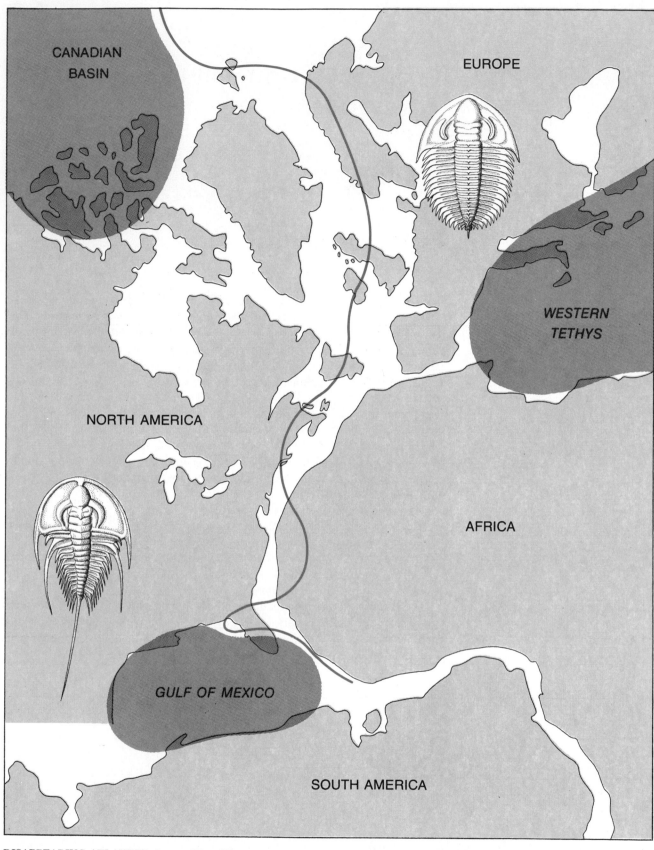

DISAPPEARING ATLANTIC of some 400 million years ago may have been reduced to two comparatively minor bodies of water: the Canadian basin and the Gulf of Mexico. The broad colored line divides regions that are connected today, such as the Scandinavian peninsula, the British Isles and parts of eastern North America. It marks the boundary between faunal realms that were distinctly different in Cambrian and early Ordovician times. A fossil typical of each realm is illustrated. The American form is the trilobite *Paedeumias transitans*; the European form is the trilobite *Holmia kjerulfi*. The difference between the two realms grew progressively less, and by the middle of the Paleozoic era it had vanished. Convergence of the two faunas suggests that what had once been a wide, deep proto-Atlantic was swallowed up along zones of subduction as American and European continental plates came together.

Continental Drift and the Fossil Record

by A. Hallam
November 1972

Similarities between the fossils found in widely separated areas led to the first theories of continental drift. Today's advocates of plate-tectonic theory are also supported by the fossil record

In the past 10 years geologists have been widely converted to plate tectonics, a concept that implies the lateral migration of the continents. If Alfred Wegener, who put forward the hypothesis of continental drift around the turn of the century, were alive today, he might be wryly amused. The workers who revitalized his conception paid comparatively little attention to the evidence in its favor provided by fossils. Yet by Wegener's own account he began to take the idea of drifting continents seriously only after learning of the fossil evidence for a former land connection between Brazil and Africa. The fossil record, rather than the much noted physical "fit" between the opposing coastlines, was what inspired him.

The fossil record is no less important to students of continental drift today. The similarities and differences between fossils in various parts of the world from Cambrian times onward are now helping paleontologists both to support the drift concept and to provide a reasonably precise timetable for a number of the key events before and after the breakup of the ancestral continent of Pangaea.

In Wegener's day there was nothing particularly novel about the idea that the various continents had been connected in various ways off and on in the distant past. Biologists and paleontologists in the 19th century and in the early 20th readily invoked such land connections to explain the strong resemblances between plants and animals on different continents. It was generally agreed that links had existed between Australia and other regions bordering the Indian Ocean until early in the Jurassic period, some 180 million years ago. The same was held to be true of a link between Africa and Brazil until early in the Cretaceous period, some 140 million years ago, and a link between Madagascar and India up to the start of the Cenozoic era, only 65 million years ago.

At the same time the orthodox explanation of these connections was that the position of the continents was fixed but that "land bridges" had spanned the considerable distances of open ocean between them. In the orthodox view these extensive bridges had later sunk without a trace [see upper illustration on page 163]. Wegener dismissed such explanations in trenchant terms. The earth's crust, he pointed out, is composed of rocks that are far less dense than the material of the earth's interior. If the floors of the oceans were paved with vast sunken bridges composed of the same thickness of light crustal material as the continental areas that lie above sea level, then gravity measurements made at sea should reveal that fact. The gravity measurements indicate the exact opposite: the underlying rock of the ocean floor is much denser than the crustal material of the continents.

The essential improbability of sunken land bridges can also be stated in terms of isostatic balance. If the low-density crustal rocks of the vanished bridges had indeed been somehow forced downward into the denser sea bottom, the bridges would tend to rise again. None of the hypothetical land bridges, however, has reemerged. This makes it necessary to assume the existence of some colossal unspecified force that continues to hold the bridges submerged. The existence of such a force seems improbable in the extreme. Unless one chose to dismiss the fossil evidence out of hand, Wegener concluded, the only feasible means of explaining intercontinental plant and animal resemblances was by the drifting of the continents themselves.

It is odd that neither paleontologists nor geophysicists paid much heed to Wegener's cogent argument. The paleontologists were almost unanimous in rejecting the notion, perhaps because they did not fully appreciate the force of Wegener's geophysical proposals. The main effect of his hypothesis on this group was that quite narrow land bridges became more popular than the embarrassingly broad avenues that had been favored at the turn of the century [see lower illustration on page 163]. Of course even when the bridges were pared down in this fashion, serious isostatic problems remained. As for the geophysicists, they largely ignored the considerable body of fossil evidence for continental drift that Wegener had assembled. Perhaps they failed to appreciate its significance or perhaps they mistrusted data of a merely qualitative kind and were suspicious of the seemingly subjective character of taxonomic assessments.

The Zoogeography of the Past

In recent years there has been a major resurgence of interest in the zoogeography of the past. This is no doubt due in part to the prospect that fossil evidence will enable paleontologists to test independently the conclusions that have been reached on purely geological and geophysical grounds about plate tectonics. Here I shall describe how some of this evidence, in particular the remains of higher animals on land and of simpler bottom-dwelling marine organisms offshore, can shed light on the making and breaking of connections between continents in the past and can help to determine how long some of these continental linkages and separations endured.

Two principal factors control the geo-

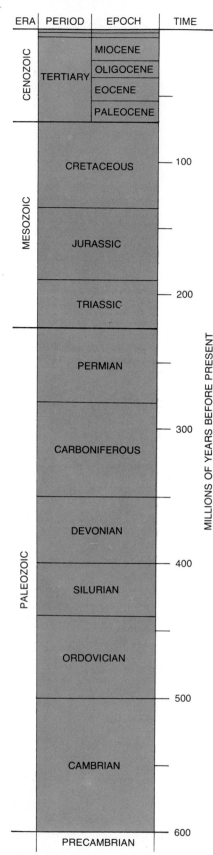

ERA	PERIOD	EPOCH	TIME
CENOZOIC	TERTIARY	MIOCENE	
		OLIGOCENE	
		EOCENE	
		PALEOCENE	
MESOZOIC	CRETACEOUS		100
	JURASSIC		
	TRIASSIC		200
PALEOZOIC	PERMIAN		
	CARBONIFEROUS		300
	DEVONIAN		
	SILURIAN		400
	ORDOVICIAN		
	CAMBRIAN		500
	PRECAMBRIAN		600

MILLIONS OF YEARS BEFORE PRESENT

THREE SUCCESSIVE ERAS occupy the 600-million-year span seen in this geological time scale. Patterns of distribution found in the fossil record of each era throw light on plate-tectonic theory of continental drift.

graphical distribution of land animals. They are on the one hand the climate and on the other various water barriers and in particular wide stretches of sea. The effectiveness of climate as a barrier is nicely illustrated by the fact that the animals that live near the poles are far less diverse than those that live in the Tropics. Polar ice, however, is not a permanent feature of the planet. In times of a more equable world climate such as prevailed during the Mesozoic era far and away the most significant deterrents to movement must have been ocean barriers.

An animal need not be exclusively terrestrial to be confined by an ocean barrier. For example, a large number of fishes cannot survive except in fresh water. Amphibians are also severely circumscribed by the sea, although frogs are better able to colonize islands by swimming across the sea than newts or salamanders. As for the probability that reptiles or mammals might successfully move across the sea by accidental rafting, the chances are obviously best for the small and the rapidly reproducing. By the same token, however, the small animals are the very ones that would first die of starvation if the rafting were prolonged. Hence even fairly narrow marine straits can be highly effective barriers to terrestrial animals.

The paleontologist George Gaylord Simpson has made a useful distinction between three kinds of dispersal route. The first, which he calls "corridors," are land connections that allow free migration of animals in both directions. The second, called "filter bridges," combine a land connection with some additional factor, such as climate, in a way that bars some prospective migrants. As an example, it seems doubtful that warm-weather-loving animals crossed the Bering Strait bridge between Asia and North America during the Pleistocene. The passage was open only when the sea level was low during the colder phases of that glacial epoch.

Simpson's third category, "sweepstakes routes," takes its name from the small proportion of winners compared with losers. The rare winners are those that survive chance rafting and succeed in colonizing isolated areas. Unlike corridors (or even filter bridges), which favor the eventual homogeneity of the faunas at both ends of the passage, sweepstakes routes lead to the development of populations that are low in diversity and ecologically unbalanced. The reason is that, in addition to the high mortality rate, chance rafting can only be possible for a

very small fraction of any continental fauna. One result of this double selectivity is that islands are often a refuge for a comparatively primitive group of animals. The tuataras of New Zealand, the sole surviving representatives of one major order of reptiles, are one example; the lemurs of Madagascar are another.

The three routes Simpson defines are the ones that are available to land animals in general and to the higher vertebrates in particular. It is obvious, however, that the same kinds of connection would also influence the dispersal of marine organisms such as bottom-dwelling invertebrates. Of course, the effects would be exactly reversed. For example, the establishment of a corridor between two landmasses would simultaneously raise a barrier between two segments of a previously homogeneous marine fauna. The disappearance of a land connection, in turn, would result in the establishment of a corridor as far as marine organisms were concerned. Analogous to the filter bridge for land animals would be the marine "filter barrier." Here a bottom-dweller on one side of an ocean basin might migrate freely to the other side as long as it could drift in tropical waters, but it would be unable to survive the rigors of such a journey in the cooler waters of the temperate zones.

Isolation and Homogenization

Bearing these considerations in mind, what does evolutionary theory predict with respect to continental drift? Clearly when a formerly unified landmass splits up, the result is genetic isolation (and hence morphological divergence) among the separated segments of a formerly homogeneous land fauna. Conversely, the suturing of two continental areas is followed by the homogenization of the corresponding faunas as there is cross-migration. The process will quite probably be accompanied by the extinction of any less well-adapted groups that may now face stronger competition.

In land areas that are unconnected two factors, parallel evolution and convergence, may produce animal species that develop a similar morphology because they occupy identical ecological niches. A well-known instance is provided by the ant bear of South America, the aardvark of Africa and the cosmopolitan pangolin. Forces of this kind are unlikely, however, to affect entire faunas.

So much for the land animals. What are the effects of continental drift on the invertebrates that inhabit the shallow ocean floor? It is obvious that a stretch of land that separates two oceans acts as a barrier to marine migration. It is less well appreciated that a wide stretch of deep ocean may be almost as effective as a land barrier in preserving the genetic isolation of the marine shelf dwellers on opposite shores. This isolation is due to the fact that such animals disperse at only one time in their life cycle: when they are newly hatched larvae that join the plankton commu-

nity at the ocean surface or close to it. A bottom dweller's larval stage is normally not long enough to enable the animal to survive a slow ocean crossing.

This matter was quantified some time ago by the Danish biologist Gunnar Thorson. He studied the larval stages of no fewer than 200 species of marine invertebrates, and he concluded that only 5 percent could survive in the plankton for more than three months. That length of time is too short to allow transoceanic

colonization except by an occasional "transport miracle." More recently Thorson has been criticized for confining his attention to organisms that inhabit cool temperate waters. New data gathered by towing fine nets through the plankton indicate that a significant number of larvae of tropical species can survive an Atlantic crossing in either direction; they drift with the equatorial surface current one way and with the subsurface countercurrent the other way. It is also true

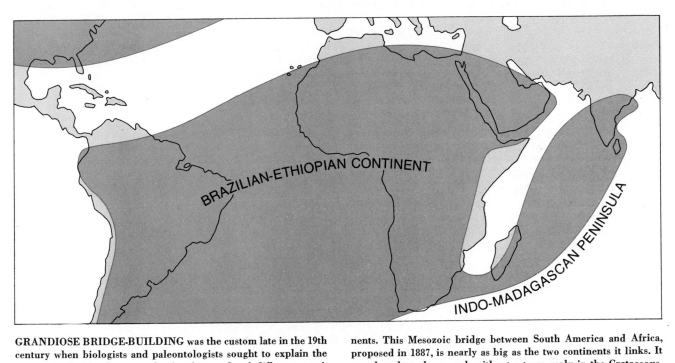

GRANDIOSE BRIDGE-BUILDING was the custom late in the 19th century when biologists and paleontologists sought to explain the strong similarities between the fossil records of different continents. This Mesozoic bridge between South America and Africa, proposed in 1887, is nearly as big as the two continents it links. It was thought to have sunk without a trace early in the Cretaceous.

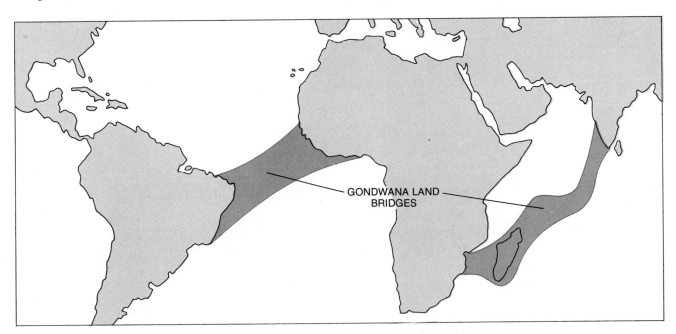

MORE MODEST BRIDGES were proposed in the 20th century as geophysical studies found no evidence in support of vast drowned intercontinental connections. This pared-down link between Brazil and West Africa and another (supposedly in existence until the end of Cretaceous times) between East Africa and India represent an effort to satisfy those who were critical of the broader bridges.

that some mollusks and some corals are found everywhere in the Tropics, a fact that gives added support to the notion that transoceanic migration is possible for long-lived larvae. At the same time it is possible that organisms with prolonged larval stages represent a late evolutionary development that has taken place only since the present Atlantic and Indian oceans began to open up.

Be this as it may, it is evident that, although Thorson's conclusions must be somewhat amended, an ocean barrier is effective in restricting the migration of a majority of the shelf-dwelling marine invertebrates. Moreover, the capacity for migration being in direct proportion to the length of the larval stage, it follows that the wider the deep-ocean barrier is, the more the faunas of the opposing shelves should differ. That is true today. There are fewer species in common among the faunas on opposite sides of the Pacific than there are on opposite sides of the Atlantic. Therefore among marine fossil faunas the degrees of similarity or difference between two coastal assemblages should allow an estimate of the amount of deep-ocean separation between the two coasts.

Given the present state of knowledge, it would be futile to seek any absolute measure of similarity between fossil faunas. If, however, one adopts a dynamic approach and deals with changes in the degree of resemblance during successive intervals of geologic time, a fair amount of progress can be made. Such an approach has several advantages. For one, animals belonging to widely separated phyla can be grouped together. For another, the specific factors that control a particular species' migration potential need not be precisely understood. Further, one can use the work of many different taxonomists, without regard to whether they are "lumpers" or "splitters" in their method of classification, as long as the work is self-consistent. Best of all, the approach takes fully into account the factor that is all-important to the geologist: the factor of time.

The Distribution of Fossils

There are four principal patterns revealed by the distribution of fossils. They are closely interrelated and I shall define them in order, accompanying each definition with one or more examples. All the examples will concern the making or breaking of land or sea connections, but not all will be the result of continental drift.

The first pattern is "convergence." The term describes an increase, as time passes, in the degree of resemblance between faunas of different regions. An example unrelated to continental drift is provided by the history of South America during the Cenozoic era, beginning some 65 million years ago. Throughout most of the era South America had a highly distinctive land fauna [see illustration below]. The fossil mammals from Cenozoic formations in Argentina were among the most spectacular finds of Darwin's voyage aboard H.M.S. Beagle. The strongly endemic nature of this fauna, comparable in this respect to the fauna of Australia today, is clear evidence that the continent was isolated for many millions of years.

At about the end of the Pliocene epoch, two million or so years ago, a drastic change took place. A land connection—the Isthmus of Panama—was established between North and South America and many New World animals

DIVERGENCE OF MAMMALS in the New World occurred during a period of millions of years when North and South America were unconnected. Mammals then unknown in the south included mastodons, tapirs, primitive camels and various carnivores. The mammals of the south included many that are now extinct. Illustrated here are (a) *Mylodon*, a giant sloth, (b) *Paedotherium*, a notoungulate, (c) *Prodolichotis*, a rodent, (d) *Macrauchenia*, an ungulate, and (e) *Plaina*, an early relative of the more successful armadillo.

that had been unable to move south now crossed the bridge. Among them were the mastodon, the tapir, primitive camels and a number of carnivores. Simultaneously many of the indigenous South American faunas became extinct; the losers included all but two genera of the many primitive marsupials that had been sheltered there and were evidently incapable of competing with the better-adapted migrants from North America. The traffic was not, however, entirely one-way. Armadillos soon extended their range northward throughout Central America and into the southern U.S. [*see illustration below*].

Simpson has estimated that before North America and South America were united perhaps 29 families of mammals lived in the area south of the Isthmus of Panama and perhaps 27 entirely different families of mammals lived to the north. After the union the faunas of both continents had 22 families of mammals in common. This is a particularly dramatic example of convergence, even though the establishment of the Panama bridge seems to have owed nothing to continental drift.

The mammalian fauna of Africa during the Cenozoic era provides another example of convergence. Up to some 25 million years ago the African fauna had a strong endemic element. The animals ancestral to the living elephants, the manatees and the hyrax were found only there. This suggests that the continent had been isolated for a substantial period.

Early in the Miocene epoch a number of mammals from Eurasia entered Africa by means of one or more land connections. The migration led to the reduction and even the extinction of some of the indigenous African fauna. At the same time the ancestral elephants crossed over to Eurasia and had soon spread around the world. Less than 25 million years later mastodons were waiting at the emerging Isthmus of Panama to enter the last major continental area in the world still barred to the proboscoids. The Miocene bridge-building that allowed a substantial degree of convergence between the faunas of Africa and Eurasia can be attributed to continental drift, specifically a northward movement of the Africa-Arabia plate.

A third example of convergence takes us all the way back to the Cambrian and early Ordovician periods, some 500 to 600 million years ago. The most spectacular marine organisms of the Cambrian, the early anthropods known as trilobites, turn out to be sharply separated into two distinct faunas; the line that divides the two faunal provinces runs through eastern North America and through the British Isles and Scandinavia [*see illustration on page 160*]. Over the next 75 million years or so, during late Ordovician and Silurian times, the two trilobite faunas tend to lose their regional distinctiveness. So do a number of other early marine invertebrates: corals, brachiopods, graptolites and conodonts. So do two groups of early freshwater fishes: primitive jawless ostracoderms belonging to the orders Anaspida and Thelodonti. Freshwater fishes are, of course, prisoners of the continental streams they inhabit. By late Silurian or early Devonian times, some 400 million years ago, only one faunal province existed in the North Atlantic region.

The paleontologist A. W. Grabau noted the difference between the trilobites of adjacent areas many years ago, and he suggested that the sharp delineation might be attributable to the former

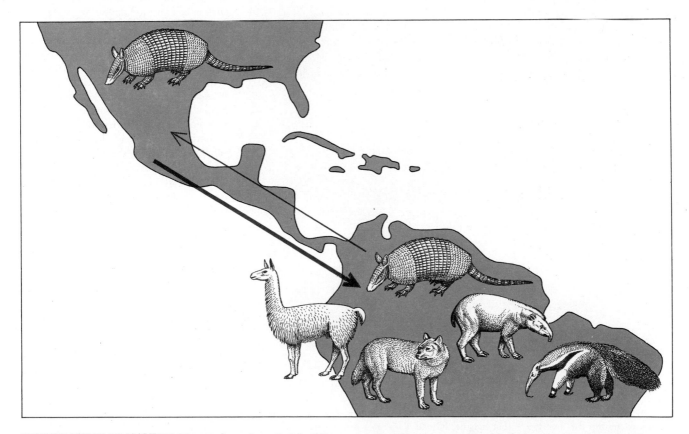

CONVERGENCE OF MAMMALS began about the end of the Pliocene epoch, after a land bridge was established between North and South America. One mammal native to South America, the armadillo, migrated northward. So many mammals formerly unknown south of Panama moved to South America, however, that the two continents soon came to have 22 families of mammals in common. At the same time many South American mammals were unable to compete with the immigrants from the north and became extinct.

existence of a deep-ocean barrier. More recently J. Tuzo Wilson of the University of Toronto followed up Grabau's suggestion and proposed that the border between the two faunal provinces marked the closure of a proto-Atlantic ocean that was eliminated by continental drift in Paleozoic times. Since then John F. Dewey of the State University of New York at Albany has developed this concept with considerable success in terms of plate tectonics and subduction: the process in which the leading edge of a drifting plate is destroyed by plunging under another plate [see "Plate Tectonics," by John F. Dewey, beginning on page 124].

Dewey's view, which is based primarily on geological grounds, envisions the loss of an ancient segment of ocean down one or more zones of subduction as an American and a European plate drifted together. Compression and subsequent uplift in the region of subduction formed the Caledonian mountain belt in northwestern Europe and the

"old" Appalachian Mountains in eastern North America. The record of faunal convergence in the interval between Cambrian and Devonian times evidently reflects the steady narrowing of the proto-Atlantic, a process that continued for tens of millions of years until most of the ancient ocean was swallowed up. Here we have an example of the fossil record supporting a reconstruction of a drift episode that has been independently inferred from geological data.

The Case of the Urals

The Caledonian belt is not the only ancient mountain range that hints of a collision between two drifting plates. The Urals, the mountains that separate the European and Asiatic parts of the U.S.S.R., have also been interpreted as a collision feature on geological grounds. Let us see whether the fossil record supports this interpretation.

In Devonian times an order of jawless freshwater fishes, cousins to the orders

that once flourished on opposite sides of the proto-Atlantic, inhabited the streams of the region that is now the European and Asiatic flanks of the Urals. These fishes are the Heterostraci. Specimens of the order from fossil formations on the European side of the Urals clearly belong to a faunal province that is distinct from the province on the Asiatic slope. It follows that, at least during the 50 million years of Devonian times, a marine barrier separated the two landmasses that now meet in the Urals.

There is negative evidence to suggest that the same barrier persisted during the succeeding interval: the Carboniferous period. Fossil deposits of Carboniferous age in the European U.S.S.R. contain the remains of amphibians and reptiles. In spite of more than a century of prospecting, however, no Carboniferous amphibians or reptiles have been found in the Asiatic U.S.S.R. Yet 50 million years later, at the beginning of the Mesozoic era, the amphibians and reptiles of Asia closely resemble those found elsewhere in the world. Evidently by that time land connections between the two regions were firmly established.

Does this array of fossil evidence, which seems very much like the evidence for a proto-Atlantic, show that the Urals are indeed the products of continental drift? Not necessarily. A further example will show why. Early in the Cenozoic era the mammals of Europe comprised a fauna that was different in many respects from the mammalian fauna of Asia. This was not, however, because two continental plates had drifted apart. Instead the regions were separated at that time by the intrusion of a long arm of shallow sea. As far as land animals are concerned, a shallow sea is quite as effective a barrier as a true deep-ocean basin. How do we know that the separation of the two regions in Paleozoic times was not a similar shallow sea?

Examination of another fossil group, the bottom-dwelling marine invertebrates, should enable us to resolve this question. If a deep-ocean basin separated the European and Asiatic parts of the U.S.S.R. in Paleozoic times, then the wider that ocean was, the more divergent the invertebrate fossils on opposite sides of the Urals should be. Unfortunately, when we apply this test, the existing data prove to be somewhat indecisive. Still, the weight of evidence seems to suggest that the Paleozoic marine gap between Europe and Asia was not very large. This fossil finding suggests that, if continental drift formed the

THREE PAIRS OF MOLLUSKS, each pair belonging to the same genus but to different species, exemplify the divergence that marine animals undergo when a barrier divides a once uniform fauna. The cowries (*top*) are of the genus *Cypraea*; the Caribbean species, *C. zebra* (*left*), has diverged from the Pacific species, *C. cervinetta* (*right*). The same is true of a second gastropod pair (*middle*), both of the genus *Strombus*, and the bivalve pair (*bottom*), both of the genus *Arca*. Divergence began when the Isthmus of Panama arose.

FOSSILS OF MESOSAURUS, a late Paleozoic reptile seen in a restoration here, are found on both sides of the South Atlantic and nowhere else in the world. If *Mesosaurus* was able to swim well enough to cross the ocean, it should have diffused far more widely. Since it did not, this example of "disjunct endemism" suggests that South America and Africa must have been joined at that time.

Urals, the ocean that disappeared as a consequence was a narrow one.

The Pattern of Divergence

We can now consider the second principal pattern of fossil distribution. This is "divergence," which is simply the reverse of convergence. To illustrate the pattern I shall use three examples drawn from the Cenozoic fossil record and one from the Mesozoic. The first is the late Cenozoic rise of the Isthmus of Panama that, as we have seen, allowed the convergence of the land faunas of North America and South America. Simultaneously the rise cut in two a marine region that until then had been inhabited by a homogeneous population of bottom-dwelling invertebrates. The consequence of this genetic isolation was divergence; during the Pleistocene epoch a number of "twin" species, the descendants of identical but isolated genera of marine invertebrates, have evolved independently on opposite sides of the isthmus.

The second example concerns the invertebrate faunas of the Tethys seaway, an ancient span of ocean that in early Cenozoic times reached all the way from the Caribbean, by way of the Mediterranean basin, to the western shores of Indonesia. Throughout this vast region in early Cenozoic times the invertebrate faunas were markedly homogeneous. Beginning about 25 million years ago, however, during the Miocene period, the homogeneity of the Tethys faunas was abruptly disturbed. Thereafter the ma-

rine invertebrates of the Indian Ocean differed sharply from those of the Mediterranean. And whereas the faunas of the Mediterranean continued in general to resemble the faunas of the Atlantic, there are indications that some groups, in particular bottom-dwelling foraminifera, also began to show divergence.

The land animals of the Cenozoic era provide the third example. During that era the fossil faunas of the various continents, the mammals in particular, differ in many respects. The most obvious differences are evident in Australia, South America and Africa—the continents of the Southern Hemisphere. During the preceding Mesozoic era, in contrast to this Cenozoic pattern of divergence, the land-dwelling animals (most of them reptiles) were quite homogeneous irrespective of their continued residence. The Cenozoic pattern of divergence is the inevitable consequence of the breakup of Pangaea in late Mesozoic times.

Two groups of late Mesozoic marine invertebrates yield a fourth example of divergence. They flourished during the Cretaceous period, when the breakup of Pangaea was well under way. When one compares certain species of bivalves and foraminifera from fossil strata of the lower Cretaceous in the Caribbean with similar organisms in the Mediterranean region, the two faunas prove to be very much alike. By late Cretaceous times, however, new genera of bivalves and foraminifera have appeared that are unique to one or the other of the formerly homogeneous regions.

This late Cretaceous divergence con-

forms with the inference, drawn from geological and geophysical data, that the final period of the Mesozoic era witnessed a progressive enlargement of the deep-ocean separation between the Mediterranean and the Caribbean. To the north of the Tropics in the newborn Atlantic a sea-shelf connection between the Old World and the New persisted, but there is little doubt that the diverging marine faunas (including the rudists, a peculiar group of reef-building bivalves) were unable to migrate in any but warm waters and were thus inhibited from crossing from one side of the Atlantic to the other through the cool shelf sea.

The Pattern of Complementarity

I term the third of the patterns of fossil distribution "complementarity" because the faunas in adjacent areas of shore and ocean shelf react to alterations of the environment in a complementary way. For example, when a land connection forms, the newly united land faunas tend to converge and the newly divided marine faunas tend to diverge, whereas the breaking of a land connection gives rise to the opposite effect. The pattern of complementarity is significant because it provides a cross-check on interpretations of the fossil record that depend exclusively on either the marine faunas or the terrestrial ones.

One additional example will suffice to define complementarity. As we have already noted, there is evidence from the Miocene of divergence among the

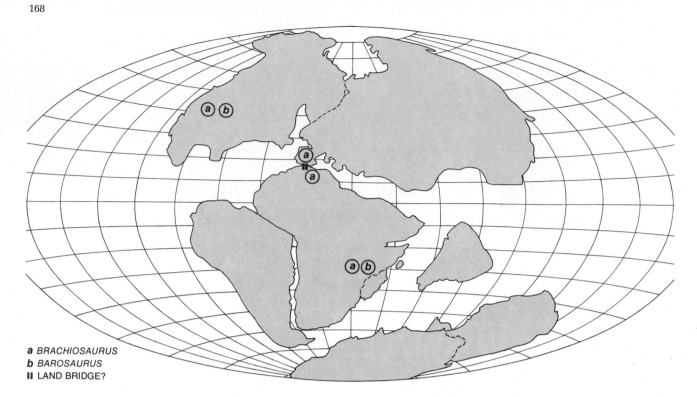

a BRACHIOSAURUS
b BAROSAURUS
ǁ LAND BRIDGE?

FOSSIL-RICH STRATA that contain the remains of two genera of late Jurassic dinosaurs are located in western North America and in East Africa respectively. Fossils of one genus (*a*) are also found in Portugal and Algeria. At that time the ancient world continent, Pangaea, had broken apart; its components are shown here at the positions calculated by Robert S. Dietz and John C. Holden of the National Oceanic and Atmospheric Administration. Unless a land bridge existed in the late Jurassic where North Africa and Spain nearly touch, the presence of identical fossils on separate continents is another example of disjunct endemism. But where the bridge should be there are oceanic rocks of Jurassic age instead, making the existence of a land connection at that time improbable.

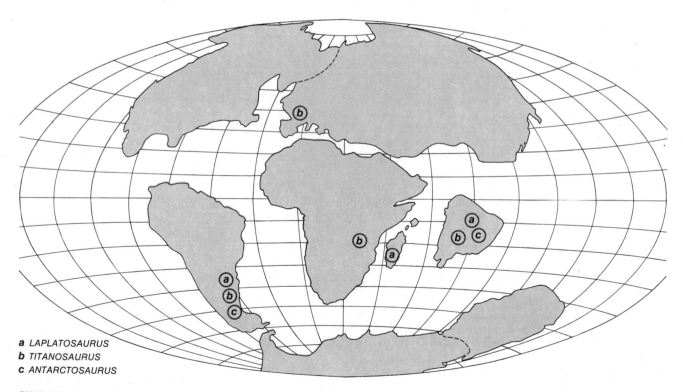

a LAPLATOSAURUS
b TITANOSAURUS
c ANTARCTOSAURUS

SIMILAR DILEMMA is posed by the disjunct endemism of three genera of late Cretaceous dinosaurs. All three genera are present in fossil formations in South America and in India. One genus (*b*) is also present in both Europe and Africa and another (*a*) is also found in Madagascar. The map shows the various continents in the positions calculated for the late Cretaceous by Dietz and Holden. By that time, students of continental drift generally agree, India had moved well away from the Africa-Arabia plate, continuing a trend that supposedly first isolated the drifting subcontinent some 100 million years earlier. Unless the dinosaur fossil identifications are mistaken, however, the land connection between India and South America could scarcely have been severed at so early a date.

marine invertebrates along the length of the Tethys seaway on the one hand and of convergence among the land mammals of Africa and Eurasia on the other. It appears that this pattern of complementarity in the Miocene fossil record signals the withdrawal of the Tethys seaway from the region of the Near East and Middle East. From the viewpoint of plate tectonics the withdrawal of the Tethys must have been a consequence of the Africa-Arabia plate impinging on Eurasia. The fossil evidence is thus in accord with the geological evidence of compressive, generally north-south earth movements and mountain uplift in southern Spain, Turkey and Iran in Miocene times.

There is also geological evidence of compressive tectonic activity in the Near East and Middle East in late Cretaceous times, some 50 million years earlier. This episode could not, however, have eliminated the Tethys seaway; the fossil record shows no matching record of complementarity in the late Cretaceous. When the story of the faunal migrations between Europe and Africa is finally known in detail, it will surely prove to be a complex narrative. It will probably not call, however, for any fundamental modification of the general picture I have presented here.

Disjunct Endemism

We come now to the fourth and last of the fossil-distribution patterns, which I call "disjunct endemism." The term describes the following situation. A group of fossil organisms is limited in its geographical distribution but nonetheless appears in two or more parts of the world that are now separated by major geographical barriers such as zones of deep ocean. The classic case in point is *Mesosaurus*, a small, snaggle-toothed reptile that lived in late Paleozoic times, some 270 million years ago. Strata that contain fossils of *Mesosaurus* are found only in Brazil and in South Africa [see illustration on page 167]. This animal, measuring some 18 inches from snout to tip of tail, was evidently aquatic. It could hardly have been able to swim very far, however, without having diffused into many parts of the world other than Brazil and South Africa. The application of Occam's razor suggests that in late Paleozoic times Brazil and South Africa were contiguous; this bit of fossil evidence in favor of continental drift was first noted many years ago.

In at least two instances that involve the ruling reptiles of the Mesozoic era, the dinosaurs, the fossil evidence seems

to require sharp revisions of the drift timetable. The dinosaurs involved are five genera of sauropods, the line that gave rise to such museum favorites as *Brontosaurus* and *Diplodocus*. Two of the genera flourished in late Jurassic times and three in the late Cretaceous, some 70 million years afterward.

The Jurassic dinosaur genera are *Brachiosaurus*, the biggest of all the sauropods, and *Barosaurus*. The remains of dinosaurs of both genera are found in the Morrison Formation of the western U.S. and in the Tendaguru fossil beds of Tanzania; *Brachiosaurus* fossils have also been found in Portugal and in Algeria. These huge herbivores seem to have occupied an ecological niche comparable to an elephant's. Although they might possibly have negotiated swamps, they would have had trouble swimming across a wide river, let alone an ocean. Their presence in both Africa and North America thus points to the existence of a land connection between these areas in late Jurassic times. The necessity for this connection in turn imposes a constraint on the timing of the oceanic separation of the northern and southern halves of Pangaea.

Now, one popular reconstruction of the continental array in Jurassic times, prepared by Robert S. Dietz and John C. Holden of the National Oceanic and Atmospheric Administration, shows a possible place of crossing between Eurasia (which was then linked to North America) and Africa. That is where Spain and North Africa nearly touch; a Jurassic land bridge here would solve the problem of the seemingly disjunct endemism of the two sauropods [see top illustration, facing page]. The existence of marine deposits of Jurassic age in the parts of Spain and North Africa that might have been joined, however, seems to rule out any such land bridge.

Another way to be rid of this supposed example of disjunct endemism is to attack the biological classifications involved. One could assert that the African species of these two dinosaur genera are actually quite different from the North American species and attribute this divergence to a break in the land connection between the two regions. By and large, however, the dinosaurs are a group that has been excessively split by taxonomists. An admitted resemblance even at the genus level is likely to signify a close genetic affinity if not an actual ability to interbreed. On balance, then, close similarities between the late Jurassic dinosaur faunas of the Tendaguru beds and of the Morrison Formation pose a problem for the continental-

drift timetable that has not yet been satisfactorily resolved.

The disjunct endemism of the three late Cretaceous sauropods—*Titanosaurus, Laplatosaurus* and *Antarctosaurus*—presents an even more clear-cut contradiction between the fossil record and the accepted timetable for continental drift. These three dinosaurs are known from fossil formations both in South America and in India. Yet the Indian subcontinent had supposedly become isolated by surrounding ocean some 100 million years before late Cretaceous times, toward the close of the Triassic [see bottom illustration, facing page]. Unless the fossil identifications are in error, such an early date for the severance of India from the rest of Gondwanaland is clearly inadmissible. Several similar arguments from the fossil record, which I need not give in detail, can be used to support the persistence of land connections (if only intermittent ones) between the Africa–South America landmass and the Australia-Antarctic landmass until quite late in the Mesozoic period.

I have tried to show here how the fossil record can contribute to plate-tectonic theory by helping to establish a more refined timetable for continental drift. Critics may say that the fossil data are often imprecise and are in addition subject to ambiguous interpretation. That is undoubtedly true in some instances. Moreover, we must beware of oversimplification and of overinterpretation of patchy evidence. We should also acknowledge that endemism cannot be explained in every instance on the basis of continental movements. Nonetheless, the general level of agreement between paleontology and the other earth sciences is sufficiently high to warrant some confidence in our fossil-based conclusions.

As time passes and the reconstruction of past plate movements becomes more precise, I believe interest in several biological questions that arise from the new view of earth history will increase. For one thing, we shall be in a better position to learn more about the rates of evolution among isolated organisms. For another, we shall be able to make findings about the relative ease of migration and colonization under different geographical circumstances. The disjunct distribution of many living animals—for example the lungfishes, the marsupials and the giant flightless birds—will be better understood. Perhaps most intriguing of all, we may acquire new insights into why many groups of plants and animals have become extinct.

15

Plate Tectonics
and Mineral Resources

INTRODUCTION

Man would have a barren existence, if he could survive at all, without the mineral wealth in the earth's crust. Ultimately these finite resources will run out because man's appetite is insatiable and nature replenishes mineral deposits exceedingly slowly. What nature has accumulated over hundreds of millions of years, man will exhaust in about a thousand years. Until a system for recycling and a technology of substitution are developed, there will be increasing pressure to find new deposits of minerals such as metals, oil, and nonmetallic ores.

Geology lacks a comprehensive theory of the origin and distribution of ore deposits. To be sure, geologists can describe terrains in which specific deposits are likely to be found, but their descriptions are based mostly on the acccumulated experience of centuries of prospecting.

If Peter Rona is correct in the main theme of this article, "Plate Tectonics and Mineral Resources," and the worldwide pattern of mineral distribution is tied to present-day and ancient plate boundaries, a powerful new tool would become available to geologists, enabling them to concentrate their mineral exploration programs in specific regions. Of particular importance is the notion that the sea floor, beyond the continental shelves, is essentially unexplored, and that these huge regions may harbor unknown deposits. The "shows" found in a few deep-sea drill holes do not discourage this possibility. Since most estimators of remaining mineral reserves have neglected the deep sea, and since most exploration programs have been planned with little or no reference to the role of plate margins, there is a real possibility that these new approaches may enable man to buy more time to find out how to recycle and substitute before his resources run out.

Plate Tectonics
and Mineral Resources

by Peter A. Rona
July 1973

*The concepts of continental drift and sea-floor
spreading provide clues to the location of economically
important minerals such as oil and metals. These clues
have already led to promising deposits*

A scientific revolution is in progress that over the past five years has already changed our understanding of the earth as profoundly as the Copernican revolution changed medieval man's understanding of the solar system. The Copernican revolution entailed a fundamental change in man's world view from an earth-centered planetary system to a sun-centered one and led to the development of modern astronomy and the exploration of space. The current scientific revolution entails a fundamental change in man's world view from a static earth to a dynamic one and presages comparable benefits. Some of the benefits may even be economic. The implications of the new global tectonics for mineral resources, particularly the mineral resources of the ocean floor, are only now beginning to emerge.

At present the only undersea mineral resources that certainly have economic value are the vast oil and gas reserves found under many continental shelves and continental slopes, gravel, sand, shells and placer deposits on the continental shelves, various other minerals buried under the continental shelves in specific relation to adjacent continental deposits, and fields of manganese nodules that blanket large areas of the deep-sea floor. Even this limited knowledge is remarkable in the light of the difficulty that was encountered in obtaining it. Consider how much we would know about the mineral deposits of the continents if our sampling procedure were limited to flying in a balloon at an altitude of up to six miles and suspending a bucket at the end of a cable to scrape up loose rocks from the surface of the land. What are the chances that we would find the major known ore bodies, which generally underlie areas of less than a square mile?

Yet this farfetched analogy accurately describes man's present capacity for sampling the sediments and rocks of the ocean bottom, utilizing a variety of coring, drilling and dredging devices lowered from ships through the water column over an area twice as large as that of the continents. Averaged over the world's oceans, the distribution of ocean-floor rocks that have been sampled to date is only about three dredge hauls per million square kilometers!

In recent years every major discovery of a hidden mineral resource has been anticipated by a theoretical vision. For example, once field geologists realized that there was a definite association between the type of sedimentary structure termed an anticline and accumulations of oil, they knew where to drill and the rate of discovery of oil deposits accelerated accordingly. In the same way the right conceptual framework can be used to extend man's limited direct knowledge of resources of the ocean basin toward a realistic appraisal of their potential. The test of the value of such a conceptual framework is how well it explains what one sees and predicts what one does not see.

The old conceptual framework of a static earth held that the continents and ocean basins were permanent features that had existed in their present form since early in the 4.5-billion-year history of the earth. Only the most accessible continental mineral deposits were discovered, largely by trial and error, with little understanding of why or where they existed. The recent change to a conceptual framework based on a dynamic-earth model, in which continents are constantly moving and ocean basins are opening and closing, is leading toward a better understanding of the global distribution of mineral deposits in both space and time.

The basis of the new conceptual framework is the theory of plate tectonics, the essentials of which have already been reported in SCIENTIFIC AMERICAN [see "Plate Tectonics," by John F. Dewey, page 124]. "Tectonics" is a geological term pertaining to earth movements. The movements in question involve the lithosphere, the rigid outer shell of the earth, which is of the order of 60 miles thick. The lithosphere, which behaves as if it were floating on an underlying plastic layer, the asthenosphere, is segmented into about six primary slabs, or plates, each of which may encompass a continent and part of an adjacent ocean basin [see top illustration on page 175].

The boundaries of the lithospheric plates are delineated by narrow earthquake zones where the plates are moving with respect to each other. Three types of boundary are recognized. One type, called a convergent plate boundary, is where two adjacent plates move together and collide or where one plate plunges downward under the other plate and is absorbed into the interior of the earth.

The second type of boundary, called a divergent plate boundary, is where two adjacent plates move apart because new lithosphere is added to each plate by the process of sea-floor spreading. The new lithosphere, which moves more or less symmetrically to each side of the divergent plate boundary, acts like a conveyor belt, carrying the continents apart in the motion that has become known as continental drift. The dual existence of convergent boundaries where lithosphere is destroyed and divergent boundaries where lithosphere is created implies that the diameter of the earth is not changing radically.

SMALL VEIN OF PURE COPPER was discovered in a core sample of sedimentary rock obtained by the Deep Sea Drilling Project some 350 miles southeast of New York City. The copper vein, the horizontal reddish structure in this longitudinal section of a piece of the original core, is about half an inch long. It was found in sediment about 65 feet above the volcanic basement rocks under the lower continental rise at a water depth of 17,000 feet.

METAL-RICH CORE, collected from the Atlantis II Deep, one of the hot-brine pools located along the axial valley of the Red Sea at a depth of about 6,600 feet below sea level, represents the most concentrated submarine metallic sulfide deposits known. The muddy sediments containing the sulfide minerals fill the Red Sea basins to a thickness estimated at between 65 and 330 feet. The deposits are saturated with (and overlain by) salty brines considered to be the hydrothermal solutions from which the sulfide minerals were precipitated. The photograph was made by David A. Ross of the Woods Hole Oceanographic Institution.

The third type of tectonic-plate boundary is the parallel plate boundary, where two adjacent plates move edge to edge along their common interface.

Hydrothermal mineral deposits, that is, mineral deposits formed by precipitation from solutions, constitute a major part of our useful metallic ores on the continents. Economically the most important types of hydrothermal deposit are the sulfides, in which various metals combine with sulfur to precipitate from the hydrothermal solution. About a year ago Frederick Sawkins, a geologist at the University of Minnesota, pointed out that most of the sulfide deposits of the world are located along present or former convergent plate boundaries where an oceanic lithospheric plate plunges under the margin of a continent (including the continental shelf) or under a chain of volcanic islands. The processes that concentrate the sulfide deposits along convergent plate boundaries, which are at present only partly understood, involve mineralizing solutions that emanate from the plunging lithospheric plate, which melts as it is absorbed into the interior of the earth.

Metallic sulfide deposits along convergent plate boundaries include the Kuroko deposits of Japan, the sulfide ore bodies of the Philippines and the deposits extending along the mountain belts of western North America and South America (the Coast Ranges, the Rockies and the Andes) and from the eastern Mediterranean region to Pakistan. Gold-bearing deposits are not sulfides but often accompany sulfide minerals. The majority of gold deposits in Alaska, Canada, the southeastern U.S., California, Venezuela, Brazil, West Africa, Rhodesia, southern India and

TROODOS MASSIF on the island of Cyprus, the site of economically important mineral deposits that originated at a divergent tectonic-plate boundary, stands out clearly as the dark-colored mountainous region in the middle of the satellite photograph on the opposite page. The photograph was made recently from an altitude of nearly 600 miles by a multispectral camera system on board the first Earth Resources Technology Satellite (ERTS I). Region is believed to be a slice of oceanic lithosphere that was formed by the process of sea-floor spreading from a submerged mid-oceanic ridge and was subsequently thrust upward.

southeastern and western Australia occur in rocks that can be associated with former convergent plate boundaries.

Divergent plate boundaries are formed by the spreading of lithospheric plates in the central portions of ocean basins. The Red Sea and the island of Cyprus in the Mediterranean Sea provide important clues to the potential of metallic sulfide deposits at divergent plate boundaries.

The Red Sea, the product of a divergent plate boundary developing between the African plate and the Eurasian plate, provides an accessible natural laboratory for the study of mineral processes associated with divergent plate boundaries. About five years ago the richest submarine metallic sulfide deposits known were found in three rather small basins along the center of the Red Sea at a depth of about 6,600 feet below sea level. The sulfide minerals are disseminated in sediments that fill the basins to a thickness estimated at between 65 and 330 feet. The top 30 feet or so of sediment, which has been explored by coring the largest of the basins, has a total dry weight of about 80 million tons, with average metal contents of 29 percent iron, 3.4 percent zinc, 1.3 per-

cent copper, .1 percent lead, .005 percent silver and .00005 percent gold. The deposits are saturated with (and overlain by) salty brines carrying the same metals in solution as those present in the sulfide deposits. The salty brines are considered to be the hydrothermal solutions from which the sulfide minerals are precipitated. It remains controversial whether the brines are being charged with minerals from volcanic sources under the Red Sea or from sediments with high copper, vanadium and zinc contents adjacent to the basins where the metallic sulfide deposits are found [see "The Red Sea Hot Brines," by Egon T. Degens and David A. Ross; SCIENTIFIC AMERICAN, April, 1970].

The Red Sea represents the earliest stage in the growth of an ocean basin: the stage where a divergent plate boundary rifts a continent in two. The most advanced growth stage of a divergent plate boundary is the mid-oceanic-ridge system, a 47,000-mile undersea mountain chain that extends through all the major ocean basins and girdles the globe. The mid-oceanic-ridge system has not been adequately sampled to determine whether or not concentrations of

metallic sulfides comparable to the Red Sea deposits are present at sites along its crest or in basins in its flanks. Measurements of the distribution of heat emanating from mid-oceanic ridges and of the chemical alteration of ridge rocks indicate that seawater forms a hydrothermal solution by penetrating fissures, dissolving minerals from rocks underlying the ridges and precipitating those minerals in concentrated deposits.

A limited amount of sampling indicates that hydrothermal processes are actively concentrating metals from volcanic sources underlying mid-oceanic ridges. Sediments on active mid-oceanic ridges are generally enriched in iron, manganese, copper, nickel, lead, chromium, cobalt, uranium and mercury, with trace amounts of vanadium, cadmium and bismuth. The concentrations typical of sediments covering widespread areas on mid-oceanic ridges are not economic, but much higher concentrations exist locally.

Metallic sulfides are found in rocks dredged from the Indian Ocean Ridge. In addition small veins of pure copper have been recovered by the Deep Sea Drilling Project at several sites. At the crest of the Ninety East Ridge near the Equator in the Indian Ocean, for example, veins of copper are found in volcanic rocks overlain by 1,440 feet of sediment at a water depth of 7,380 feet. Some 350 miles southeast of New York City a small vein of pure copper and clusters of copper crystals have been discovered in sediment about 65 feet above the volcanic basement rocks under the lower continental rise at a water depth of 17,000 feet [see top illustration on page 172].

A specimen of manganese 1.7 inches thick recently dredged from a water depth of about 12,000 feet in the median valley of the Mid-Atlantic Ridge by the Trans-Atlantic Geotraverse of the National Oceanic and Atmospheric Administration has particular significance. The composition, form and thickness of this manganese sample, which accumulated at a rate about 100 times faster than the manganese in nodules, indicates a hydrothermal origin and demonstrates that hydrothermal mineral deposits are actively accumulating at certain divergent plate boundaries in ocean basins. Because the sea floor is supposed to originate by spreading from mid-oceanic ridges, a mineral deposit on a mid-oceanic ridge would be expected to extend in a linear zone from the ridge across the ocean basin to the adjacent continental margin if the depositional process is a

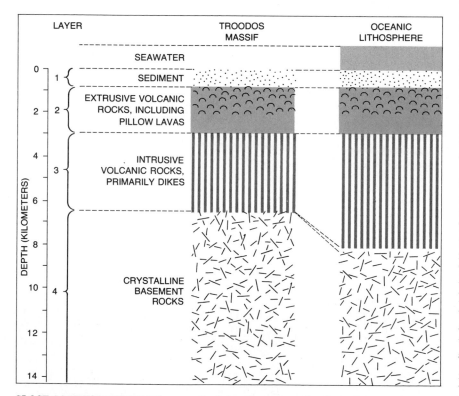

CLOSE CORRESPONDENCE between the layered sequence of rocks in the Troodos Massif (*left*) and that of the oceanic lithosphere (*right*) is evident from this comparison. The geological structure of the Troodos Massif was determined directly from rock outcrops; the structure of the oceanic lithosphere was determined indirectly by seismic-refraction techniques. The sulfide ore bodies of the Troodos Massif are in the upper portion of layer made up of extrusive volcanic rocks. Pillow shapes form when volcanic lava cools on the sea floor.

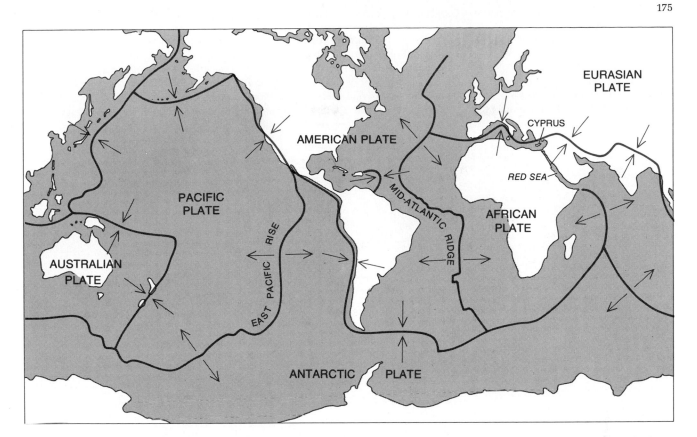

SIX PRINCIPAL TECTONIC PLATES of the lithosphere, the rigid outer shell of the earth, are delineated by the heavy color lines on this world map. The paired arrows indicate whether a plate boundary is convergent or divergent (*see illustration below*).

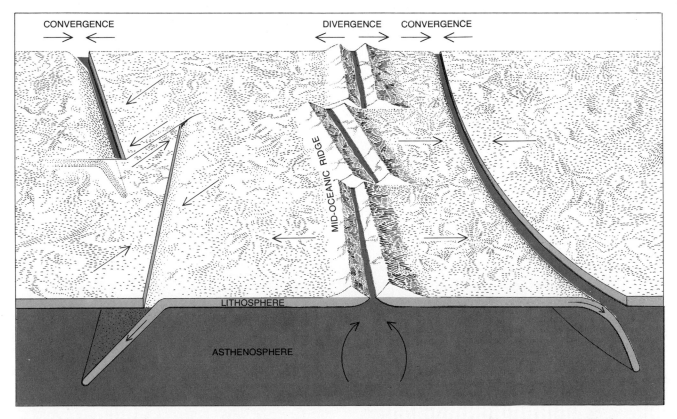

TWO TYPES OF PLATE BOUNDARY are illustrated schematically in this block diagram. The 60-mile-thick lithospheric plates move outward like conveyor belts from the mid-oceanic ridges (divergent plate boundaries) and plunge downward under the deep-sea trenches (convergent plate boundaries). The third major type of plate boundary, not shown here, is the parallel plate boundary.

continuous one [see illustration on facing page].

At this point in man's exploration of the oceans it would seem to be too much to expect that it would be possible to make detailed observations on an eco-nomically important metallic sulfide de-posit that originated at a divergent plate boundary on a submerged mid-oceanic ridge. Yet such a deposit is known and has been extensively studied. The Troo-dos Massif on the island of Cyprus is interpreted as being a slice of oceanic lithosphere that was formed by the proc-ess of sea-floor spreading from a mid-oceanic ridge and was subsequently thrust upward to its present position [see illustration on page 173]. The composi-tion and layered sequence of rocks that constitute the Troodos Massif are the

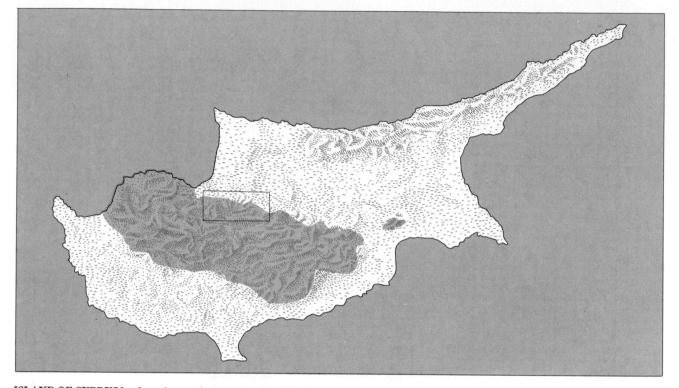

ISLAND OF CYPRUS has been famous for its mineral wealth since Phoenician times. The principal ore bodies are in the uppermost volcanic layers of the Troodos Massif, the total extent of which is indicated by the dark-colored area. The hatched area repre-sents sediments, including alluvium. A geological map of a portion of the Troodos igneous complex (small rectangle) is shown below.

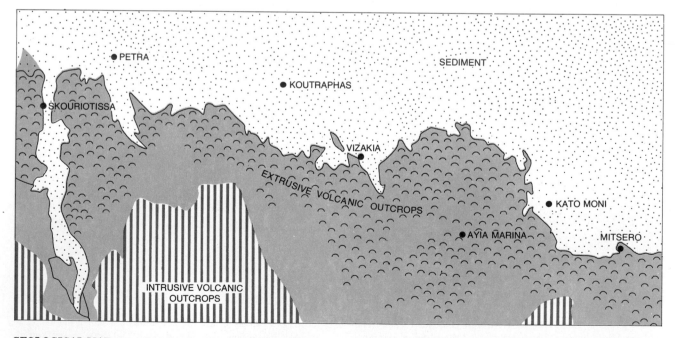

GEOLOGICAL MAP of a region that lies along the northern fringe of the Troodos Massif is based on studies that were undertaken by the Geological Survey of Cyprus. The map shows outcrops of extru-sive volcanic rock that incorporate bodies of metallic sulfide ore.

same as those known to underlie the seabed.

Cyprus has long been famous for its mineral wealth. The mining of copper (for which the island is named) was an important industry in Roman and even in Phoenician times. The brilliant green stains of copper sulfides on ancient mine tailings have attracted modern prospectors. Between 1965 and 1970 the average annual exports amounted to about a million tons each of iron pyrites, chromite and gypsum, about 150,000 tons of copper pyrites and 100,000 tons of copper concentrates. The estimated value of the mineral products exported from Cyprus in 1970 amounted to $30 million.

The principal ore bodies are in the uppermost volcanic layers of the Troodos Massif. It has been uncertain whether the Troodos sulfide-ore bodies originated before the upthrust of the Troodos Massif or afterward. In the first instance the ore bodies would be representative of the seabed. In the second the ore bodies would be attributed to special conditions unrelated to the seabed. The sulfide deposits are clearly related to the volcanic rocks in which they occur. Recent studies reveal that iron-rich and manganese-rich sediments interlayered with the volcanic rocks and associated with the ore bodies of the Troodos Massif are chemically identical with those metal-enriched sediments found on active mid-oceanic ridges, indicating that both the sediments and the ore bodies were formed on the sea floor by hydrothermal processes.

The Troodos ore bodies may provide the first firm evidence on the nature of metallic sulfide deposits in ocean basins. The Skouriotissa ore body, for example, is roughly elliptical in plan view, measures approximately 2,000 feet long by 600 feet wide and is lens-shaped in cross section. Its estimated mass is six million tons. The average composition of the ore is 2.25 percent copper (ranging to greater than 5 percent), 48 percent sulfur and 43 percent iron.

The Mavrovouni ore body is also roughly elliptical in plan view, measures approximately 1,000 feet long by 600 feet wide and forms a lens that attains a thickness of 800 feet in cross section. Its estimated mass is greater than 15 million tons. The average composition of the ore is 4.2 percent copper, 48 percent sulfur, 43 percent iron, .4 percent zinc, .25 ounce per ton gold and .25 ounce per ton silver.

Sediments underlying the Skouriotissa ore body, presumably a disintegration product of the pyrite in the ore body,

contain 2.12 ounces of gold per ton and 12.96 ounces of silver per ton. Exposed patches of metallic oxides indicate the presence of the ore bodies under the mountainous surface of the Troodos Massif. The Skouriotissa ore body is exploited by underground shafts and the Mavrovouni ore body by strip-mining.

What kind of target for exploration would a Troodos ore body make if it were submerged under thousands of feet of water on the crest or flank of a mid-oceanic ridge? It is unlikely that any of

the present exploration methods would be capable of detecting the ore body. The resolution of present geophysical exploration methods will have to be improved in order to detect such an ore body under the sea. Both the exploration methods and the engineering development involved will be costly.

The prerequisites for the accumulation of petroleum consist of a source of organic matter to generate the petroleum, a natural reservoir to contain it

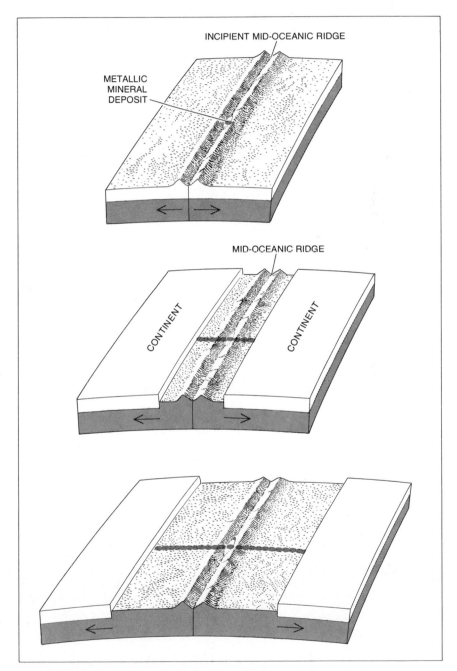

HYDROTHERMAL MINERAL DEPOSIT (*color*) formed in a hot-brine pool on the axis of a mid-oceanic ridge would be expected to extend in a linear zone from the ridge across the ocean basin to the adjacent continental margins as the ocean basin progressively widens (*from top to bottom*) as a consequence of sea-floor spreading from the mid-oceanic ridge.

and a trap to concentrate its fluid and gas constituents. Petroleum is hydrocarbons derived from the remains of plants and animals. As the progenitor of the petroleum, the organic matter must accumulate in an environment where it is preserved. The preservation of organic matter is favored by an environment that is toxic to life (so that the organic matter is not consumed as food) and deficient in oxygen (so that the hydrocarbon is not decomposed). How do conditions favorable to the accumulation of petroleum relate to convergent and divergent plate boundaries?

Convergent plate boundaries where the oceanic portion of a lithospheric plate plunges under the margin of a continent are characterized by the presence of a deep-sea trench running along their length. A system of deep-sea trenches runs along the entire western margin of North America and South America where the Pacific lithosphere is plunging under the continents. In addition to a deep-sea trench, chains of volcanic islands are present along some convergent plate boundaries; they are located between the trench and the continent. There are many such chains of volcanic islands at the western margin of the Pacific, including the Aleutians, the Kuriles, Japan, the Ryukyus, the Philippines and Indonesia. Other such chains are the Marianas, the South Sandwich

Islands and the West Indies. The island chains divide an ocean basin into smaller basins partially enclosed between the islands and the adjacent continent; such basins include the Bering Sea, the Sea of Okhotsk, the Sea of Japan, the Yellow Sea, the East China Sea and the South China Sea.

Both the marginal trenches and the volcanic-island chains create a habitat that is favorable for the accumulation of petroleum in several respects. First, the trenches and island chains act as barriers that catch sediment and organic matter from the continent and the ocean basin. Second, the shape of the trenches and the small ocean basins acts to restrict the circulation of the ocean, so that oxygen is not replenished in the seawater and organic matter is preserved. Third, the accumulation of sediments and the geological structures that develop as a result of the deformation of the sediments by tectonic forces provide reservoirs and traps for the accumulation of petroleum. According to Hollis D. Hedberg of Princeton University, "these marginal semienclosed basins constitute some of the most promising areas in the world for petroleum accumulation."

The development of divergent plate boundaries may also create a habitat favorable for the accumulation of oil, a finding that would open immense possibilities for petroleum resources in the

deep ocean basin. When a divergent plate boundary develops under a continent, the continent is rifted in two and the continental fragments are carried apart on a conveyor belt of new lithosphere generated at the divergent plate boundary. As the two continental fragments move apart, a sea forms between them. The surrounding continents act as barriers to restrict the circulation of the sea. As a result organic matter is preserved and, if the evaporation of the seawater exceeds its replenishment, layers of rock salt are deposited along with the organic matter. As the continental fragments continue to move apart and to subside along with the adjacent sea floor,

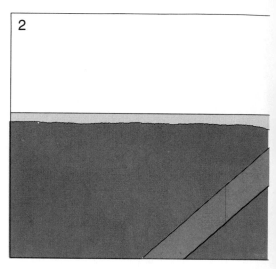

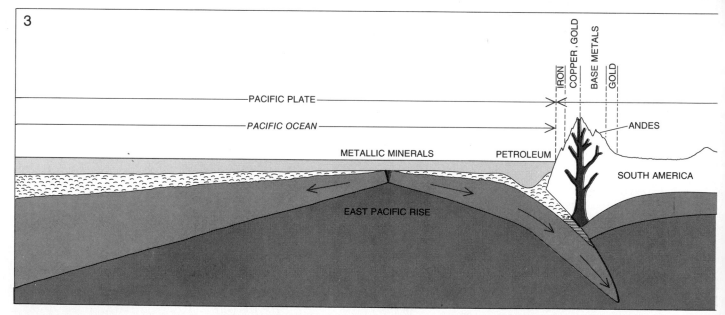

ROLE OF PLATE BOUNDARIES in the accumulation of mineral deposits is exemplified in this sequence of cross-sectional views of the development of the South Atlantic Ocean. The position of Africa is assumed to be stationary throughout the sequence of cross sections. In stage *1* a single ancestral continent, called Pangaea, is rifted into two continents (South America and Africa) about a divergent plate boundary. In stage *2* the oceanic crust created by the process of sea-floor spreading from the divergent plate boundary (a precursor of the Mid-Atlantic Ridge) rafts South America westward and is compensated for by the consumption of oceanic crust at a trench (a convergent plate boundary) that develops to the west of South America. Thick layers of rock salt, organic matter and metallic minerals accumulate in the Atlantic Sea during this early stage of continental drift. In stage *3* continued sea-floor spreading

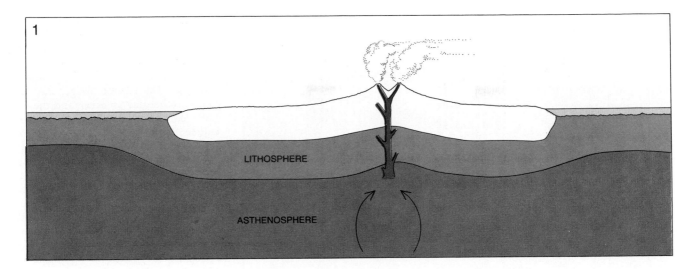

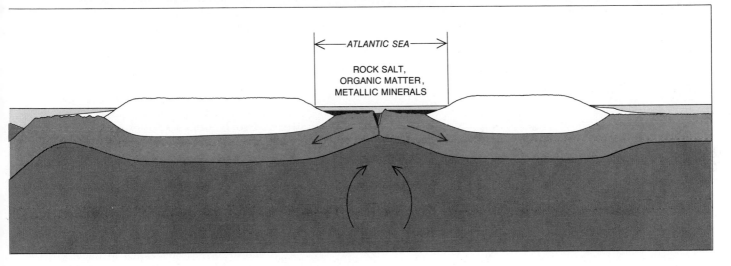

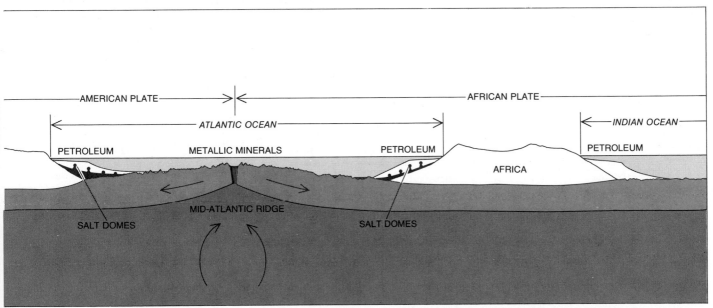

from the Mid-Atlantic Ridge widens the Atlantic into an ocean, rafts South America westward over the trench, reversing the inclination of the trench and producing the Andes mountain chain as a consequence of the deformation that develops at the convergent plate boundary along the western margin of South America. Metallic minerals that are melted from the Pacific plate as it plunges under South America ascend through the overlying crustal layers and are deposited in them to form the metal-bearing provinces of the Andes. Meanwhile in the Atlantic Ocean metallic minerals continue to accumulate about the Mid-Atlantic Ridge. Salt originating in the thick layers of rock salt that have been buried under the sediments of the continental margins rises in large, dome-shaped masses that act to trap the oil and gas that are generated from the organic matter that was preserved in the former Atlantic Sea.

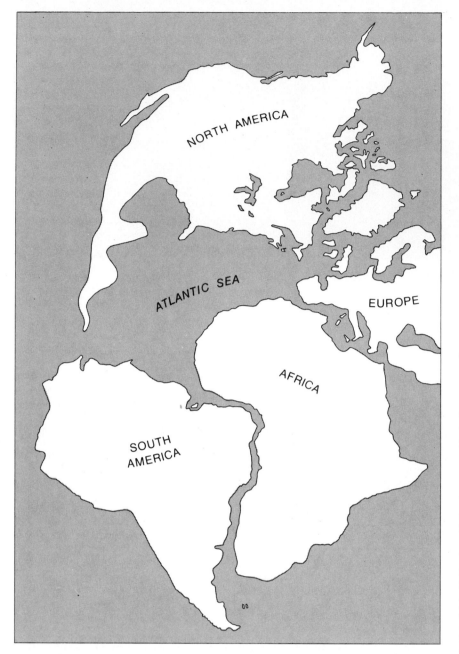

AT AN EARLY STAGE of continental drift the Atlantic was a sea with its circulation restricted by the surrounding continents. As in the present Red Sea, conditions in the Atlantic Sea favored the preservation of organic matter and the deposition of rock salt, leading to the formation of petroleum accumulations under the present continental margins.

the restricted sea becomes an open
ocean. The layers of organic matter and
salt are buried under sediments. The
organic matter subsequently develops
into petroleum (by processes that are
only partly understood) and the salt
forms into dome-shaped masses that act
to trap the petroleum.

The Red Sea is an example of a restricted sea formed at an early stage of
development of the divergent plate
boundary along which Arabia is rifting
from Africa. Layers of rock salt up to
17,000 feet thick and organic muds have

been found under it. Along both the
eastern and western margins of the
North Atlantic and the South Atlantic
apparent salt domes have been discovered extending seaward from continental
shelves to continental rises in water
depths of up to 16,500 feet. The occurrence of these salt domes in the deep
Atlantic indicates that at an early stage
of continental drift the Atlantic was a
sea with its circulation restricted by the
surrounding continents in their positions
at that time [see illustration above].

Like the present Red Sea, conditions

in the Atlantic Sea favored the preservation of organic matter and the deposition of rock salt. As the Atlantic widened
in response to the symmetric creation of
new lithosphere by sea-floor spreading
from the Mid-Atlantic Ridge, the Atlantic Sea became an ocean and the organic matter and salt were buried under
sediments, forming the present margins
of the Atlantic Ocean. It is reasonable
to expect that petroleum accumulations
will extend seaward under the continental shelf, the continental slope and
the continental rise to water depths of
about 18,000 feet along large portions
of both the eastern and western margins
of the North Atlantic and South Atlantic.
Petroleum may likewise be found in
other ocean basins that have grown
through the stage of a restricted sea by
sea-floor spreading.

In short, the patterns of mineral distribution that are emerging from the
conceptual framework provided by the
new global tectonics will clearly help
to guide man's search for new mineral
deposits. Hydrothermal processes have
concentrated the majority of known
metallic sulfide ore bodies along convergent lithospheric plate boundaries
originally at continental margins. Hydrothermal processes are also active at divergent plate boundaries from initial
stages (represented by the metallic sulfide deposits accumulating in the Red
Sea) to advanced stages (represented by
the metal concentration in sediments on
mid-oceanic ridges and by possible
metallic sulfide deposits of the Troodos
Massif type). The Troodos Massif metallic sulfide ore bodies provide an actual
example of the type of deposits that can
be expected in sea-floor rock generated
by mid-oceanic ridges. The confirmation and economic evaluation of metallic sulfide deposits of the Troodos Massif
type in ocean basins await technological
advances in marine exploration methods.

With regard to petroleum, convergent
plate boundaries create conditions that
form accumulations in small ocean basins and deep-sea trenches marginal to
continents. Divergent plate boundaries,
on the other hand, create conditions that
favor the development of oil accumulations extending from the continental
shelf into the deep ocean basin under
the continental rise.

The global patterns of mineral distribution that are emerging from such
models can be expected to accelerate
the discovery of resources not only on
the seabed but also on the continents.

IV

GEOLOGY

Much geological talk in the past few years has been about plate tectonics and associated phenomena, but most geologists continue to work in more established fields, some of which, however, may be affected by the new theories. Though they may be working on problems that have been with us for many years, geologists still have fresh and imaginative approaches. Stimulus frequently comes from advances in related sciences. Thus Norman Newell's article on reefs trades upon the increasing knowledge of macroevolution in geological time and puts that information together with what plate tectonics has revealed about continental movements. Kenneth Hsü's article on the Mediterranean Sea, set against the backdrop of movements of the Eurasian and African plates, discusses findings from the Deep Sea Drilling Project's sediment cores, which were produced in a completely new kind of sampling.

The invention of new approaches, in terms of mathematical probability, to physical problems was a welcome tool for Leopold and Langbein to use in geomorphology. River meanders have been noted for as long as geology has been a science but there is a lot of life left in the subject, particularly when new perceptions of the nature of the phenomenon are brought to it. In this, as in the following article by McNeil on laterites, plate tectonic theory is absent, for the incidence of meanders and laterites has little to do with larger movements of the earth. The article on laterites expresses a growing concern of geologists and soil scientists with social problems arising from the natural environment, in this case agriculture, a concern that makes us look harder at problems that once seemed not so important. Finally, the space exploration program has stimulated geological thought in many ways. We have included an article by Bruce Murray on Mars because it illustrates so well how geological ways of looking at the earth can be extended—with suitable modifications—to other planets where many of the subjects, from volcanism to geomorphology, are the traditional ones of geology, and susceptible to the same general ways of scientific thinking that have informed us so well about our own planet.

16

The Evolution of Reefs

INTRODUCTION

The title—but not the content—of this article, "The Evolution of Reefs," might have been chosen by Charles Darwin in 1842 when he published the results of the voyage of the *Beagle*. In that report he outlined what is still a reasonable hypothesis for the origin of Pacific coral reefs. The study of reefs has come a long way since Darwin's time and has preoccupied geologists, paleontologists, biologists, and oceanographers, all of whom have been fascinated by this distinctive ecological phenomenon. For a long time scientists were puzzled by these colonies of shallow water organisms and how they managed to build up from the deep sea floor. Darwin's theory of their evolution from originally volcanic islands remains the strongest contender. As scientists recognized ancient reefs in the limestones of the geological column, their attention shifted first to the paleontologic exploration of the many species of invertebrates found in them and then to the stratigraphic and sedimentary framework of the reef as an ecological entity. Since World War II, stimulated by the desire to understand ancient reefs, geologists and paleontologists started to explore modern reefs and it has become a standard joke among envious geologists in other specialities that men such as Norman Newell do their winter field work in Bermuda, Bikini, or the Bahamas.

Norman Newell was in fact one of the prime movers in the drive to understand the biology and geology of modern reefs and other tracts of limestone deposits and relate them to ancient reefs. He brings to the study a broad understanding of paleontology and geological history as well as of modern processes. In this article he again is leading us to a new stage in our thinking about reefs, this time about the pattern of invertebrate evolution through geological time in relation to the reef environment. Long aware of the evolutionary changes in the types of organisms that create the wave-resistant reef structure and the associated reef dwellers, Newell now considers that the plate tectonics model of the earth can provide new information. Shifts in temperature zones and marine transgressions on the continents that may be linked to plate convergences and divergences can be related to evolutionary pressures on reef organisms. Perhaps, he says, this can help explain what the fossil record shows us, a history, during geological time, of four major evolutionary collapses of reef communities.

The Evolution of Reefs

by Norman D. Newell
June 1972

*The community of plants and animals that builds
tropical reefs is descended from an ecosystem of two
billion years ago. The changes in this community
reflect major events in the history of the earth*

To a mariner a reef is a hazard to navigation. To a skin diver it is a richly populated underwater maze. To a naturalist a reef is a living thing, a complex association of plants and animals that build and maintain their own special environment and are themselves responsible for the massive accumulation of limestone that gives the reef its body. The principal plants of the reef community are lime-secreting algae of many kinds, including some whose stony growths can easily be mistaken for corals. The chief animal reef-builders today are the corals, but many other marine invertebrates are important members of the reef community.

This association of plants and animals in the tropical waters of the world is the most complex of all ocean ecological systems; as we shall see, it is also the oldest ecosystem in earth history. Its closest terrestrial counterpart, in terms of organization and diversity, is the tropical rain forest. Both settings evoke an image of exceptional fertility and exuberant biomass. Both are dependent on light in much the same way; the sunlight filters down through a stratified canopy, and the associations at each successive level consist of organisms whose needs match the available illumination and prevailing conditions of shelter. There is even a parallel between the birds of the rain forest and the crabs and fishes of the reef. Both play the part of lords and tenants, yet their true role in the history and destiny of the community is essentially passive.

It is a common belief that a reef consists principally of a rigid framework composed of the cemented skeletons of corals and algae. In reality more than nine-tenths of a typical reef consists of fine, sandy detritus, stabilized by the plants and animals cemented or otherwise anchored to its surface. Physical and biochemical processes that are little understood quickly convert this stabilized detritus into limestone. The remains of dead reef organisms make a substantial contribution to the detritus. This major component of the reef, however, has a fabric quite different from the upward-growing lattice of stony algal deposits and intertwined coral skeletons that forms the reef core.

The Reef Community

The interaction of growth and erosion gives the reef an open and cavernous fabric that in an ecological sense is almost infinitely stratified and subdivided. In the dimly lit bottom waters at the reef margin, rarely more than 200 feet below the surface, caves and overhanging ledges provide shelter for the plants and animals that thrive at low levels of illumination. From the bottom to the surface is a succession of reef-borers, cavern dwellers, predators and detritus-feeders, each living at its preferred or obligatory depth, that includes representatives of nearly every animal phylum. Near and at the surface the sunlit, oxygen-rich and turbulent waters provide an environment that contributes to a high rate of calcium metabolism among the myriads of reef-builders active there.

The most familiar of the reef animals, the corals, are minute polyps that belong to the phylum Coelenterata. The polyps live in symbiosis with zooxanthellae, microscopic one-celled plants embedded in the animals' tissue, where they are nourished by nitrogenous animal wastes and, through photosynthesis, add oxygen to the surrounding water. Experiments show that the zooxanthellae promote the calcium metabolism of the corals. The corals themselves are carnivores; they feed mainly on small crustaceans and the larvae of other reef animals.

The limestone-secreting algae—blue-green, green and red—are the principal food base of the reef community, just as the plant life ashore nourishes terrestrial herbivores. The algae are distributed across the reef in both horizontal and vertical zones. The blue-green algae are most common in the shallows of the tidal flat, an area where red algae are absent. The green algae are predominantly back-reef organisms and the reds are mostly reef and fore-reef inhabitants [*see illustration on page 193*].

The other important members of the reef community are all animals. Next in significance to the corals as reef-builders are several limestone-secreting families of sponges, members of the phylum Porifera. The phylum Protozoa is represented by a host of foraminifera species whose small limy skeletons add to the deposits in and around reefs. Several species of microscopic colonial animals of the phylum Bryozoa also contribute their limestone secretions, as do the spiny sea urchins and elegant sea lilies of the phylum Echinodermata, the bivalves of the phylum Brachiopoda and such representatives of the phylum Mollusca as clams and oysters, all of whose accumulated skeletons and shells contribute to the reef limestones.

Many organisms in the community do not add significantly to the reef structure; some burrowers and borers are even destructive. The marine worms that inhabit the reef are soft-bodied and thus incapable of contributing to the reef mass. The hard parts of such reef dwellers as crabs and fishes are systematically consumed by scavengers. A few fragments may escape, but except for such passive and minor contributions to the reef detritus these organisms are not reef-builders.

The reef community is adapted to a low-stress environment characterized by the absence of significant seasonal change. The mean winter temperature of the water where reefs grow is between 27 and 29 degrees Celsius and the difference between summer and winter monthly mean temperatures is three degrees C. or less. The water is clear (so that the penetration of light is at a maxi-

mum), agitated (so that it is rich in oxygen) and of normal salinity. Even under these ideal circumstances many reef organisms (for example corals) do not grow at depths greater than 65 feet. This adaptation to freedom from stress makes the community remarkably sensitive to environmental change.

The fossil record documents hundreds of episodes of sweeping mass extinction, some continent-wide and others worldwide. These times of ecological disruption have simultaneously affected such disparate organisms as ammonites at sea and dinosaurs ashore, the plants on land and the protozoans afloat among the

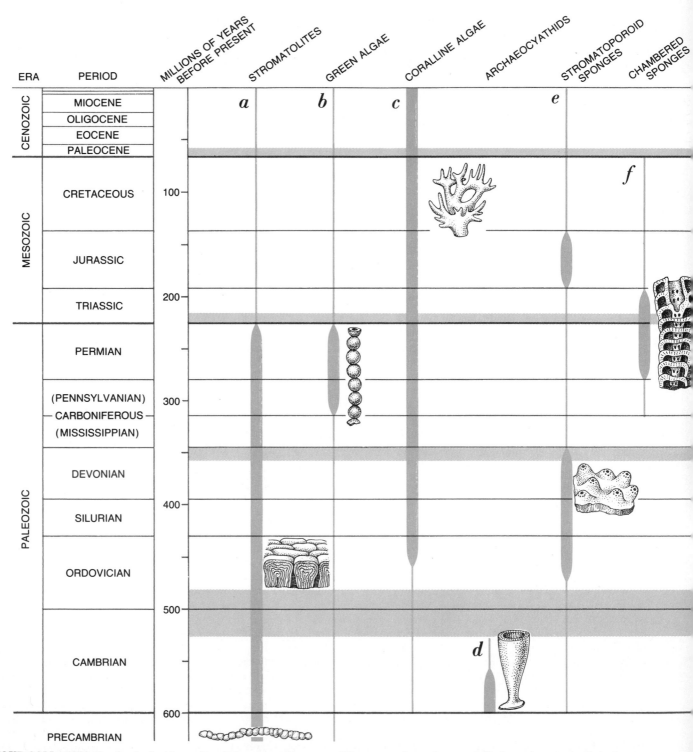

FOUR COLLAPSES (*bands of color*) have altered the composition of the reef community since the initial association between plant and animal reef-builders was established nearly 600 million years ago. This was when a group of spongelike animals, the archaeocyathids (*d*), appeared among the very much older reef-forming algal stromatolites (*a*) at the start of the Paleozoic era. In less than 70

million years the archaeocyathids became extinct; their demise marks the first community collapse. A successor community arose in mid-Ordovician times. Its members included coralline algae (*c*); the first corals, tabulate (*h*) and rugose (*i*); stromatoporoid sponges (*e*), and communal bryozoans (*m*). The group flourished almost to the end of the Devonian period, some 350 million years ago, the

oceanic plankton. The causative phenomena underlying the disruptions must therefore be unlike the ordinary, Darwinian causes of extinction—natural selection and unequal competition—that tend to affect species individually and not en masse.

For generations, in a reaction to the biblical doctrine of catastrophism that dominated 18th-century geology, scholars have viewed the apparent lack of continuity in fossil successions with skepticism. The breaks in the record, they proposed, were attributable to inadequate collections or to accidents of fossil preservation. At the same time, certain pioneers—T. C. Chamberlin and A. W. Grabau in the U.S. and Hans Stille in Germany—saw the breaks in fossil continuity as reflections of real events and sought a logical explanation for them. These men were eloquent proponents of a theory of rhythmic pulsations within the earth: diastrophic move-

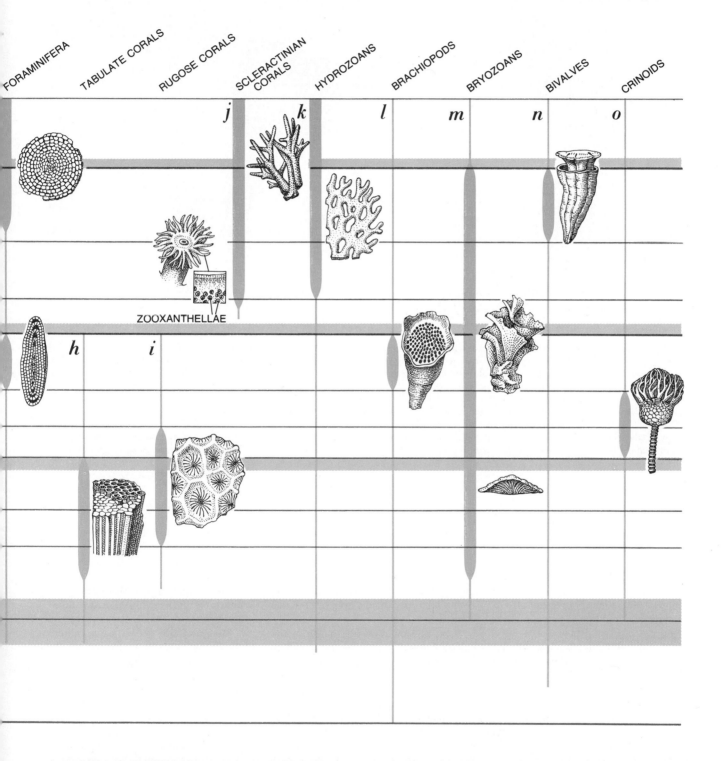

time of the second collapse. Its successor, some 13 million years later, contained a new sponge group (f) and increased numbers of green algae (b), foraminifera (g), brachiopods (l) and crinoids (o). These reef-builders flourished until the end of the Paleozoic era and the third collapse. The next resurgence occupied most of the Mesozoic era. It was marked by the appearance of modern corals (j) and a dramatic upsurge of a mollusk group, the rudists (n), that became extinct at the time of the fourth community collapse some 65 million years ago. Draining of shallow seas during the Cenozoic era produced cooler climates and led to formation of the Antarctic ice cap. Both developments have been factors in restricting the successor community in diversity and distribution.

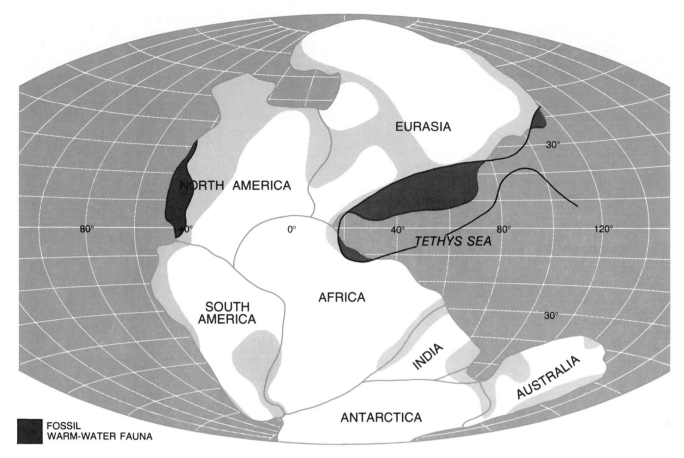

EURASIA

30°

NORTH AMERICA

80°   0°   40°   80°   TETHYS SEA   120°

SOUTH
AMERICA   AFRICA

30°

INDIA

AUSTRALIA

ANTARCTICA

FOSSIL
WARM-WATER FAUNA

TWO FACTORS that have affected world geography and climate are the movement of the continental plates and their greater or lesser invasion by shallow seas. The status of both factors during three key intervals in earth history is shown schematically here and on the opposite page. Near the end of the Paleozoic era (*a*) the continental plates had gathered into a single land area. Many of the reef-building species were then pantropical in distribution. By the end of the Mesozoic era (*b*) sea-floor spreading had separated the continental plates. The Atlantic Ocean had become enough of a barrier to the migration of reef organisms between the Old and New World to allow the evolution of species unique to each region. At the end of the Mesozoic era the shallow seas encroaching on the continents were drained completely. Early in the Cenozoic era (*c*) the shallow seas had been reestablished in certain zones, but the total continental area above water was larger than in the Mesozoic. The change resulted in a trend toward greater seasonal extremes of climate; the distribution of tropical and subtropical organisms, however, remained quite broad, as palm-tree fossils show.

ments that had been accompanied by significant fluctuations in sea level and consequent disruptive changes in climate and environment.

The lack of any demonstrable physical mechanism that might have produced such simultaneous worldwide geological revolutions kept a majority of geologists and paleontologists from accepting the proposed theory of the origins of environmental cycles. Today, however, we have in plate tectonics a demonstrated mechanism for changes in sea level and shifts in the relative extent of land and sea such as Chamberlin and his colleagues were unable to provide [see "Plate Tectonics," by John F. Dewey, beginning on page 124]. Significant changes in the volume of water contained in the major ocean basins, produced by such plate-tectonic phenomena as alterations in the rate of

lava welling up along the deep-ocean ridges or in the rate of set-floor spreading, have resulted sometimes in the emergence and sometimes in the flooding of vast continental areas. The changing proportions of land and sea meant that global weather patterns alternated between mild maritime climates and harsh continental ones.

Now, reef-building first began in the earth's tropical seas at least two billion years ago. As we have seen, the modern reef community is so narrowly adapted to its environment as to be very sensitive to change. It seems only logical to expect that the same was true in the past, and that changes in reef communities of earlier times would faithfully reflect the various rearrangements of the earth's land masses and ocean basins that students of plate tectonics are now documenting. The expectation is justified;

whereas many details of earth history will be clarified only by future geological and paleontological research, the record of fossil reef communities accurately delineates a number of the main catastrophic episodes.

As one might expect, the oldest of all types of reef is the simplest. Algae alone, without associated animals, are responsible for limestone reef deposits, billions of years old, that were formed in the seas of middle and late Precambrian times.

The Precambrian algae produced extensive accumulations of a distinctively laminated limestone that are found, flanked by aprons of reef debris, in rock formations around the world. Geologists call these characteristic limestone masses stromatolites, from the Greek for "flat" and "stone." The microscopic organisms that built the stromatolites are rarely

b

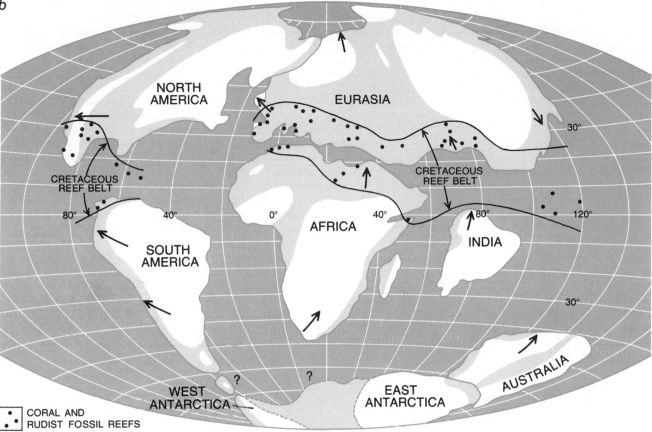

CORAL AND
RUDIST FOSSIL REEFS

c

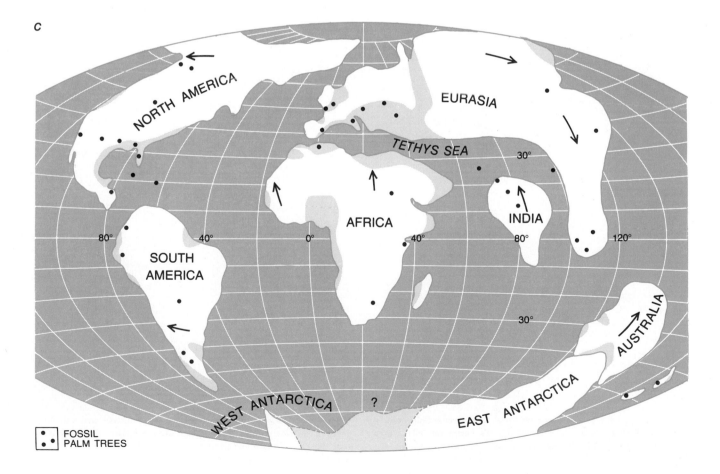

FOSSIL
PALM TREES

preserved as fossils, but they must surely have been similar to the filamentous blue-green algae that form similar masses of limestone today.

The accomplishments of these Precambrian algal reef-builders were not inconsiderable: individual colonies grew upward for tens of feet. They did so by trapping detrital grains of calcium carbonate and perhaps by precipitating some of the lime themselves. The resulting fossil bodies take the form of trunklike columns or hemispherical mounds.

As I outline the evolution of the reef community the reader will note that I often speak of first and last appearances. This does not mean, of course, that the organism being discussed was either instantly created or instantly destroyed. Each had an extensive evolutionary heritage behind it when some chance circumstance provided it with an appropriate ecological niche. Similarly, the decline and extinction of many major groups of organisms can be traced over periods of millions of years, although in numerous instances the time involved was too short to be measured by the methods now available.

Enter the Animals

The long Precambrian interval ended some 600 million years ago. The opening period of the Paleozoic era, the Cambrian, saw the first establishment of a reef community. The stromatolites' first partners were a diverse group of stony, spongelike animals named archaeocyathids (from the Greek for "ancient" and "cup"). In early Cambrian times these stony animals rooted themselves along the stromatolite reefs, grouped together in low thickets or scattered like shrubs in a meadow. It is not hard to imagine that the vacant spaces within and between these colonies provided shelter for the numerous bottom-feeding trilobites that inhabited the Cambrian seas. Not every reef harbored archaeocyathids, however; some reefs of early and middle Cambrian times are composed only of stromatolites.

By the end of the middle Cambrian, some 540 million years ago, the archaeocyathids had vanished. No single cause for this extinction, the first of the four major disasters to overtake the reef community, can be identified. One imaginable cause—competition from another reef-building animal—can be ruled out completely. The seas remained empty of any kind of reef-building animal throughout the balance of the Cambrian and until the middle of the Ordovician period, some 60 million years later. All reefs built during this long interval were the work of blue-green algae alone.

Fossil formations in the Lake Champlain area of New York, a region that lay under tropical seas in middle Ordovician times some 480 million years ago, contain the first evidence of a renewed association between reef-building plants and animals. The community that now arose was a rather complex one. Stromatolites continued to flourish and a second kind

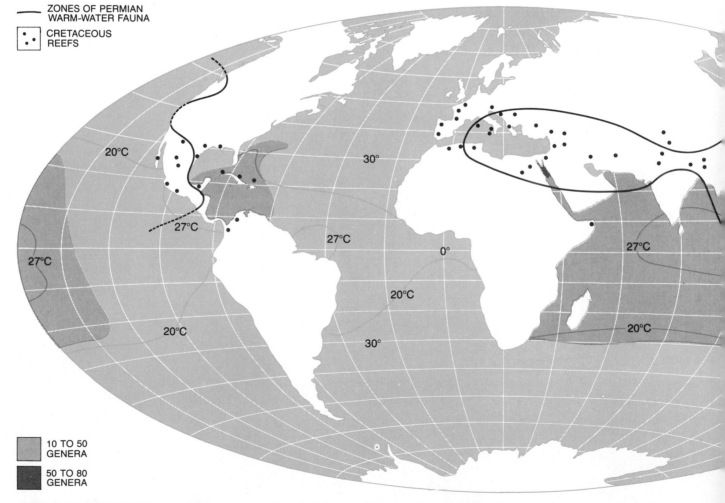

ZONES OF PERMIAN WARM-WATER FAUNA

CRETACEOUS REEFS

10 TO 50 GENERA

50 TO 80 GENERA

REEF-BUILDERS TODAY are confined to a narrow belt, mainly between 30 degrees north and 30 degrees south latitude (*light gray*), and even within this belt the greatest number of species are found where the minimum average water temperature is 27 degrees Celsius. Paleozoic and Mesozoic reef fossils, however, are found in areas far outside today's limits (*black lines and dots*). This suggests that the true Equator in those times lay well to the north of the equators shown on the maps on the preceding pages. The asym-

of plant life appeared: the coralline red alga *Solenopora,* a direct progenitor of the modern coralline algae. Colonial bryozoans, previously insignificant in the fossil record, now assumed an important role in the expanding reef community. Animal newcomers included a group of stony sponges, the stromatoporoids, some shaped like encrusting plates and others hemispherical or shrublike in form. These calcareous sponges were to play a major role in the community for millions of years. The most significant new animals, however, in light of subsequent developments, were certain stony coelenterates: the first of the corals. The intimate collaboration between algae and corals, apparently unknown before middle Ordovician times, has continued (albeit with notable fluctuations) to the present day.

These new arrivals and the other corals that appeared during the Paleozoic era were mainly of two types. In one type the successive stages of each polyp's upward growth were recorded as a

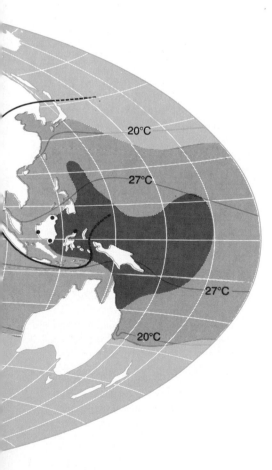

metry of the fossil-reef belt with respect to the present belt suggests that much of a once wider array has been engulfed by subduction along continental-plate boundaries.

series of parallel floors that subdivide the stony tube sheltering the animal; these organisms are called tabulate corals. The conical or cylindrical stone tube that sheltered the second type has conspicuous external growth wrinkles on its surface; these corals are called rugose.

A little more than 350 million years ago, near the end of the Devonian period, worldwide environmental changes caused a number of mass extinctions. Among the victims were many previously prominent marine organisms, including several groups in the reef community. That community now underwent a major retrenchment. Up to this point the tripartite association between algae, sponges and corals that first appeared in Ordovician times had proliferated for 130 million years without significant disturbance. The environmental alterations that nearly wiped out the reef community at the end of that long and successful period of radiation remain unidentified, although one can conjecture that a change from a mild maritime climate to a harsh continental one probably played a part. In any event the episode was severe enough so that only scarce and greatly impoverished reef communities, consisting in the main of algal stromatolites, survived during the next 13 million years. Not until well after the beginning of the Carboniferous period was there a community resurgence.

Some 115 million years passed between the revival of the reef community in Carboniferous times and the end of the Paleozoic era; the interval includes most of the Mississippian and all of the Pennsylvanian (the two subdivisions of the Carboniferous) and the closing period of the Paleozoic, the Permian. The revitalized assemblage that radiated in the tropical seas during this time continued to include stromatolites, numerous bryozoans and brachiopods and a dwindling number of rugose corals. Except for these organisms, however, the community bore no great resemblance to its predecessor of the middle Paleozoic. Both the stromatoporoid sponges and the tabulate corals are either absent from Carboniferous and Permian reef deposits or are present only in insignificant numbers.

Two new groups of calcareous green algae, the dasycladaceans and codiaceans, now attained quantitative importance in the reef assemblage. As if to match the decline of the stromatoporoid sponges, a second poriferan group—the calcareous, chambered sphinctozoan sponges—entered the fossil record. At the same time a group of echinoderms—the crinoids, or sea lilies—assumed a

larger role in the reef community. As the Paleozoic era drew to a close, the crinoids and the brachiopods achieved their greatest diversity; their skeletons preserved in Permian reef formations number in the thousands of species.

The Third Collapse

Half of the known taxonomic families of animals, both terrestrial and marine, and a large number of terrestrial plants suffered extinction at the end of the Paleozoic era. The alteration in environment that occurred then, some 225 million years ago, had consequences far severer than those of Devonian times. In the reef community the second successful radiation—based principally on a new tripartite association involving algae, bryozoans and sphinctozoan sponges—came to an end; reefs are unknown anywhere in the world for the first 10 million years of the Mesozoic era.

What was the cause of this vast debacle? There is little enough concrete information, but analogy with later and better-understood events encourages the conjecture that once again unfavorable changes in climate and habitat were major factors. In late Paleozoic times all the continents had come together to form a single vast land mass: Pangaea. Continental ice sheets appeared in the southern part of this supercontinent, a region known as Gondwana, in Carboniferous and early Permian times. These glaciers are concrete evidence of cooler climates; paleomagnetic studies show that the glaciated areas were then all located near the South Pole. Any relation between these early Permian ice sheets and the widespread late-Permian extinctions, however, is not yet evident. What is probably more significant is evidence that, at least during a brief interval, all the shallow seas that had invaded continental areas were completely drained at the end of the Paleozoic era. Serious climatic consequences must have resulted from the disappearance of a mild, primarily maritime environment.

In late Paleozoic times a wide tropical seaway, the Tethys, almost circled the globe. The only barrier to the Tethys Sea was formed by the combined land masses of North America and western Europe, which were then connected. The tongue of the Tethys that eventually became the western Mediterranean constituted one end of the seaway. The opposite end invaded the west coast of North America so deeply that great Permian reefs arose in what is now Texas.

The Mediterranean extremity of the Tethys Sea was the setting for a significant development when, after 10 million years of eclipse, the reef community was once again revitalized. There, in mid-Triassic times, a new group of corals, the scleractinians, made its appearance. The scleractinians were the progenitors of the more than 20 families of corals living in the reef community today. At first the new coral families, six in all, were represented in only a few scattered reef patches found today in Germany, in the southern Alps and in Corsica and Sicily. Even by late Triassic times, some 200 million years ago, the new corals were still subordinate as reef-builders to the calcareous algae.

The Mesozoic Community

During the 130 million years or so of Jurassic and Cretaceous times the reef community once again thrived in many parts of the world. The stromatoporoid sponges, all but extinct since the community collapse of Devonian times, returned to a position of some importance during the Jurassic. The new coral families steadily increased in diversity and reached an all-time peak in the waters bordering Mediterranean Europe during the Cretaceous period. In that one region there flourished approximately 100 genera of scleractinians; this is a greater number than can be found worldwide today. The reef community in this extremity of the Tethys Sea was also rich in other reef organisms. It included the two groups of sponges and such reef-builders as sea urchins, foraminifera and various mollusks. In addition a hitherto minor group of coralline red algae, the lithothamnions, now began to play an increasingly important role. By this time the stromatolites, important reef-builders throughout the Paleozoic, were no longer conspicuous members of the reef community.

Early in the Cretaceous period, some 135 million years ago, there was an unfavorable interval: reefs are unknown in the fossil record for some 20 million years thereafter. This pause merely set the stage, however, for a further efflorescence. Both in the Mediterranean area and in the waters of the tropical New World, hitherto unknown or insignificant coral families appeared. The present Atlantic Ocean was just beginning to form. Regional differences between reef communities in the Old World and the New that appeared at this time testify to the growing effectiveness of the Atlantic deeps as a barrier to the ready migration of reef organisms.

An extraordinary evolutionary per-formance was now played. Certain previously obscure molluscan members of the reef community, bivalves known as rudists, abruptly came into prominence as primary reef-builders. The next 60 million years saw a phenomenal rudist radiation that brought these bivalves to the point of challenging the corals as the dominant reef animals. Along the sheltered landward margins of many fringing barrier reefs rudists largely supplanted the corals. Their cylindrical and conical shells were cemented into tightly packed aggregates that physically resembled corals, and many of the aggregates grew upward in imitation of the coral growth pattern. Before the end of the Cretaceous period some 65 million years ago the rudists had attained major status in the reef community. Then at the close of the Cretaceous they quite abruptly died out everywhere.

Ever since the pioneer days of geology in the 18th century the end of the Cretaceous period has been known as a great period of extinctions. Nearly a third of all the families of animals known in late Cretaceous times were no longer alive at the beginning of the Cenozoic era. The reef community was not exempt; in addition to the rudists, two-thirds of the known genera of corals died out at this time. Forms of marine life other than reef organisms were also hard hit. The ammonites, long a major molluscan group, had suffered a decline during the final 10 million years of the Cretaceous; by the end of the period they were all extinguished. The belemnites, another major group of mollusks, declined sharply, as did the inoceramids, a diverse and abundant group of clams that had previously flourished worldwide. Among the foraminifera, the free-floating, planktonic groups suffered in particular.

The environmental changes that devastated life at sea also took their toll of land animals. Perhaps the most spectacular instance of extinction ashore involved the group that had been the dominant higher animals during most of the Mesozoic era: the dinosaurs. Of the 115-odd genera of dinosaurs found in late Cretaceous fossil deposits, none survived the end of the period. The concurrent breakdown of so many varied communities of organisms clearly suggests a single common cause.

What was the source of this biological crisis? We have now come sufficiently close to the present to have a better grasp of the evidence. Throughout almost all of Mesozoic times life both on land and in the sea seems to have been remarkably cosmopolitan. A broad belt of equable climate extended widely in both directions from the Equator and there was no very evident segregation of organisms into climatic zones. The earth has always been predominantly a water world; the total land area may never have exceeded the present 30 percent of the planet's surface and has often been as little as 18 percent. In the late Cretaceous almost two-thirds of today's land area was submerged under shallow, continent-invading seas. It appears that during times of extensive inundation such as this one there was nothing like the blustery global circulation of air and strong ocean currents of today.

Paleoclimatology reveals the contrast between contemporary and Cretaceous conditions. By measuring the proportions of different isotopes of oxygen present in the carbonate of fossil foraminifera and mollusks it is possible to calculate the temperature of the water when the animals were alive. Today the temperature of deep-ocean water is about three degrees C. Cesare Emiliani of the University of Miami has shown that early in the Miocene epoch, some 20 million years ago, the bottom temperature of the deep ocean was about seven degrees C. He finds that in the Oligocene, 10 million years earlier, the temperature was about 11 degrees and that in late Cretaceous times, 75 million years ago, it was 14 degrees. He suggests that the onset of cooling may have been lethal to dinosaurs. In any event it is clear that cold bottom water has been accumulating in the deep-ocean basins since early in the Cenozoic era.

The accentuated seasonal oscillations in temperature and rainfall at the end of the Cretaceous period, a trend that evidently started when the shallow continental seas began to drain away into the deepening ocean basins, have been credited by Emiliani and also by Daniel I. Axelrod of the University of California at Davis and Harry P. Bailey of the University of California at Riverside with the simultaneous reduction in numbers or outright extinction of many other animal and plant species at that time. Genetically adapted to a remarkably equable world climate over millions of years, many of these Mesozoic organisms were ill-prepared for the extensive emergence of land from under the warm, shallow continental seas and the climate of accentuated seasonality that followed. Perhaps we should be surprised not that so many Cretaceous organisms died but that so many managed to adapt to the new conditions and thus survived.

Near the end of the Paleocene epoch, some 10 million years after the great

FOSSIL BARRIER REEF was built in upper Devonian times by a community of marine plants and animals living in the warm seas that covered part of Australia more than 350 million years ago. Ex-posed by later uplift and erosion, the reef now forms a belt of jagged highlands, known as the Napier Range, in Western Austra-lia. A stream has cut a canyon (*foreground*) through the reef rock.

YOUNGER FOSSIL REEF, built some 250 million years ago in the Permian period, forms a rim of rock 400 miles long surrounding the Delaware Basin (*right*) on the border between Texas and New Mexico. Most of the reef is buried under later deposits, but one exposure forms this 40-mile stretch of the Guadalupe Mountains. El Capitan (*foreground*), a part of the reef front, is 4,000 feet high.

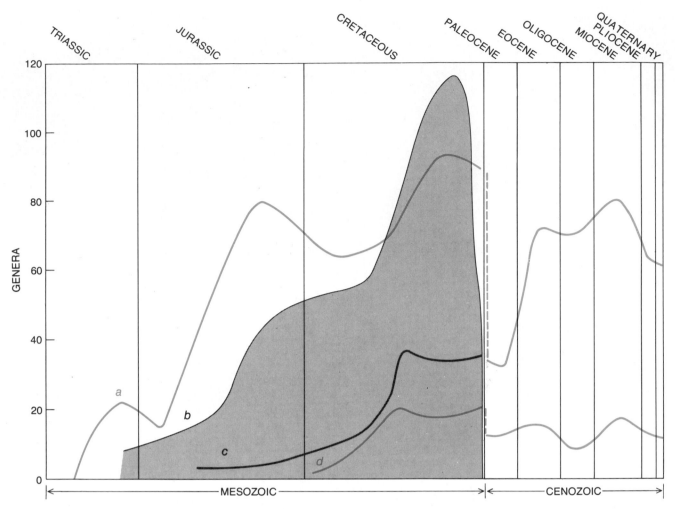

MESOZOIC INCREASE in the diversity of ocean and land animal genera reached a peak in the Cretaceous period. It was followed both by extinctions and by severe reductions in the number of genera that chanced to survive. At sea the explosively successful rudists (c) and ashore the long-dominant dinosaurs (b) became extinct. Corals (a) and globigerine foraminifera (d) fell in numbers.

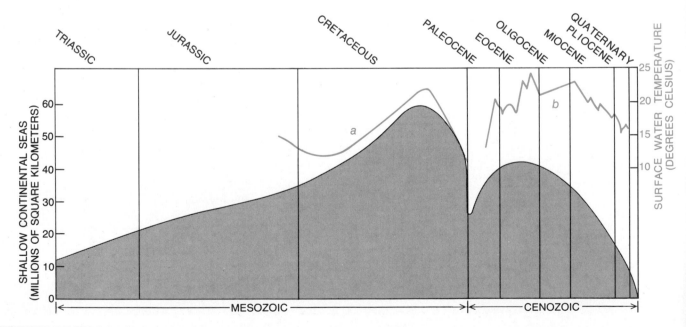

SHRINKING OF SEAS that had invaded large continental areas (gray) began near the end of the Mesozoic era and accelerated in the Miocene epoch. Oxygen isotopes in belemnite (a) and forami- nifera (b) skeletons show that the ocean has become progressively cooler since early in Miocene times (color). A relation seems to exist between a larger land area and greater seasonal extremes in climate.

Cretaceous collapse, a reef community sans rudists reappeared in the tropical seas. The following epoch, the Eocene, saw a new radiation of the scleractinian corals. Several genera that now appeared worldwide were unknown earlier in the fossil record; many of them are still living today.

The Cenozoic Decline

A sharp reduction in coral diversity that began in late Eocene times and lasted throughout the Oligocene epoch seems to reflect a continued increase in seasonality of climate and a substantial lowering of mean temperatures over large areas of the globe. Nonetheless, communities built around a bipartite association of corals and coralline algae continued to build extensive reefs in the Gulf of Mexico and the Caribbean area, in southern Europe and in Southeast Asia. Continued sea-floor spreading and deepening of the Atlantic basin, which enhanced the effectiveness of the Atlantic as a barrier to migrating corals, are evidenced in increasing differences between Caribbean and European coral species in late Oligocene fossil deposits.

By now the earth had been free from continental glaciers for almost 200 million years, but a change in the making was to have profound consequences for world climate: the Antarctic ice cap was becoming established. Fossil plant remains and foraminifera from nearby deep-sea deposits both testify that even in early Miocene times the climate in much of the Antarctic continent was not much different from the blander days of

the early Cenozoic, when palm trees grew from Alaska to Patagonia. At this time, moreover, Antarctica was some distance away from its present polar position. Nonetheless, mountain glaciers had begun to appear there millions of years earlier, in Eocene and Oligocene times. Sands of that age, produced by glacial streams and then rafted out to sea on shelf ice, are found in offshore deep-sea cores. The cooling trend was well established before the end of the Miocene epoch. In the Jones Mountains of western Antarctica lava flows of late Miocene age overlie consolidated glacial deposits and extensive areas of glacially scoured bedrock.

With the formation of the Antarctic ice cap some 15 to 20 million years ago a factor came into being that strongly influences world weather patterns to this day. The ice cap energizes the world weather system. In the broad reaches of open ocean surrounding Antarctica the surface water is aerated and cooled until it is too heavy to remain at the surface. The cold water sinks and spreads out along the sea floor, following the topography of the bottom. The result is a gravity circulation of cold water from Antarctica into the world's ocean basins, with a consequent lowering of the mean ocean temperature and cooling of the overlying atmosphere. The energetic interactions of the atmosphere above with the ocean and land below, in turn, strongly influence global wind patterns and worldwide weather. Today's climate is the product of a long cooling trend, marked by ever greater seasonal extremes; the trend became strongly ac-

centuated when the Antarctic ice cap came into being late in the Miocene epoch.

This event and others in the Cenozoic era are faithfully recorded in terms of changes in the reef community. For example, in spite of the development of new barriers to the migration of reef organisms during the Mesozoic era, such as the Atlantic deep, the reef community had remained predominantly cosmopolitan up to the close of the Cretaceous. By Miocene times, however, what had once been essentially a single pantropical community was effectively divided into two distinct biogeographic provinces: the Indo-Pacific province in the Old World and the Atlantic province in the New World.

In the Old World the increasingly unfavorable climate had eliminated the reef community in European waters. It was during Miocene times, when Australia reached its present tropical position, that the Old World reef-builders first began to colonize the shallows of the Australian shelf; in terms of maximum diversity, the headquarters of the Indo-Pacific province today lies in the Australasian region where seasonal contrasts in water temperature are minimal.

Like the Atlantic deep, the deep waters of the eastern Pacific formed a generally effective barrier to the migration of reef organisms from the Indo-Pacific province into the hospitable tropical waters along the west coast of the Americas, principally around Panama. In Miocene times this Pacific coastal pocket was still connected to the reef-rich Caribbean, the headquarters of the At-

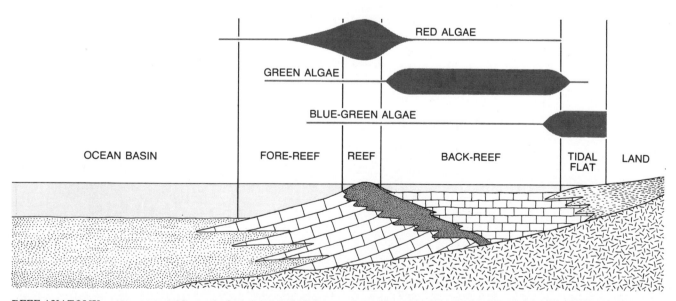

REEF ANATOMY centers on a rigid core (*color*) composed of the cemented skeletons of algae and corals. In this diagram growth of the core began at right and continued upward and outward. Most of the bulk of the reef consists of wide areas of stabilized detritus that are continuously being converted into the limestone that comprises the fore-reef and back-reef. The sandlike detritus is stabilized by the growth of the animals and plants in the reef community. The zones occupied by different algae are indicated here.

lantic province. Contact between the Atlantic province and the small Pacific enclave continued until the two areas were separated during the Pliocene epoch by the uplifting of the Isthmus of Panama.

The Pliocene saw a further contraction of the world tropics. The community of reef-builders gradually retreated to its present limits, generally south of 35 degrees north latitude and north of 32 degrees south latitude. Rather than serving as a center for new radiations, this tropical belt essentially became a haven. The epoch that followed, the Pleistocene, was marked by wide fluctuations in sea level and sharp alternations in climate that accompanied a protracted series of glacial advances and retreats. Oddly enough, the reef community was scarcely affected by such ups and downs; the main reasons for this

apparent paradox seem to be that neither the total area of the deep tropical seas nor their surface water temperature underwent much change during the Pleistocene ice ages.

Is a fifth collapse in store for the most complex of ocean ecosystems? It would be foolhardy to respond with a flat yes or no. If, however, the past is prologue, the answer to a collateral question seems clear. The question is: Will a reef community in any event survive? The most significant lesson that geological history teaches about this complex association of organisms is that, in spite of the narrowness of its adaptation, it is remarkably hardy. At the end of the long interlude that followed each of four successive collapses, the reef community entered a new period of vigorous expansion. Moreover, without exception each revitalized aggregate of reef-builders in-

cluded newcomers in its ranks.

The conclusion is inescapable that even during the times most unfavorable to the reef ecosystem the world's tropical oceans must have had substantial refuge areas. In these safe havens many of the threatened organisms managed to adapt and survive while others seem to have crossed some evolutionary threshold that had prevented their earlier appearance in the community. Today the Atlantic and Indo-Pacific provinces are two such refuge areas.

As long as the cooling trend that began in Cretaceous times does not entirely destroy these tropical refuge areas, one long-range conclusion seems firm. Any collapse of the present reef community will surely be followed by an eventual recovery. The oldest and most durable of the earth's ecosystems cannot easily be extirpated.

17

When the Mediterranean Dried Up

INTRODUCTION

Science advances in jumps, both when new ideas are born and when new instruments are invented that allow us to observe things never seen before. Such a new instrument is the *Glomar Challenger,* the deep-sea drilling ship of the National Science Foundation's Joint Oceanographic Institutions for Deep Earth Sampling program (JOIDES). For the first time geologists were able to penetrate the sea floor more than a few tens of meters and to bring up on deck the sediment and rock that lay below the surface. Hitherto the nature of the subsurface could only be vaguely inferred from a combination of reasoning from general principles and poorly diagnostic data on seismic velocities. Now we have the actual materials so that we can identify the mineralogy, the chemical composition, and the fossils that give us the geological age of the deposit. The surprises have been great. One early upset was the discovery that layers thought from seismic profiles to be volcanic ash beds turned out to be chert (silica) beds—a lithology no one had predicted. The first hole to be drilled, in the deep of the Gulf of Mexico, went into the cap of a salt dome. When the *Glomar Challenger* went to the Mediterranean, a series of startling finds were made—of coarse gravels unusual in deep-sea deposits, then of anhydrite, indicating evaporite conditions, and finally of rock salt, indicating extensive conditions of advanced evaporation.

Kenneth J. Hsü, one of the chief scientists of that part of the Deep Sea Drilling Project, is one of those versatile geologists who have worked in many areas of geology, ranging from Alpine tectonics, through the study of sedimentary structures of turbidity current deposits, to sedimentary geochemistry and petrology. Perhaps it was this combination of experience, perhaps it was that his kind of mind was at the right place at the right time, and no doubt it was his work with the other chief scientist, William B. F. Ryan of Lamont-Doherty Geological Observatory, that led to the outrageous—but hard to dispute—hypothesis of this article, "When the Mediterranean Dried Up."

A generation ago the thought of such an unusual—even catastrophic—event as the drying up of the Mediterranean would have been rejected out of hand as violating the principle of uniformitarianism: the idea that geological processes had always operated in the past as they do today. Now we understand uniformitarianism in a broader way as the tenet that chemical and physical processes have always operated according to the same basic laws. We now explicitly recognize that the earth did gradually evolve through many states and that geologic processes may have operated at different rates and with different consequences in the past. So we are now able to think about the unusual geological event, such as the one Hsü describes and argues for with compelling evidence.

NIGHT VIEW FROM THE BRIDGE of the *Glomar Challenger* shows, straddling the middle of the ship, the lower legs of the 142-foot drilling derrick, the vessel's most conspicuous structural feature. Beyond the derrick, on the foredeck, is the automatic pipe-racker, where 24,000 feet of drill pipe can be stored. The length of pipe suspended vertically above the derrick floor is the top end of a 15,000-foot drill string, which at the time the photograph was made was in the process of boring through the floor of the Mediterranean Sea 80 miles west of Sardinia. As the photograph suggests, such drilling and coring operations are carried out 24 hours a day. At present the *Glomar Challenger* is the only drilling ship capable of operating in the open ocean. It uses a dynamic positioning system to keep virtually stationary over the borehole, even in a stormy sea. The ship, which is owned by Global Marine Inc. and is operated by the Scripps Institution of Oceanography under a contract with the National Science Foundation, is named after the world's first full-time oceanographic research vessel, H.M.S. *Challenger*, which was launched in December, 1872, exactly 100 years ago.

When the Mediterranean Dried Up

by Kenneth J. Hsü
December 1972

Evidence acquired on a recent cruise by the deep-sea drilling vessel Glomar Challenger has revealed that six million years ago the Mediterranean basin was a desert 10,000 feet deep

Six million years ago a biological revolution swept across the Mediterranean Sea. The ancient marine fauna of the Mediterranean, descendants of mixed races from the Atlantic and Indian oceans, effected an unorganized mass exodus to find a refuge west of Gibraltar. Those that remained were soon to face annihilation, except for some hardy species that could tolerate the deteriorating environment. Thus ended the Miocene epoch, the less recent of the two dynasties that preceded our own Quaternary period. With the dawn of the Pliocene, or more recent, epoch the refugees returned, bringing with them new species from the Atlantic. They are the ancestors of the present marine fauna of the Mediterranean. This dramatic event, as recorded by fossils in certain sands and marls of Italy, did not escape the attention of Sir Charles Lyell, one of the founders of geology. The end of the revolution, signaled by the establishment of a new faunal dynasty, was chosen by Lyell in 1833 as the historical datum dividing the Miocene and Pliocene epochs. What was the cause of this revolution?

Near the end of the 19th century a deep gorge buried under the plain of Valence in southern France was discovered during a search for ground water. The gorge was cut into hard granite to a depth of hundreds of feet below sea level. Filling the gorge are Pliocene oceanic sediments, which in turn are covered by the sands and gravels of the Rhône river. When the gorge was first discovered, it was found to extend for some 15 miles between Lyons and Valence. Eventually the buried channel was traced for more than 100 miles downstream to La Camargue in the Rhône delta, where the valley was reached by drilling 3,000 feet below the surface. Obviously the modern Rhône is a lazy weakling compared with its ancestor, which sculptured a system of gorges almost comparable in size to the Grand Canyon of the Colorado. What caused the deep incision of the Rhône?

In 1961 the American oceanographic-research vessel *Chain* sailed to the Mediterranean with newly developed seismic equipment to explore the sea floor. The CSP, or continuous seismic-profiling, device aboard could be considered a super echo-sounder. Besides recording bottom reflections (the acoustic signals bounced back from the sea bottom), the instrument picked up signals transmitted by sound waves that were able to penetrate beyond the bottom and be reflected by hard layers hundreds of feet below. The new tool made possible a new discovery: it was found that the Mediterranean floor is underlain by an array of pillar-like structures, each a few miles in diameter and hundreds or thousands of feet high, protruding into the beds of sediments [*see top illustration on next page*]. Geophysicists were familiar with structures of this type; they looked very much like salt domes.

Salt domes are formed after rock salt from a deeply buried mother bed has forced its way upward into overlying sediments. Salt domes are common, for example, along the U.S. Gulf Coast, where many oil fields have been located around the domes. To find salt in coastal sediments is not unexpected, because geologists have long thought that the rock-salt formations were precipitated in coastal salinas, or lagoons. It was entirely unexpected, however, that salt domes would be discovered under the abyssal plains of the Mediterranean. Where could the salt have come from? Or are those structures indeed salt domes?

The exploration continued in the 1960's. William B. F. Ryan of the La-

mont-Doherty Geological Observatory, a participant in four cruises to the Mediterranean, and others working in the area were soon impressed by the presence of a strong acoustic reflector everywhere in the Mediterranean [*see bottom illustration on next page*]. This reflector, the M reflector, is commonly found a few hundred feet below the sea bottom, and its geometry closely simulates the bottom topography.

A layer that could send back distinct echoes must be very hard. Yet ocean sediments are commonly soft oozes made up of minute skeletons of the small organisms called foraminifera and nannoplankton. What could this hard layer be? Why should such a hard rock be down there under the Mediterranean?

To solve these and many other puzzles a group of investigators constituting the Mediterranean Advisory Panel of the Joint Oceanographic Institutions for Deep Earth Sampling program (JOIDES) in 1969 recommended to the Deep Sea Drilling Project (which is funded by the National Science Foundation and administered by the Scripps Institution of Oceanography) that the deep-sea drilling vessel *Glomar Challenger* be sent to the Mediterranean. The proposal was approved, and a two-month cruise in the fall of 1970 yielded the surprising answer to our mysteries: The biological revolution of the Mediterranean, the deep incision of the Rhône and the oceanic salt are all silent witnesses to an event six million years ago when the Mediterranean was almost completely dry. The hard-rock layer serving as the strong acoustic reflector is composed of the inorganic residues left behind by the desiccated Mediterranean: minerals known as evaporites.

The *Glomar Challenger* left Lisbon on August 13, 1970, for the 13th cruise

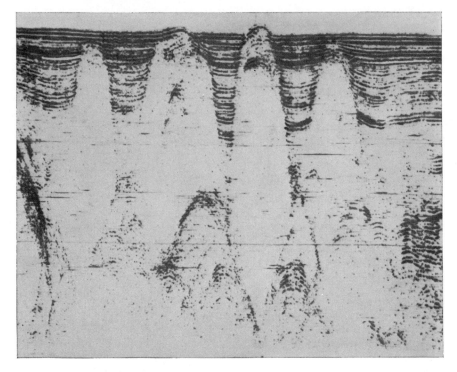

PILLAR-LIKE STRUCTURES, believed to be salt domes, are evident under the sea floor in this continuous seismic profile of a 10-mile-wide section of the Balearic abyssal plain in the western Mediterranean. The continuous-seismic-profiling device records not only reflections from the sea bottom but also signals transmitted by sound waves that penetrate beyond the bottom and are reflected by hard layers below. Some of the domes protrude as knolls above the sea floor; others are completely buried. The discovery of these pillar-like structures was the first hint that vast salt deposits are located under the Mediterranean floor.

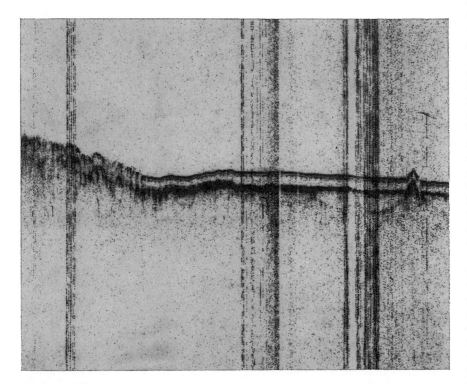

M REFLECTOR, a strong acoustic reflecting layer, underlies much of the Mediterranean floor. The relief of this layer (*lower dark contour*) closely simulates the bottom topography (*upper dark contour*). Drilling has shown that the M reflector corresponds to the top of an extensive underground evaporite formation consisting of inorganic residues precipitated from brines when the Mediterranean was isolated from the Atlantic Ocean some six million years ago. This continuous seismic profile was obtained by the *Glomar Challenger* as she worked her way northeastward (*right to left*) from a point southeast of Sicily toward Crete.

of the Deep Sea Drilling Project. The vessel is unique because of its ability to keep virtually stationary in a stormy sea. Guided by a system of radio beacons and computers, the four thrusters of the vessel can position her above the drill hole within a circle with a diameter of about 100 feet. Ryan and I were co-chief scientists on this cruise, heading an international team of 20 investigators and technicians. On the evening of August 23 we arrived at a spot some 100 miles southeast of Barcelona. After positioning our vessel above a buried submarine volcano we were ready to tackle the problem of identifying the M reflector.

The drill string was lowered to the bottom at a depth of 6,000 feet, and we bored some 600 feet into the sediments. According to our seismic record, we should have been hitting our goal and reaching the top of the hard layer. Since we had penetrated the oldest Pliocene sediments, the hard layer should date back to the late Miocene: some six million years ago. At the critical juncture, however, while we all waited eagerly for the next rock sample to come up, we ran into trouble: the core barrel got stuck inside the drill string. There was nothing to do but to haul all the drill pipes back on deck and pull the core barrel out of the pipe at the end of the string. Furthermore, we were advised to find another drill site; we could not expect to dig deeper at this spot.

Neither Ryan nor I slept that night as we supervised the drilling details. Now that the core barrel had come up, it was found to be buried in a tube full of sand. The sand was brought in a bucket to our shipboard laboratory. All morning long on the 24th Ryan and I busied ourselves sorting pea gravel out of the sand; we were too tired to work and too keyed up to sleep. We needed the menial labor to ease the tension. As the morning wore on, however, we became more and more amazed by what we saw.

Gravels are rare in the oceans. Submarine slumping can generate an underwater current, known as a turbidity current, that can transport coastal sands and gravels to deep abyssal plains. One would expect gravels of this type to be composed of many different kinds of rock, derived from erosion on land. As our pickings accumulated, however, we noticed that our gravel was made up of only three different types of rock: oceanic basalt, hardened oceanic ooze and gypsum. In addition there was an unusual dwarf fauna of small shells and snails. We found no quartz, no feldspars, no granites, no rhyolites, no gneisses, no

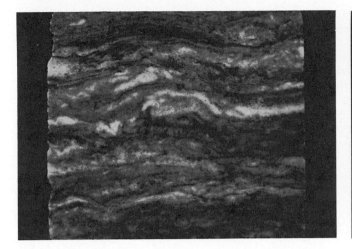

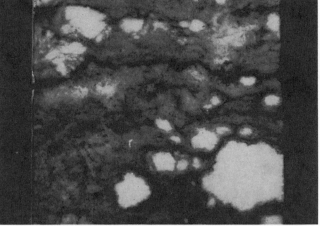

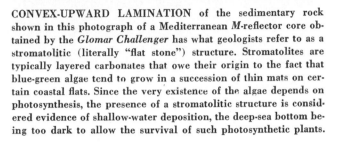

CONVEX-UPWARD LAMINATION of the sedimentary rock shown in this photograph of a Mediterranean *M*-reflector core obtained by the *Glomar Challenger* has what geologists refer to as a stromatolitic (literally "flat stone") structure. Stromatolites are typically layered carbonates that owe their origin to the fact that blue-green algae tend to grow in a succession of thin mats on certain coastal flats. Since the very existence of the algae depends on photosynthesis, the presence of a stromatolitic structure is considered evidence of shallow-water deposition, the deep-sea bottom being too dark to allow the survival of such photosynthetic plants.

LIGHT-COLORED NODULES evident in this Mediterranean rock core are composed of anhydrite: a high-temperature form of calcium sulfate that can only be precipitated from brines at temperatures higher than 35 degrees Celsius. At lower temperatures the low-temperature form of the sulfate, gypsum, is precipitated. Anhydrite is formed today almost exclusively in the sedimentary layers of the hot and arid coastal deserts called sabkhas, where it is precipitated in the pore spaces of sediments near the ground-water table. The high-temperature form typically grows as irregular nodules, similar to the limy concretions one finds in arid soils.

schists, no quartzites, no sandstones. In fact, we found no trace of anything that could be identified as coming from the nearby continent. The gravels could not have been brought out here by turbidity currents from the Spanish coast. What was the meaning of this unusual gravel?

Toward the evening of the 24th the drill bit was brought in. Caught in the teeth of the bit, which had been stuck near the top of the hard layer, were fine aggregates of anhydrite, an anhydrous calcium sulfate. Anhydrite is a common mineral in evaporites, which represent the inorganic residues left behind by evaporated brines, and ancient evaporites are commonly lithified (converted into rock). Now we had the first solution to our mysteries: The *M* reflector, the hard layer under the Mediterranean, is a late Miocene evaporite.

This solution only deepened the mystery. Evaporites should be the deposits of coastal lagoons or of deserts. Why should we find an evaporite formation under the Mediterranean at depths of thousands of feet below sea level?

The gravel provided the clue. The fragments of gravel could not have come from land, only from a dried-up ocean. Was it possible, then, that the Mediterranean had been isolated from the Atlantic and had changed into a desert basin during the late Miocene?

One could imagine the gradual shrinkage of the Mediterranean and the increasing salinity of its waters, with the death of all normal marine animals except for some dwarf species of clams and snails tolerant of supersaline conditions. The inland sea would eventually be changed into a salt lake, like the Dead Sea, where the brine would be dense enough to precipitate gypsum. Continued evaporation would eventually have laid bare the Mediterranean bottom. The submarine volcano would be converted into a volcanic mountain, and the oceanic ooze on its flank would become lithified. Streams draining such a desiccated ocean bottom would produce an unusual gravel such as the one we had found.

It seemed preposterous to make up such a story on the flimsy bit of evidence we had. The Mediterranean abyssal plains are more than 10,000 feet deep and the basin holds almost a million cubic miles of water. It was unthinkable that this beautiful blue ocean should disappear and be replaced by a series of Dead Seas and Death Valleys.

Or was it so unthinkable? A few facts and figures showed that it would be rather easy to dry up the Mediterranean. The climate of the Mediterranean region is arid. The annual evaporation loss is approximately 1,000 cubic miles. Only a tenth of the loss is compensated by rainfall and by the influx of fresh water from rivers. The Mediterranean manages to maintain its normal marine salinity through an exchange of water masses with the Atlantic. If the Strait of

Gibraltar were to be closed today, the annual evaporative loss could not be compensated, and the Mediterranean would be dried up in about 1,000 years.

Still, was it necessary to invoke such a drastic explanation? Many geologists had speculated that salts could be precipitated from a deep brine pool when the salt content in the brine exceeds the saturation concentration. The distribution of the evaporite layer indicated to us that this sedimentary formation was deposited in a deep Mediterranean basin. Is it not possible that the basin was filled to the brim with brines? True, a deep-water theory could not explain the genesis of our unusual gravel, but there should be more convincing evidence for desiccation.

The Arabic word sabkha is used in the Arabian Gulf countries to denote arid desert flats, particularly that part of a desert coastal plain situated slightly above the high-tide level. Sabkhas became an object of considerable interest to geologists soon after it was realized that certain types of ancient rock formation are practically identical with the sabkha sediments; both are characterized by the presence of nodular anhydrite and stromatolitic dolomite.

Anhydrite is a variety of calcium sulfate formed at high temperatures. At temperatures below 35 degrees Celsius in the presence of a brine that is saturated with sodium chloride (NaCl) anhydrite would be hydrated to form gypsum ($CaSO_4 \cdot 2H_2O$). (The hydration temperature would be higher if the brine were less salty). Since deep brine pools rarely exceed 35 degrees C. in bottom temperatures, anhydrite is formed today almost exclusively as a mineral in sabkha sediments. Since it is precipitated in the minute pore space of sediments near the ground-water table,

it tends to grow as irregular nodules rather like the concretions one finds in arid soils.

Stromatolite (literally "flat stone") is a laminated carbonate that owes its genesis to the growth of algae [see "The Evolution of Reefs," by Norman D. Newell, beginning on p. 183]. A dense growth of blue-green algae forms a thin mat on coastal flats. After a severe

storm the mat may be buried under a thin cover of sediments, but the algal growth persists and a new mat is constructed. This alternation ultimately results in the laminated rock called stromatolite. Stromatolite is common under the sabkhas of the Arabian peninsula, where nodular anhydrite also grows.

To return to our narrative, we sailed

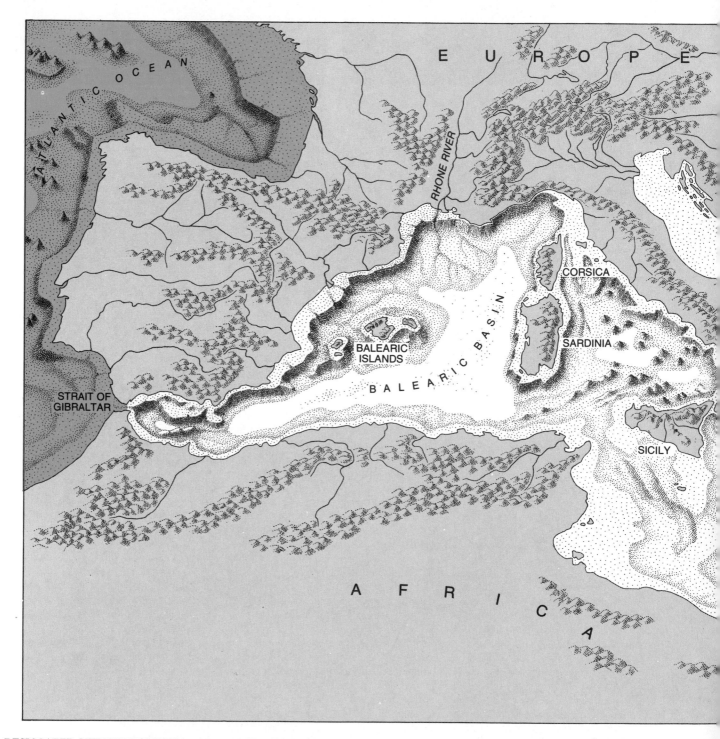

DESICCATED MEDITERRANEAN is represented by this panoramic drawing of the modern submarine topography of the Mediterranean basin. The Mediterranean must have looked something like this approximately six million years ago, when the basin was a great interior desert lying 10,000 feet below sea level. The Balearic abyssal plain was then a salt lake where evaporite minerals, including rock salt, were precipitated. Meanwhile nodular anhydrite grew in the soils on the lake shore. Gravels and variegated silts were

our vessel on August 27 to the Balearic Sea south of Majorca. Our drill site was positioned slightly north of the abyssal plain of the Balearic basin; our precision depth recorder registered 1,417 fathoms. The drill penetrated 1,000 feet of soft oozes before reaching the top of the hard layer. The core came up, and it was a late Miocene evaporite, as we expected. What was surprising, at least

to those who advocated deep-water salt deposition, was the sampling of nodular anhydrite and stromatolitic dolomite [*see illustrations on page 199*].

Stromatolite cannot form in deep water because the growth of algae requires sunlight. Moreover, one could not expect the bottom of a deep-water Mediterranean to ever get as hot as the 35 degrees C. (95 degrees Fahrenheit)

needed to precipitate anhydrite. Later a detailed petrological investigation by G. M. Friedman of Rensselaer Polytechnic Institute confirmed our prognosis that the Mediterranean evaporites were deposited on a desert flat; such sediments could not possibly have been deposited in several thousand feet of water.

More sophisticated methods of research led to the same conclusion. For

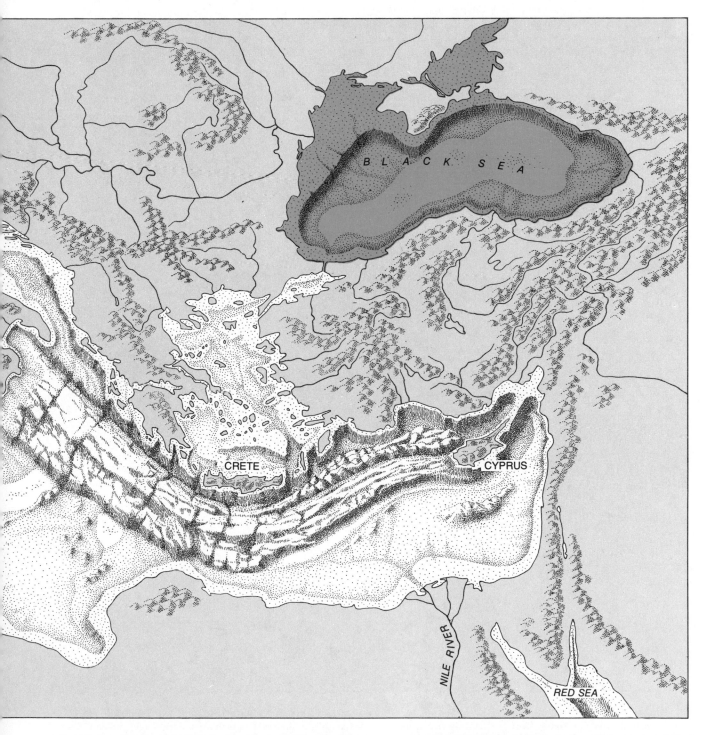

deposited around the edge of the basin at the foot of the steep slope, which is now the continental slope. At the end of the Miocene epoch, some 5.5 million years ago, an opening was breached at the Strait of Gibraltar. The inrushing water of the Atlantic constituted a

great waterfall, which probably had a discharge rate 100 times greater than Victoria Falls. Within a few thousand years the desiccated Mediterranean would be filled to the brim, and deep marine sediments would again be deposited on the Balearic abyssal plain.

example, we know that the oxygen in sulfates and carbonates consists of two isotopes: oxygen 16, the common isotope, and oxygen 18, the heavy isotope. The ratio of oxygen 18 to oxygen 16 in a sample could reveal its genesis. Evaporites precipitated from evaporated seawater have a narrow range of isotopic compositions. In contrast, those deposited on playas, or desert lakes, have a wide range of values. Analyses of our samples by R. M. Lloyd of the Shell Oil Company laboratory in Houston showed a high variability of isotopic composition and thus provided additional confirmation for their playa origin.

Thus the anhydrite was deposited on a desert flat, but what a desert flat it must have been! The flat is now buried 9,000 feet below sea level. Do we have any unequivocal evidence that the floor of the Mediterranean basin was so very deep?

In fact, we do. Maria Cita of the University of Milan, one of our shipboard paleontologists, studied the microfossils in the marine sediments above and below the anhydrite. She told us that the sediments are normal deep-ocean ones. Clearly the basin was deep and submerged under marine waters when it was open to the Atlantic, but it turned into a deep hole when the floodgate was shut and the basin dried up. Because we discovered several oceanic sediments interbedded with the evaporites, we concluded that the floodgate swung open and shut repeatedly during an interval of about a million years.

After six weeks of drilling we were able to confirm that the M reflector is an evaporite everywhere. Nonetheless, we had sampled only evaporative carbonates and sulfates and not the rock salt. Some doubted that a piece of salt could ever be brought on deck before it had been pulverized and dissolved by the drilling. We were aware of the fact that we might be searching in the wrong places. The distribution of saline minerals in desert playas shows a bull's-eye pattern [see illustration on page 204]. The less soluble carbonates and sulfates, being the first to precipitate, are found at the edge of a salt pan, whereas the more soluble rock salt is usually deposited in the more central, deepest part of a basin, where the last bitter waters were concentrated. Our earlier drill sites had been on the peripheral, slightly elevated parts of the basin. To find rock salt we had to search under the abyssal plain. It was now October and we were heading home. A last borehole was spudded on the abyssal plain some 80 miles west of Sardinia. After drilling through 1,100 feet of soft oozes we hit pay dirt. The driller brought in a cylinder of shining, transparent crystals. Their bitter taste left us with no doubt that we had found rock salt 10,000 feet below sea level.

Interbedded in the salt layers are some windblown silts. This aeolian detritus includes land-formed quartz as

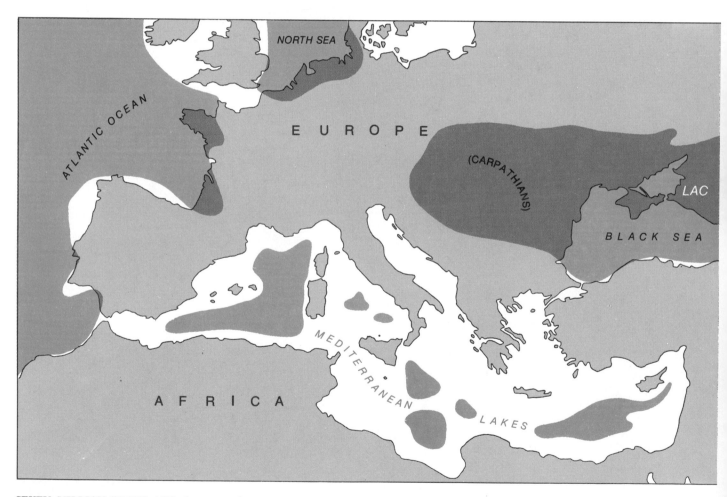

SEVEN MILLION YEARS AGO the geography of Europe was quite different from what it is today. Most of northeastern Europe was covered by a very large fresh-water-to-brackish-water lake that extended from the vicinity of Vienna eastward to beyond the Aral Sea. This great inland water body was named the Lac Mer by the French geologist Maurice Gignoux. At that time the Mediterranean was already separated from the Atlantic. Just before the uplift of the Carpathian Mountains, which took place roughly seven million years ago, the Lac Mer drained into the Mediterranean, supplying fresh and brackish waters to form a series of large inland lakes there. Laminated diatomites were deposited in such lakes. Eventually the Carpathians rose and formed a barrier, depriving the Medi-

well as the broken skeletons of foraminifera. Those tiny marine creatures flourished in the Mediterranean during the intervals when marine conditions prevailed. After the sea was isolated and desiccated the dead skeletons were blown across the desert flat by dust storms and laid down on the shore of a salt lake, to be buried eventually in rock salt. Bearing testimony to the alternate wetting and drying of the playa are nodular anhydrite and salt-filled mud cracks.

Examined under the microscope, the rock salt showed evidence of repeated solution and recrystallization, much like the salt in the modern coastal salinas of Lower California or in parts of Death Valley. The analogy to Death Valley can be carried a step further. We sampled red and green floodplain silts and well-rounded arroyo gravels from a nearby site. Those were carried to the base of an exposed continental slope by flash floods from the mountains of Sardinia

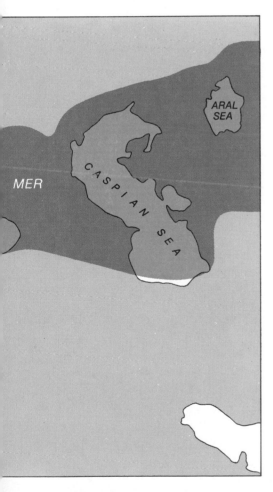

terranean of its water supply from the Lac Mer and turning the entire Mediterranean basin into a vast desert. The large fresh-water and brackish-water lakes were reduced to shallow lakes and playas, where salt and various other evaporites were precipitated.

and were deposited as alluvial fans fringing the salt pan. The similarity ends when one recalls that whereas Death Valley lies nearly at sea level, the floors of the Mediterranean desert basins were some 10,000 feet lower.

Our drilling had barely scratched the top of a huge salt deposit. The Mediterranean salt should be 5,000 or 6,000 feet thick, according to geophysical surveys. This estimate is probably not too far off the mark, since we were told that late Miocene salt formations are present in Sicily and are several thousand feet thick. We now believe the Sicilian evaporite represents a segment of the Mediterranean sea bottom that was pushed up by mountain-building movements a few million years ago.

After the salt came the deluge. The evidence is unmistakable from our drill cores. We obtained the geological record from three drill sites (in the Balearic, Tyrrhenian and Ionian basins), showing that the separate parts of the Mediterranean were simultaneously flooded and submerged under deep marine waters at the end of the Miocene epoch some 5.5 million years ago. The first deposit is a dark gray marl five inches thick, deposited when the basin was being filled up, followed by a white ooze with local patches of red ooze.

One can picture the desiccated Mediterranean as a giant bathtub, with the Strait of Gibraltar as the faucet. Seawater roared in from the Atlantic through the strait in a gigantic waterfall. If the falls had delivered 1,000 cubic miles of seawater per year (equivalent to 30 million gallons per second, 10 times the discharge of Victoria Falls), the volume would not have been sufficient to replace the evaporative loss. In order to keep the infilling sea from getting too salty for even such a hardy microfauna as the one found in the dark gray marl the influx would have had to exceed evaporation by a factor of 10. Cascading at a rate of 10,000 cubic miles per year, the Gibraltar falls would have been 100 times bigger than Victoria Falls and 1,000 times more so than Niagara. Even with such an impressive influx, more than 100 years would have been required to fill the empty bathtub. What a spectacle it must have been for the African ape-men, if any were lured by the thunderous roar.

By the time the first Pliocene white ooze was deposited the Mediterranean must have been filled to the brim. Then the present system of exchange with the Atlantic would have been in operation. The white ooze is a typical oceanic sediment, made up almost entirely of the

skeletons of microfossils and nannofossils. In addition to floating creatures bottom-dwelling organisms were found. William E. B. Benson of the National Science Foundation has examined the marine ostracods and Orville L. Bandy of the University of Southern California has studied the benthic foraminifera. They have both concluded that at that time the bottom of the Mediterranean was either colder or deeper than it is today. Those new immigrants from the Atlantic crawled through the deep gash at the western end of the Mediterranean. Eventually the cold-water bottom fauna died out when the Strait of Gibraltar was sufficiently shoaled to prevent the inflow of deep Atlantic waters, a condition that has persisted until today.

As we drilled through the evaporite formation we encountered an unusual laminated sediment. In addition to organic materials and minute crystals of dolomite the sediment contains fossil diatoms, whose skeletons consist of silica (SiO_2). Herbert Stradner of the University of Vienna, another of our shipboard paleontologists, recognized that some of the diatoms could only have lived in bodies of fresh or brackish water. This identification was later confirmed by Marta Hajós of the Hungarian Academy of Sciences; she found not only floating species but also bottom-dwelling ones. Later, at an eastern Mediterranean drilling site south of Crete, we sampled a late Miocene ostracod fauna. As Arredo Decima of the University of Palermo, an expert on ostracods, told us after he had examined our specimens, these tiny creatures also could have lived only on the bottom of brackish-water lakes.

We were puzzled by this discovery. Where could all that fresh water have come from? There was absolutely no reason to assume that the Mediterranean climate had been much more humid six or seven million years ago, nor could we find any evidence for sudden and drastic changes in precipitation or in evaporation that would have converted great lakes into Death Valleys. The mystery was resolved when we talked to our Austrian and Balkan colleagues, who told us about the Lac Mer. Apparently a large part of eastern Europe was covered by fresh or brackish waters during the late Miocene and the Pliocene. At one time a giant lake, called by the French Lac Mer, extended from Vienna to the Urals and the Aral Sea; its last descendants are the Caspian Sea and Black Sea of today. This body of water, collecting all the excess precipitation from the then wet and cold northeastern

Europe, was draining into the Mediterranean during the earlier part of the late Miocene, some seven or eight million years ago. Shortly thereafter tremendous earth movements led to the uplift of the Carpathian Mountains and a radical reorganization of the drainage system. The Lac Mer now found an outlet to the north. The eastern faucet to the Mediterranean was turned off. Cut off from its major supply of fresh water, the arid Mediterranean suffered the fate of desiccation.

From the eastern Mediterranean boreholes we obtained a series of middle Miocene cores. Aided by the outcrop section on Sicily, we were able to reconstruct the history of the Mediterranean during the past 15 or 20 million years. The Mediterranean was once a broad seaway linking the Indian and Atlantic oceans. With the collision of the African and Asiatic continents and the advent of mountain-building in the Middle East, the connection to the Indian Ocean was severed. Meanwhile Africa was also advancing toward Europe, and communication to the Atlantic was maintained by way of two narrow straits, the Betic in southern Spain and the Riphian in North Africa. We saw evidence in our cores of the gradual deterioration of the Mediterranean environment, the advancing stagnation of its waters, the inevitable extinction of its bottom-dwellers, the struggle for existence by its swimming and floating population and the evolution of a hardy race that could survive widely changing salinities. We saw a change from an inland sea to a series of great lakes, and we saw their desiccation and the complete extermination of the fauna and flora at the bottom of the Miocene Death Valleys, 10,000 feet below sea level. We saw also the deluge, the establishment of a new faunal dynasty and the changing marine population leading up to the population of the present.

The discovery of the Mediterranean desert was made possible by deep-sea drilling. The first people acquainted with the discovery were the shipboard investigators who were selected for qualifications other than their mastery of local geology. Hence the full impact of the discovery was only appreciated after the *Glomar Challenger* returned to Lisbon and the drill results were communicated to the public by press conferences in Paris and New York.

If the Mediterranean had indeed been

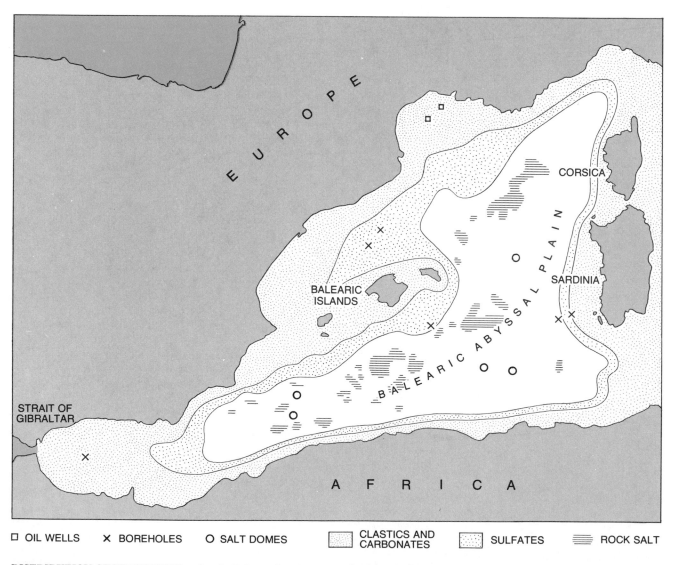

□ OIL WELLS   ✕ BOREHOLES   ○ SALT DOMES   CLASTICS AND CARBONATES   SULFATES   ☰ ROCK SALT

DISTRIBUTION OF EVAPORITES under the Balearic Sea shows the characteristic bull's-eye pattern one would expect in a completely enclosed basin. The less soluble carbonates and sulfates, being the first to precipitate, are found around the periphery of the basin; anhydrite and gypsum are found in a narrow ring just inside this outermost region, whereas rock salt, being the most soluble, is present only under the central, deepest part of the abyssal plain, where the last bitter waters must have been concentrated.

emptied, the coastal plains of the surrounding lands would have become high
plateaus, and islands would have been
lofty peaks. The first response to a lowering of the water level would be rejuve-
nation of streams: a marked increase in
their down-cutting power. The buried
Rhône gorge, first discovered 80 years
ago, should thus be one of the many surrounding the Mediterranean. Where are
the other buried gorges?

Soon after we returned to port Ryan
received a letter from a Russian geologist, I. S. Chumakov, who had learned
of our findings through an article in *The
New York Times*. Chumakov was one
of the specialists sent by the U.S.S.R. to
Aswan in Egypt to help build the famous
high dam. In an effort to find hard rock
for the dam's foundation 15 boreholes
were drilled. To the Russians' amaze-
ment they discovered a narrow, deep
gorge under the Nile valley, cut 700 feet
below the sea level into hard granite
[*see illustration on next page*]. The valley was drowned some 5.5 million years
ago and filled with Pliocene marine
muds, which are covered by the Nile
alluvium. Aswan is 750 miles upstream
from the Mediterranean coast. In the
Nile delta boreholes more than 1,000
feet deep were not able to reach the
bottom of the old Nile canyon. Chumakov estimated that the depth of the incision there might reach 5,000 feet, and
he visualized a buried Grand Canyon
under the sands and silts of the Nile
delta.

Chumakov was not the only one who
had been puzzled. Oil geologists exploring in Libya had also had their share of
surprises. First, their seismograms would
register anomalies; there were linear
features underground transmitting seismic waves at abnormally high velocities.
Drilling into the anomalies revealed that
they are buried channels incised 1,300
feet below sea level. The geologic record tells the same story: vigorous down-
cutting by streams and sudden flooding
by marine waters at the beginning of
the Pliocene. Frank T. Barr and his co-
workers of the Oasis Oil Company, based
at Tripoli in Libya, concluded in a report
that the Mediterranean must have been
thousands of feet below its present level
when the channels were cut. They could
not get their manuscript published in a
scientific journal, since no one would
accept such an outrageous interpretation.

Still other buried gorges and channels have been found in Algeria, Israel,
Syria and other Mediterranean countries. Rivers emptying into a desiccating
Mediterranean not only would have in-

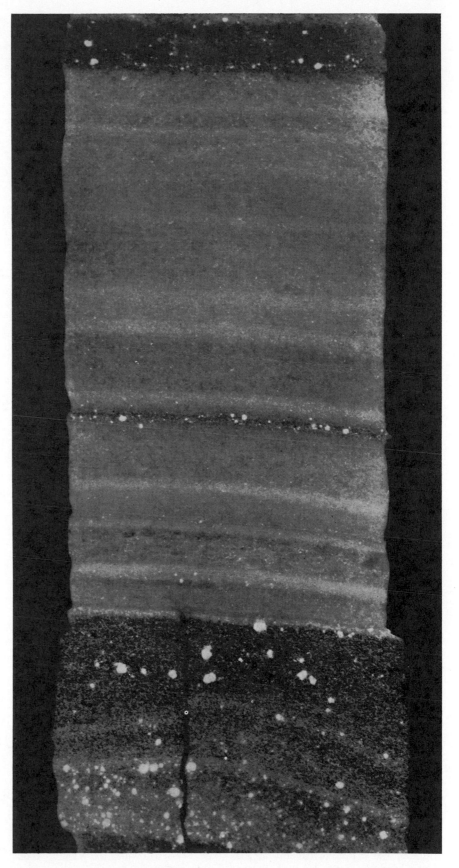

DEEP-SEA SALT CORE, the first of its kind ever obtained, was retrieved from a borehole
drilled by the *Glomar Challenger* through some 1,100 feet of soft oozes underlying the Balearic abyssal plain about 80 miles west of Sardinia. The sea bottom in this region is about
10,000 feet deep. The vertical crack in the lower part of the core is believed to be a desiccation crack, further evidence that the Mediterranean was dry down to 10,000-foot level.

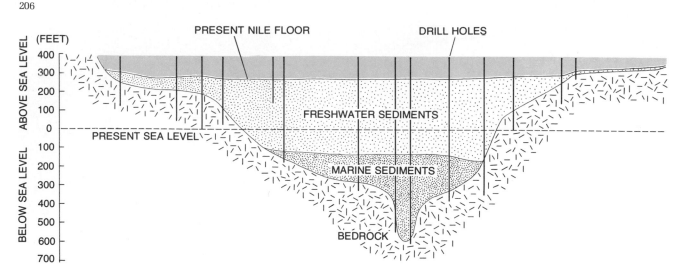

DEEP GORGE under the upper Nile valley near Aswan in Egypt was discovered by a team of Russian geologists while drilling test boreholes preparatory to the construction of the Aswan high dam. The narrow, deep central portion of the canyon cuts some 700 feet below the present sea level. The incision was apparently carved when the Mediterranean was dry. When the Mediterranean was flooded again at the beginning of the Pliocene epoch, the canyon was also drowned and marine sediments filled its lower section.

cised deeply on land but also would have continued on down across the exposed continental shelf and the continental slope to the flat bottom of the abyssal plain that was turning into a playa. In the course of a lecture at Yale University devoted to a discussion of the results of our expedition a student in the audience called my attention to such an eventuality and asked if such channels exist.

They do indeed. Extensive oceanographic surveys have been carried out by the French in the Balearic basin. The late Jacques Bourcart of the University of Paris reported in 1950 the discovery of numerous submarine canyons indenting the continental margins off the coast of France, Corsica, Sardinia and North Africa. The canyons are typically river-cut and are filled with alluvial gravels. Most of them can be related to a river on land and can be traced to a depth of 6,000 or 8,000 feet at the edge of the abyssal plain. They too were drowned by the early Pliocene deluge. Similar canyons have since been found in all parts of the Mediterranean. Their genesis had never been satisfactorily explained until it was realized that the Mediterranean was desiccated six million years ago.

The key opens the door to the solution of other mysteries. For example, one can now begin to understand the origin of the extensive caverns in the circum-Mediterranean lands and the peculiar karst topography of Yugoslavia, where sinkholes and pinnacles abound. One can also provide an answer to the long-standing question of why ground-water circulation once

penetrated 10,000 feet below sea level in a mid-ocean island such as Malta. Not only the geomorphic changes but also the biological ones were catastrophic. Giuliano Ruggieri of the University of Palermo, an authority on the evolution of the shallow marine fauna of the Mediterranean, recently sent me a reprint of an article he wrote in 1955, in which he surmised that the Mediterranean must have been desiccated in order to explain the biological revolution six million years ago. Lyell's historical datum marked the return of marine waters to this inland basin.

To remove all the water from the Mediterranean and pile it elsewhere would raise the sea level of the world ocean by 35 feet, an event that would drown many of our coastal cities. The magnitude of the negative load on the desiccated Mediterranean was comparable to the weight of the Fennoscandian ice sheet on Europe during the last ice age. The resubmergence of the Mediterranean must have led to the subsidence of the basin and the uplift of surrounding lands. Oceanographers and geologists began to see such evidence in their records. The presence of a hot desert where the Mediterranean is now should have had a heavy climatic impact. Indeed, European paleobotanists have noticed a change toward warm aridity in central Europe during the late Miocene, when the Vienna woods were changed into steppes and when palms grew in Switzerland. With the return of marine waters to the Mediterranean in the Pliocene epoch the central European climate again became wet and cold and deteriorated gradual-

ly into the ice age. It is also interesting to note that the Arctic polar ice cap began to build up in the late Miocene. Was this a coincidence, or was the initiation of glaciation triggered by the drying up of the Mediterranean?

Was the disappearance of a large inland sea a unique event in the geological history? Probably not. The existence of large saline deposits indicates that there have been other desiccated oceans. The famous Zechstein salts of northern Europe may have been the residues of an inland sea that dried up 250 million years ago. The giant salt and potash deposit of Alberta and Saskatchewan, some 350 million years old, may have had a similar origin. In fact, the discovery that a small ocean basin can be converted into desert has led us to reexamine the entire problem of salt genesis. Geologists used to worry about the occurrence of oceanic salt deposits under the Gulf of Mexico, under the South Atlantic off the coast of the Congo and Angola and under the North Atlantic off the coast of Nova Scotia. We can now postulate that those salts were formed when the Gulf and the Atlantic were isolated inland seas undergoing desiccation.

It must of course seem somewhat farfetched to imagine the Mediterranean as a deep, dry, hot hell. We ourselves were reluctant to come to that conclusion until all other explanations had failed; the facts left us with no alternative. As Sherlock Holmes once remarked: "It is an old maxim of mine that when you have excluded the impossible, whatever remains, however improbable, must be the truth."

18

River Meanders

INTRODUCTION

Though geologists and engineers have studied rivers and their meander belts for a great many years, they rarely offered rational explanations for the meandering. "Why does a river meander?" was a question usually asked only by ignorant freshmen and usually answered by a teleological and circular argument such as that it has to in order to keep its longitudinal profile the way it is. Some years ago a new approach was made to the study of the dynamics of rivers—the authors of this article, "River Meanders," were leaders of that effort—involving a knowledge of hydraulics, observations of natural streams, and experimental studies in laboratory flumes. The river came to be seen as a large and complex system governed by a group of physical laws relating the fluid dynamics of water and sediment transport to the shape of the river channel and its longitudinal profile.

In order to understand a physical state we have to devise suitable parameters for measurement. Students of streams had to define and measure such quantities as meander wave length and radius of curvature. One might say that it was in the perceptive choice of quantities to measure that scientists such as Leopold and Langbein made the beginnings of discoveries about how rivers work. The analysis had to wait on the development of the mathematics that describes the probability of a given outcome in a series of random trials—in particular, the mathematics of random walks. Given this background, analysts of river geometry were able to formulate some general principles of the shapes of meanders. Then they could proceed to seek some underlying physical explanation for the shape and use it to predict the general behavior of streams.

The work of Leopold and Langbein, their colleagues, and other fluvial morphologists in the past twenty years did not follow such a neat pattern of discovery. And the work is by no means finished, for dozens of important questions remain and new applications have been found. Now, for instance, planetologists want estimates of the reasonability of distinctive morphologies on Mars, seen in photographs taken by *Mariner 9*, as the products of rivers. And on earth, where we will be increasingly harnessing rivers for water power and irrigation, we will need to understand more so that we can predict better how streams respond to engineering by man.

River Meanders

by Luna B. Leopold and W. B. Langbein
June 1966

The striking geometric regularity of a winding river is no accident. Meanders appear to be the form in which a river does the least work in turning; hence they are the most probable form a river can take

Is there such a thing as a straight river? Almost anyone can think of a river that is more or less straight for a certain distance, but it is unlikely that the straight portion is either very straight or very long. In fact, it is almost certain that the distance any river is straight does not exceed 10 times its width at that point.

The sinuosity of river channels is clearly apparent in maps and aerial photographs, where the successive curves of a river often appear to have a certain regularity. In many instances the repeating pattern of curves is so pronounced that it is the most distinctive characteristic of the river. Such curves are called meanders, after a winding stream in Turkey known in ancient Greek times as the Maiandros and today as the Menderes. The nearly geometric regularity of river meanders has attracted the interest of geologists for many years, and at the U.S. Geological Survey we have devoted considerable study to the problem of understanding the general mechanism that underlies the phenomenon. In brief, we have found that meanders are not mere accidents of nature but the form in which a river does the least work in turning, and hence are the most probable form a river can take.

Regular Forms from Random Processes

Nature of course provides many opportunities for a river to change direction. Local irregularities in the bounding medium as well as the chance emplacement of boulders, fallen trees, blocks of sod, plugs of clay and other obstacles can and do divert many rivers from a straight course. Although local irregularities are a sufficient reason for a river's not being straight, however, they are not a necessary reason. For one thing, such irregularities cannot account for the rather consistent geometry of meanders. Moreover, laboratory studies indicate that streams meander even in "ideal," or highly regular, mediums [*see illustration on page 212*].

That the irregularity of the medium has little to do with the formation of meanders is further demonstrated by the fact that meandering streams have been observed in several naturally homogeneous mediums. Two examples are ocean currents (notably the Gulf Stream) and water channels on the surface of a glacier. The meanders in both cases are as regular and irregular as river meanders.

The fact that local irregularities cannot account for the existence of river meanders does not rule out other random processes as a possible explanation. Chance may be involved in subtler and more continuous ways, for example in turbulent flow, in the manner in which the riverbed and banks are formed, or in the interaction of the flow and the bed. As it turns out, chance operating at this level can explain the formation of regular meanders. It is a paradox of nature that such random processes can produce regular forms, and that regular processes often produce random forms.

Meanders commonly form in alluvium (water-deposited material, usually unconsolidated), but even when they occur in other mediums they are invariably formed by a continuous process of erosion, transportation and deposition of the material that composes the medium. In every case material is eroded from the concave portion of a meander, transported downstream and deposited on the convex portion, or bar, of a meander. The material is often deposited on the same side of the stream from which it was eroded. The conditions in which meanders will be formed in rivers can be stated rather simply, albeit only in a general way: Meanders will usually appear wherever the river traverses a gentle slope in a medium consisting of fine-grained material that is easily eroded and transported but has sufficient cohesiveness to provide firm banks.

A given series of meanders tends to have a constant ratio between the wavelength of the curve and the radius of curvature. The appearance of regularity depends in part on how constant this ratio is. In the two drawings on page 210 the value of this ratio for the meander that looks rather like a sine wave (*top*) is five for the wavelength to one for the radius; the more tightly looped meander (*bottom*) has a corresponding value of three to one. A sample of 50 typical meanders on many different rivers and streams has yielded an average value for this ratio of about 4.7 to one. Another property that is used to describe meanders is sinuosity, or tightness of bend, which is expressed as the ratio of the length of the channel in a given curve to the wavelength of the curve. For the large majority of meandering rivers the value of this ratio ranges between 1.3 to one and four to one.

Close inspection of the photographs

ENTRENCHED MEANDERS of the Colorado River in southern Utah were photographed from a height of about 3,000 feet. The meanders were probably formed on the surface of a gently sloping floodplain at about the time the entire Colorado Plateau began to rise at least a million years ago. The meanders later became more developed as river cut deep into layers of sediment. Mean downstream direction is toward right.

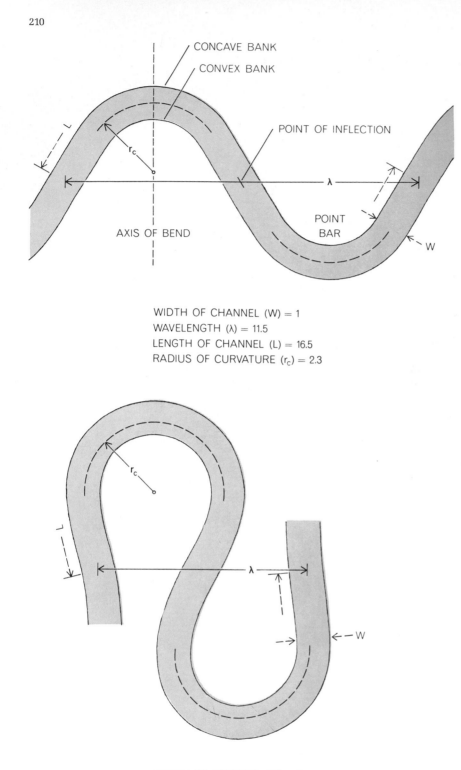

CONCAVE BANK

CONVEX BANK

POINT OF INFLECTION

r_c

AXIS OF BEND

POINT BAR

λ

W

L

WIDTH OF CHANNEL (W) = 1
WAVELENGTH (λ) = 11.5
LENGTH OF CHANNEL (L) = 16.5
RADIUS OF CURVATURE (r_c) = 2.3

r_c

L

λ

W

WIDTH OF CHANNEL (W) = 1
WAVELENGTH (λ) = 6.9
LENGTH OF CHANNEL (L) = 24.8
RADIUS OF CURVATURE (r_c) = 2.3

PROPERTIES used to describe river meanders are indicated for two typical meander curves. A series of meanders has a regular appearance on a map whenever there tends to be a constant ratio between the wavelength (λ) of the curve and its radius of curvature (r_c). The value of this ratio for the meander that looks rather like a sine wave (*top*) is five to one; the more tightly looped meander (*bottom*) has a corresponding value of three to one. An average value for this ratio is about 4.7 to one. Sinuosity, or tightness of bend, is expressed as the ratio of the length of the channel (*L*) in a given curve to the wavelength of curve. The value of this ratio for the top curve is 1.4 to one and for the bottom curve 3.6 to one. On the average the value of this ratio ranges between 1.3 to one and four to one.

and maps that accompany this article will show that typical river meanders do not exactly follow any of the familiar curves of elementary geometry. The portion of the meander near the axis of bend (the center of the curve) does resemble the arc of a circle, but only approximately. Neither is the curve of a meander quite a sine wave. Generally the circular segment in the bend is too long to be well described by a sine wave. The straight segment at the point of inflection—the point where the curvature of the channel changes direction—prevents a meander from being simply a series of connected semicircles.

Sine-generated Curves

We first recognized the principal characteristics of the actual curve traced out by a typical river meander in the course of a mathematical analysis aimed at generating meander-like curves by means of "random walk" techniques. A random walk is a path described by successive moves on a surface (for example a sheet of graph paper); each move is generally a fixed unit of distance, but the direction of any move is determined by some random process (for example the turn of a card, the throw of a die or the sequence of a table of random numbers). Depending on the purpose of the experiment, there is usually at least one constraint placed on the direction of the move. In our random-walk study one of the constraints we adopted was that the path was to begin at some point *A* and end at some other point *B* in a given number of steps. In other words, the end points and the length of the path were fixed but the path itself was "free."

The mathematics involved in finding the average, or most probable, path taken by a random walk of fixed length had been worked out in 1951 by Hermann von Schelling of the General Electric Company. The exact solution is expressed by an elliptic integral, but in our case a sufficiently accurate approximation states that the most probable geometry for a river is one in which the angular direction of the channel at any point with respect to the mean down-valley direction is a sine function of the distance measured along the channel [see *illustration on opposite page*].

The curve that is traced out by this most probable random walk between two points in a river valley we named a "sine-generated" curve. As it happens,

shape of real river meanders [*see illustration on next page*]. At the axis of bend the channel is directed in the mean down-valley direction and the angle of deflection is zero, whereas at the point of inflection the angle of deflection reaches a maximum value.

A sine-generated curve differs from a sine curve, from a series of connected semicircles or from any other familiar geometric curve in that it has the smallest variation of the changes of direction. This means that when the changes in direction are tabulated for a given distance along several hypothetical meanders, the sums of the squares of these changes will be less for a sine-generated curve than for any other regular curve of the same length. This operation was performed for four different curves of the same length, wavelength and sinuosity—a parabolic curve, a sine curve, a circular curve and a sine-generated curve—in the illustration on page 213. When the squares of the changes in direction were measured in degrees over 10 equally spaced intervals for each curve, the resulting values were: parabolic curve, 5,210; sine curve, 5,200; circular curve, 4,840; sine-generated curve, 3,940.

Curve of Minimum Total Work

Another property closely associated with the fact that a sine-generated curve minimizes the sum of the squares of the changes in direction is that it is also the curve of minimum total work in bending. This property can be demonstrated by bending a thin strip of spring steel into various configurations by holding the strip firmly at two points and allowing the length between the fixed points to assume an unconstrained shape [*see top illustration on pages 214 and 215*]. The strip will naturally avoid any concentration of bending and will assume a shape in which the bend is as uniform as possible. In effect the strip will assume a shape that minimizes total work, since the work done in each element of length is proportional to the square of its angular deflection. The shapes assumed by the strip are sine-generated curves and indeed are good models of river meanders.

A catastrophic example of a sine-generated curve on a much larger scale was provided by the wreck of a Southern Railway freight train near Greenville, S.C., on May 31, 1965 [*see bottom illustration on page 215*]. Thirty adjacent flatcars carried as their load 700-foot sections of track rail chained in a bundle to the car beds. The train, pulled by five locomotives, collided with a bulldozer and was derailed. The violent compressive strain folded the train-load of rails into a drastically foreshortened snakelike configuration. The elastic properties of the steel rails tended to minimize total bending exactly as in the case of the spring-steel strip, and as a result the wrecked train assumed the shape of a sine-generated curve that distributed the bending as uniformly as possible. This example is particularly appropriate to our discussion of river meanders because, like river meanders, the bent rails deviate in a random way from the perfect symmetry of a sine-generated curve while preserving its essential form.

The Shaping Mechanism

The mechanism for changing the course of a river channel is contained

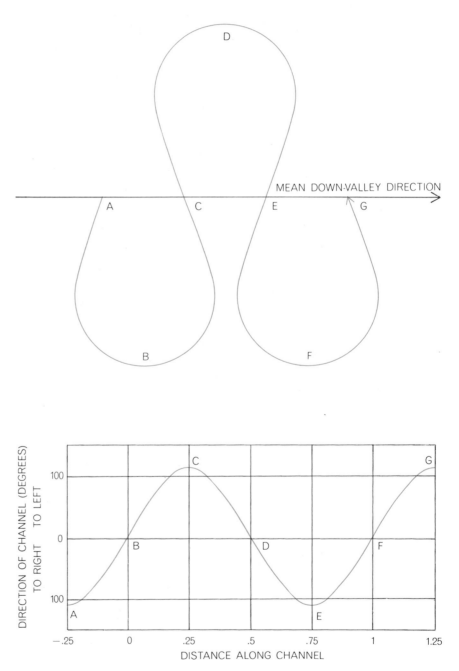

SINE-GENERATED CURVE (*top*) closely approximates the shape of real river meanders. This means that the angular direction of the channel at any point with respect to the mean down-valley direction (*toward the right*) is a sine function of the distance measured along the channel (*graph at bottom*). At the axis of each bend (*B, D and F*) the channel is directed in the mean down-valley direction and the angle of deflection is zero, whereas at each point of inflection (*A, C, E and G*) the angle of deflection reaches a maximum value.

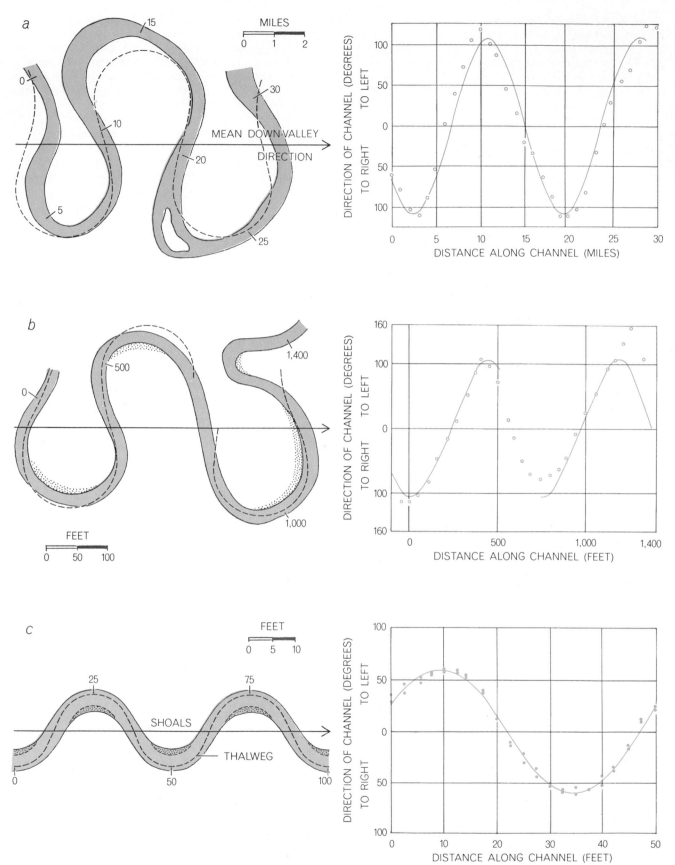

MAPS at left depict segments of two typical meandering streams, the Mississippi River near Greenville, Miss. (*a*), and Blackrock Creek in Wyoming (*b*), as well as a segment of an experimental meander formed in a homogeneous medium in the laboratory (*c*). Measurements of the angular direction of the channels with respect to the mean downstream direction were made at regular inter- vals along the center lines of the two natural meanders and along the thalweg, or deepest part of the channel, of the experimental meander. When these measurements were plotted against the dis- tance of each channel, the resulting curves closely approximated sine waves (*right*). The corresponding sine-generated curves are superposed on their respective channel maps (***broken black curves***).

this curve closely approximates the in the ability of water to erode, transport and deposit the material of the river's medium. Especially on a curve, the velocity gradient against the channel bank sets up local eddies that concentrate the expenditure of energy and localize erosion. An idealized flow pattern in a typical meander is shown in the top illustration on page 216. The left side of the illustration indicates the velocity vectors at various points for five cross sections along the curve. As the cross sections indicate, the depth of the channel changes systematically along the curve, the shallowest section being at the point of inflection and the deepest section at the axis of bend. At the same time the cross-sectional shape itself changes; it is symmetrical across the channel just downstream from the point of inflection and most asymmetrical at the axis of bend, the deeper section being always nearer the concave bank. The velocity vectors show a normal decrease in velocity with depth except at the axis of bend and near the concave bank, where the highest velocity at any point in the meander occurs somewhat below the surface of the water.

The right side of the same illustration shows the streamlines of flow at the surface of the meander. The maximum-velocity streamline is in the middle of the channel just downstream from the point of inflection; it crosses toward the concave bank at the axis of bend and continues to hug the concave bank past the next point of inflection. Riverboatmen navigating upstream on a large river face the problem that the deepest water, which they usually prefer, tends to coincide with the streamline of highest velocity. Their solution is to follow the thalweg (the deepest part of the river, from the German for "valley way") where it crosses over the center line of the channel as the channel changes its direction of curvature but to cut as close to the convex bank as possible in order to avoid the highest velocity near the concave bank. This practice led to the use of the term "crossover" as a synonym for the point of inflection.

The lack of identity between the maximum-velocity streamline and the center line of the channel arises from the centrifugal force exerted on the water as it flows around the curve. The centrifugal force is larger on the faster-moving water near the surface than on the slower-moving water near the bed. Thus in a meander the surface water is deflected toward the concave bank, requiring the bed water to move toward the convex bank. A circulatory system

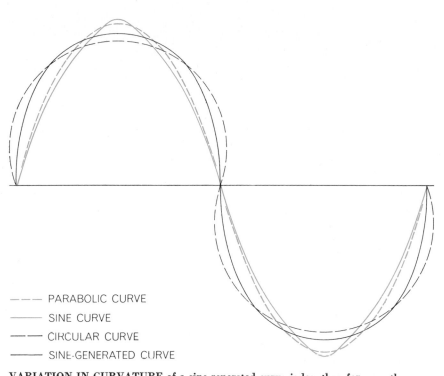

--- PARABOLIC CURVE
— SINE CURVE
— CIRCULAR CURVE
— SINE-GENERATED CURVE

VARIATION IN CURVATURE of a sine-generated curve is less than for any other regular geometric curve. This means that when the changes in direction are tabulated for small distances along several hypothetical meanders, the sums of the squares of the changes in direction will be less for a sine-generated curve than for any other curve. The changes in direction were measured in degrees over 10 equally spaced intervals for each of the four curves depicted here. When the squares of these changes were summed, the following values were obtained: parabolic curve, 5,210; sine curve, 5,200; circular curve, 4,840; sine-generated curve, 3,940. The four curves are equal in length, wavelength and sinuosity.

is set up in the cross-sectional plane, with surface water plunging toward the bed near the concave bank and bed water rising toward the surface near the convex bank. This circulation, together with the general downstream motion, gives each discrete element of water a roughly helical path that reverses its direction of rotation with each successive meander. As a result of this helical motion of water, material eroded from the concave bank tends to be swept toward the convex bank, where it is deposited, forming what is called a point bar.

Erosion of the concave banks and deposition on the convex banks tends to make meander curves move laterally across the river valley. Because of the randomness of the entire process, the channel as a whole does not move steadily in any one direction, but the combined lateral migration of the meanders over a period of many years results in the river channel's occupying every possible position between the valley walls. The deposition on the point bars, combined with the successive occupation by the river of all possible positions, results in the formation of the familiar broad, flat floor of river valleys—

the "floodplain" of the river. The construction of a floodplain by the lateral movement of a single meander can be observed even in the course of a few years; this is demonstrated in the bottom illustration on page 216, which is made up of four successive cross sections surveyed between 1953 and 1964 on Watts Branch, a small tributary of the Potomac River near Washington.

The overall geometry of a meandering river is an important factor in determining the rate at which its banks will be eroded. In general the banks are eroded at a rate that is proportional to the degree with which the river channel is bent. Any curve other than a sine-generated curve would tend to concentrate bank erosion locally or, by increasing the total angular bending, would add to the total erosion. Thus the sine-generated curve assumed by most meandering rivers tends to minimize total erosion.

Riffles and Pools

In the light of the preceding discussion it is possible to examine some of the hydraulic properties of meanders in greater detail. If a river channel is re-

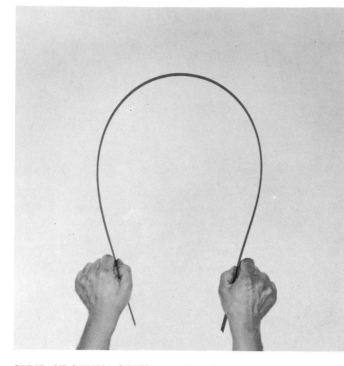
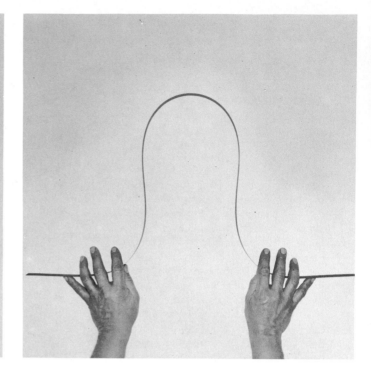

STRIP OF SPRING STEEL is used to demonstrate that a sine-generated curve is the curve of minimum total work. The strip is bent into various configurations by holding it firmly at two points and allowing the length between the fixed points to assume an un-

garded as being in a steady state, the form it assumes should be such as to avoid concentrating variations in *any* property at the expense of another property.

For example, variations in depth and velocity are inherent in all river channels, whether they are straight or curved. Even a reach, or length of channel, that is quite straight has a more or less uneven bed that consists of alternating deeps and shallows. Although this is not so obvious in a period of high flow, it becomes quite apparent at low flow, when the shallow sections tend to ripple in the sunlight as water backs up behind each hump in the bed before pouring over its downstream slope. To a trout fisherman this fast reach is known as a riffle. Alternating with the riffles are deeps, which the fisherman would call pools, through which the water flows slower and more smoothly.

The alternation of riffles and pools in a trout stream at low flow is noteworthy for another reason. The humps in the stream bed that give rise to the riffles tend to be located alternately on each side of the stream [see top illustration on page 217]. As a consequence the stream at low flow seems to follow a course that wanders successively from one side of the channel to the other, in a manner having an obvious similarity to meandering.

The analogy between this temporary sinuosity and full-scale meandering is strengthened by the fact that the riffles occur at roughly equal intervals along the channel. Moreover, the spacing of the riffles is correlated with the width of the channel. Successive riffles are located at intervals equal to about five to seven times the local channel width, or roughly twice the wavelength of a typical meander. This surprisingly consistent ratio seems even more remarkable when one realizes that each meander contains two riffles, one at each point of inflection. This observation led us to hypothesize that the same mechanism that causes meanders must also be at work in straight channels, and that a detailed study of the form and the hydraulic properties of two segments of channel that differ only in their degree of curvature might shed some light on the formation of meanders.

Obtaining Meander Profiles

In order to test this hypothesis it was necessary to obtain accurate data for all the pertinent hydraulic factors: depth, velocity, water-surface profile and bed profile. For several years we had attempted to measure such factors in small rivers near Washington just after every heavy rainstorm, when there was a rapid increase in streamflow. The water level changed so quickly in such storms, however, that there was never enough time to measure all the hydrau-

lic factors in detail through a succession of two riffles and an intervening pool. Then in 1959 we tried another strategy: we decided to measure a small stream in Wyoming, named Baldwin Creek, in early June, a period of maximum runoff from melting snow. Measurements were made in two places, a meandering reach and a straight reach, that were comparable in all outward aspects except sinuosity. The stream was about 20 feet wide and was nearly overflowing its banks, so that we could just barely walk in it wearing chest-high rubber waders.

Robert M. Myrick, an engineer with the Geological Survey, and one of us (Leopold) began a series of measurements in the midafternoon of June 19, surveying water-surface and bed profiles with a level and a rod, and making velocity and depth measurements with a current meter and a rod. When darkness came, we lighted lanterns and continued our measurements. At about daybreak we slept for a few hours and then resumed the survey, grateful that the melting snow had kept the stream at a steady high flow for such a long time.

Several days later we were able to sit down under a tree and plot the profiles, velocities and depths on graph paper. What emerged was a quite unexpected contrast between meandering reach and straight reach [see bottom illustration on page 217]. The slope of the water surface in the meandering reach

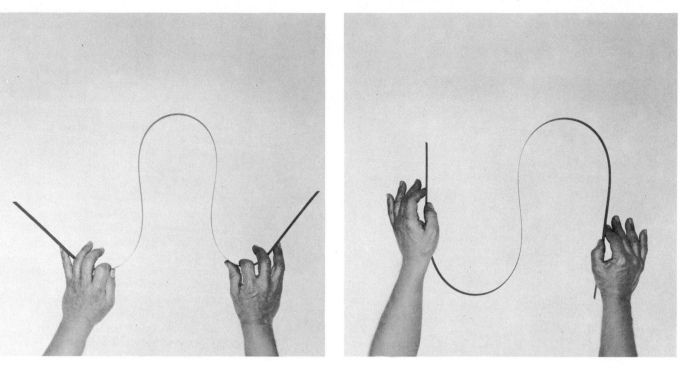

constrained shape. The strip will naturally avoid any concentration of bending and will assume a shape in which the bend is as uniform as possible. In each of the four cases shown here this shape is a sine-generated curve and indeed a good model of a river meander.

was clearly steeper than that in the straight reach; moreover, the water-surface profile of the meandering reach was nearly a straight sloping line, whereas the straight reach had a stepped profile, steep over the riffle bars and comparatively flat over the intervening pools.

What did this mean? It was as if the river had, to use somewhat anthropomorphic terms, chosen to cut a meander curve in order to achieve a more uniform water-surface profile. This suggested that the river had chosen the curved path in order to achieve the objective of uniform energy loss for each unit of distance along the channel, but had paid a price in terms of the larger total energy loss inherent in a curved path.

Conclusions

These data provided the key to further research, which ultimately resulted in several conclusions. First, it appears that a meandering channel more

CATASTROPHIC EXAMPLE of a sine-generated curve on a much larger scale was provided by the wreck of a Southern Railway freight train near Greenville, S.C., on May 31, 1965. Thirty adjacent flatcars carried as their load 700-foot sections of track rails chained in a bundle to the car beds. The train, pulled by five locomotives, collided with a bulldozer and was derailed. The violent compressive strain folded the trainload of rails into the drastically foreshortened configuration shown in this aerial photograph. The elastic properties of the steel rails tended to minimize total bending exactly as in the case of the spring-steel strip shown at top of these two pages, and the wrecked train assumed the shape of a sine-generated curve that distributed the bending as uniformly as possible.

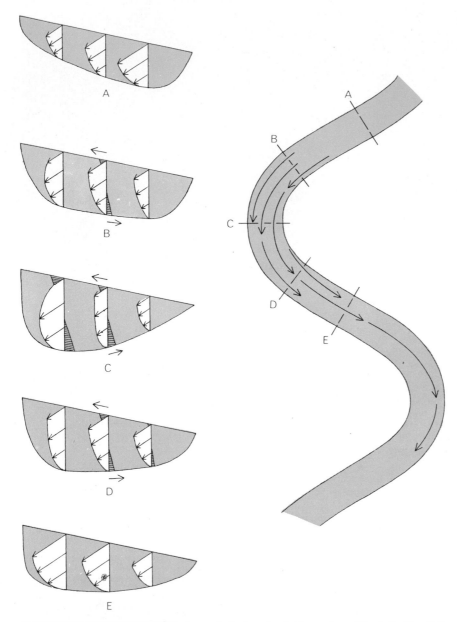

IDEALIZED FLOW PATTERN of a typical meander is shown here. The left side of the illustration indicates the velocity vectors in a downstream direction for five cross sections across the curve; the lateral component of the velocity is indicated by the triangular hatched areas. The right side of the illustration shows the streamlines at the surface of the meander.

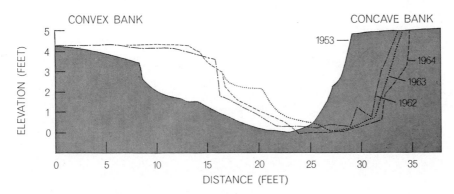

LATERAL MIGRATION of a typical meander is demonstrated in this drawing, made up of four successive cross sections surveyed between 1953 and 1964 on Watts Branch, a small tributary of the Potomac River near Washington. The lateral migration of meanders by the erosion of the concave banks and deposition on the convex banks over many years results in a river channel's occupying every possible position between the valley walls.

closely approaches uniformity in the rate of work over the various irregularities of the riverbed than a straight channel does. Of course the slope of the water surface is, with a slight correction for velocity, an accurate indicator of the rate at which energy is lost in the form of frictional heat along the length of the stream. Therefore a uniform longitudinal water-surface slope signifies a uniform expenditure of energy for each unit of distance along the channel.

A meander attains a more uniform rate of energy loss by the introduction of a form of energy loss not present in a straight reach, namely the curved path. It is evident that work is required to change the direction of a flowing liquid. Thus the slope of the water surface should increase wherever a curve is encountered by a river. In a meander it is at the deep pools, where the water-surface slope would be less steep than the average, that the introduction of a curve inserts enough energy loss to steepen the slope, thereby tending to make the slope for each unit of river length nearly the same. Accordingly the alternation of straight shallow reaches with curved deep reaches in a meander appears to be the closest possible approach to a configuration that results in uniform energy expenditure.

It is now possible to say something about the development of meandering in rivers. Although one can construct in a laboratory an initially straight channel that will in time develop a meandering pattern, a real meandering river should not be thought of as having an "origin." Instead we think of a river as having a heritage. When a continent first emerges from the ocean, small rills must form almost immediately; thereafter they change progressively in response to the interaction of uplift and other processes, including irregularities in the hardness of the rock.

Today the continuous changes that occur in rivers are primarily wrought by the erosion and deposition of sedimentary material. As we have seen, rivers tend to avoid concentrating these processes in any one place. Hence any irregularity in the slope of a river—for example a waterfall or a lake—is temporary on a geological time scale; the hydraulic forces at work in the river tend to eliminate such concentrations of change.

The formation of meander curves of a particular shape is an instance of this adjustment process. The meandering form is the most probable result of the processes that on the one hand tend to

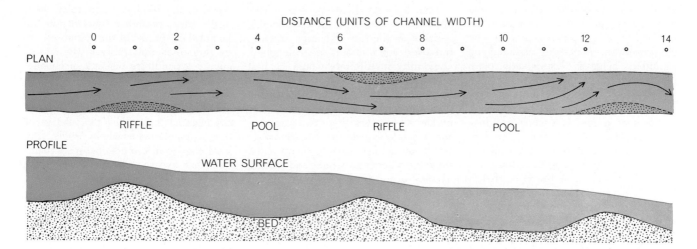

DISTANCE (UNITS OF CHANNEL WIDTH)

PLAN

RIFFLE   POOL   RIFFLE   POOL

PROFILE

WATER SURFACE

BED

STRAIGHT REACH of a river has a more or less uneven bed that consists of alternating deeps and shallows, known to trout fishermen as riffles and pools. The humps in the stream bed that give rise to the riffles tend to be located alternately on each side of the stream at intervals roughly equal to five to seven times the local stream width. As a consequence the stream at low flow seems to follow a course that wanders from one side of the channel to the other, in a manner having an obvious similarity to meandering.

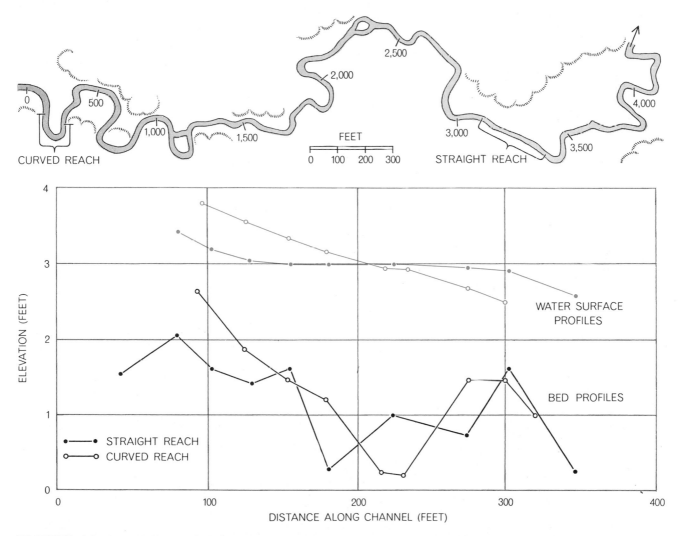

PROFILES of the water surface and bed of a small stream in Wyoming named Baldwin Creek were obtained by one of the authors (Leopold) and a colleague in 1959 during a period of maximum runoff from melting snow. Measurements were made in two places, a meandering reach and a straight reach, that were comparable in all outward aspects except sinuosity (*map at top*). What emerged was a quite unexpected contrast between the two reaches (*bottom*). The slope of the water surface in the meandering reach was clearly steeper than that in the straight reach; moreover, the water-surface profile of the meandering reach was nearly a straight sloping line, whereas the straight reach had a stepped profile, steep over the riffle bars and comparatively flat over the intervening pools.

eliminate concentrations of energy loss and on the other tend to reduce the total energy loss to a minimum rate. The sine-generated curvature assumed by meanders achieves these ends more satisfactorily than any other shape.

The same tendencies operate through the erosion-deposition mechanism both in the river system as a whole and in a given segment of the river. The tendency toward uniform power expenditure in the entire river leads toward a longitudinal profile of the river that is highly concave, inasmuch as uniformity in the rate of work per unit of length of channel would be achieved by concentrating the steepest slopes near the headwaters, where the tributaries and hence discharges are small. The longitudinal concavity of the river's profile also minimizes work in the system as a whole.

Such a longitudinal concave profile, however, would lead to considerable variation in the rate of energy expenditure over each unit area of channel bed. Uniformity in this rate would be best achieved by a longitudinal profile that was nearly straight rather than by one that was highly concave. Actual river profiles lie between these two extremes, and meanders must be considered in both contexts: first, as they occur within the river system as a whole, and second, as they occur in a given segment of channel.

In the context of the entire river system a meander will occur where the material constituting the banks is comparatively uniform. This will be more likely to take place downstream in a floodplain area than upstream in a headwater area. To the extent that the meandering pattern tends to lengthen the downstream reaches more than those upstream, it promotes concavity in the longitudinal profile of the system, thereby promoting uniformity in the rate of energy expenditure per unit of channel length.

In the local context of a given segment of channel the average slope of the channel is fixed by the relation of that segment to the whole profile. Any local change in the channel must maintain that average slope. Between any two points on a valley floor, however, a variety of paths are possible, any one of which would maintain the same slope and hence the same length. The typical meander shape is assumed because, in the absence of any other constraints, the sine-generated curve is the most probable path of a fixed length between two fixed points.

19

Lateritic Soils

INTRODUCTION

The dependence of agriculture on the quality of the soil is becoming one of the major concerns of a world increasingly hungry for wheat, rice, and the other staples of diet and this article, "Lateritic Soils," by Mary McNeil reflects that concern. Early in 1974 members of the United Nations Food and Agriculture Organization were predicting the virtual disappearance of reserve food supplies even in the agriculturally rich countries. The "green revolution" and agriculture in general depend heavily on large fertilizer supplies, which need great amounts of energy for their production. The poorer the grade of our finite reserves of phosphate rock, the raw material of fertilizer, the more energy will be required to make good fertilizer. Nowhere is the agricultural bind more extreme than in the tropical belts of lateritic soils , for these are the poorest soils and they have the largest populations to feed.

Modern geochemical and mineralogical studies of soils by geologists and soils scientists have done much to clarify the relation between climate, geological factors such as slope and bed-rock type, and soil type. Soil-forming processes are seen as a series of progressive chemical reactions between rain water that infiltrates the soil and the mineral components of the parent rock. The soil waters are helped in their attack on rocks by bacteria and by dissolved carbonic acid derived both from the carbon dioxide in the atmosphere and from respiration by plant roots. The result is a slowly downward-moving series of soil layers altered by the loss of easily soluble components, and retaining a residue of less soluble materials. Laterites are an extreme example of this process, having advanced so far, because of high rainfall and lush vegetation, that only the most insoluble elements, iron, aluminum, and titanium remain in the top layers.

Laterite soils are fragile. If agriculture is just slightly too intensive it will destroy them. Unusual climatic seasons may render them totally barren. Once made useless for agricultural purposes, laterite soils may take many years, even generations, to recover. Yet these skimpy soils are the basis of food for an enormous population in Asia, Africa, and South America. The problems of linking modern geochemistry, soil science, agronomy, and the anthropology of traditional agricultural practices of tropical countries will be central to maintenance of the world's food supply.

Lateritic Soils

by Mary McNeil
November 1964

*Rich in minerals and poor in organic matter, these
soils are common in the Tropics. They present an
obstacle to increasing the production of food in many
of the underdeveloped nations*

In the year 1860 a French naturalist named Henri Mouhot, struggling through the jungles of Indochina in search of plant species, came on a clearing in the tangled wilderness about 150 miles north of Pnompenh, the present capital of Cambodia. Looking up at the sky, he was astonished to see, standing out above the treetops, the sculptured towers of an ancient temple. It proved to be one of the remains of a "lost" civilization that had ruled the area from the ninth to the 16th century. Near the temple was a great walled city, Angkor Thom, with other superb edifices. The wooden parts of these structures had long since rotted away, but the walls, floors, stairs, towers and works of sculpture still stood virtually untouched by time. They were built of sandstone and the extraordinarily durable material known as laterite.

Laterite is a mineral-rich earth that, when exposed to air, turns into a brick-like form of rock (its name comes from the Latin word for brick). It has been an important building material since prehistoric times. Ancient roads were constructed of laterite, and it is still used for highways in parts of southeast Asia and Africa. In modern Thailand many public buildings are built at least in part of laterite. And many communities in India and Africa still rely on laterite, as early man did, as a source of iron.

Paradoxically this interesting and useful material may have been one of the principal reasons for the disappearance of the Khmer civilization that built the city of Angkor Thom. Laterites and lateritic soils are disastrous handicaps to agriculture. Today they are known to be major obstacles to the development of many of the underdeveloped countries. Because of laterite, attempts to grow more food in the Tropics may turn much of that region of the earth into wasteland. Laterite is a grave danger to projects such as the flood-control program for the Mekong Valley of southeast Asia [see "The Mekong River Plan," by Gilbert F. White; SCIENTIFIC AMERICAN, April, 1963]. Unless the laterite problem is dealt with, flood control might actually reduce, instead of improve, the food productivity of such areas.

Let us look more closely at this material. It has been a subject of much controversy among geologists and soil scientists. The two groups define laterite somewhat differently. The geologist thinks of laterites primarily as rock or earth aggregates with a high content of iron, aluminum, nickel or manganese. The soil scientist is concerned with the minerals as components of the soil, particularly with the way their role in the soil is affected by weathering and leaching. This article will discuss mainly the "laterization" of soil.

A lateritic soil is rich in iron and aluminum, low in silica and chemically acidic. It is usually red or yellow—a reflection of its high iron and/or aluminum content. Laterized soils occur most commonly in the tropical belt between the latitudes of 30 degrees North and 30 degrees South. High temperature and heavy rainfall, at least during part of the year, are basic causes of laterization.

The nature of a lateritic soil can be seen in its profile, or cross section. All soils consist of distinct layers, and soil scientists generally describe them in terms of three main "horizons," labeled *A*, *B* and *C*. In ordinary soils the top layer, horizon *A*, usually contains considerable amounts of organic material, silica, bases and undecomposed minerals; horizon *B* holds an accumulation of material that has leached down from *A*, and horizon *C* is composed of transitional parent rock that is in the process of breaking down into soil through physical and chemical weathering. Lateritic soil shows a radically different picture [*see bottom illustration on page 224*]. Most of the organic material has been broken down and leached out of horizon *A*; the silica and bases also are leached away, and the layer is largely depleted of potassium, calcium, phosphorus and other elements required by plants. The result is that the *A* horizon is composed in large part of oxides of iron, aluminum and other minerals. Below this, the *B* horizon is often either thin or completely missing, and the *C* horizon also may have failed to develop. The soil is so porous that most of the decomposed material has been washed away.

Laterization is a function of the soil climate, which in turn is closely related to the atmospheric climate. The thorough leaching of the soil is primarily due to heavy rainfall, but other tropical conditions play their part. The dampness and high temperature combine to produce a luxuriant growth of bacteria, insects, earthworms and other organisms that break down the organic material and also aerate the soil. The oxygen of the air, permeating this porous soil, oxidizes its iron and aluminum. (It is fortunate that laterization is almost entirely limited to the high-rainfall Tropics and even there is held in check by protective vegetation; if it were not, the earth's atmosphere would soon be denuded of oxygen. All the oxygen in the atmosphere would be used up if only a small percentage of the ferrous iron estimated to be in the earth's rocks were oxidized to ferric iron.)

As a result of laterization vast areas of the earth's soil have been converted into deposits of bauxite (aluminum ore) and into hematite and limonite ores of iron. At the same time laterization has also operated to reduce the high-rainfall tropical regions to near-desert conditions from the standpoint of the agricultural quality of their soil. At first thought this generalization may seem unbelievable. Are not the lush jungles, rain forests and savannas of the Tropics plain signs of the fertility of their soil? Actually this lushness is deceptive; it is created only by the abundance of moisture and belies an essential poverty of the soil. Even soil scientists have not found it easy to accept this conclu-

sion, but there is now abundant proof of it. That the tropical forests and grasslands cover some of the earth's most inhospitable and unproductive soils has been demonstrated by attempts to wrest cultivated crops from them. Put to the plow, these lands yield an amazingly small return and soon become completely infertile, as we shall see. Indeed, once a lateritic field has been laid bare to the air, it may even harden into stony laterite such as the brick of the temples of Angkor Thom.

Soil laterization has been taking place

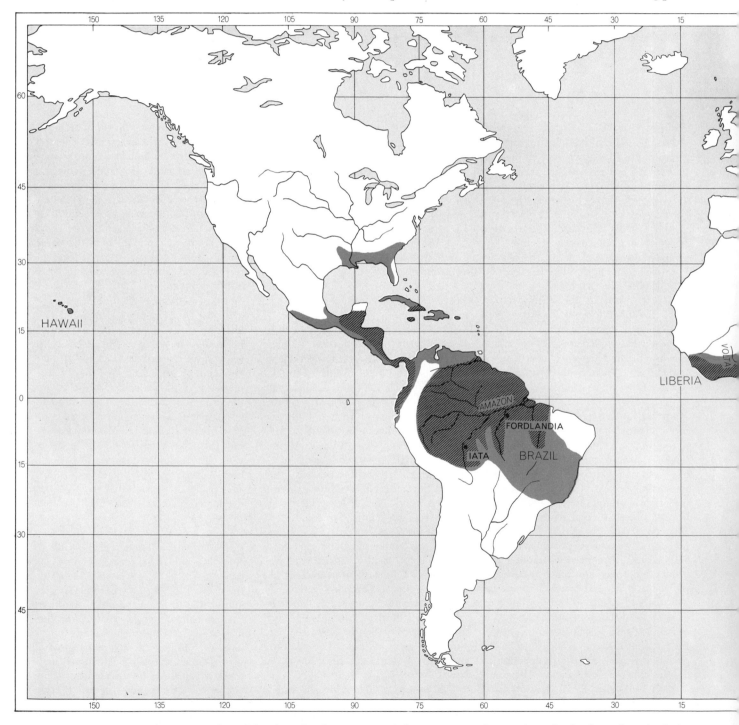

EXTENT OF LATERIZATION is indicated by the colored areas on this map. Laterization is mostly confined to tropical and sub-tropical regions. Rain forests, shown by hatching, deter hardening of laterite by insulating the soil somewhat from the effects of tropi-

at the earth's surface since the Paleozoic era at least, and probably throughout geologic time. In the Amazon basin of Brazil there are accumulations of lateritic soil 70 feet thick. Deposits such as these are truly fossil soils that provide a record of the soil's various stages of evolution and of the changes in climate, vegetation, topography and geologic processes that took place during their history. In the Brazilian profile we can see all the phases of development of laterite, from its origin as a soil from the parent rock to its final transformation into the vast deposits of bauxite, manganese, iron and new rock that now cover about 1,000 square miles of the basin.

Laterite deposits have been found as far north as Ireland, indicating that it once had a more tropical climate. In many areas of the world—South America, Central America, Africa, Australia, India, southeast Asia—the strata of exposed hillsides show layers of laterite capping various types of underlying rock (igneous, metamorphic and sedimentary). In most cases it looks as if the lateritic soil or laterite once covered a great plain or basin and sections of the deposit were later raised by uplifts of the earth's surface that formed hills and plateaus.

Whatever the details of its history may be, it is clear that the laterization of the soil throughout the tropical belt is still taking place and that the intervention of man now threatens to accelerate the process on a large scale. The ambitious plans to increase food production in the Tropics to meet the pressure of

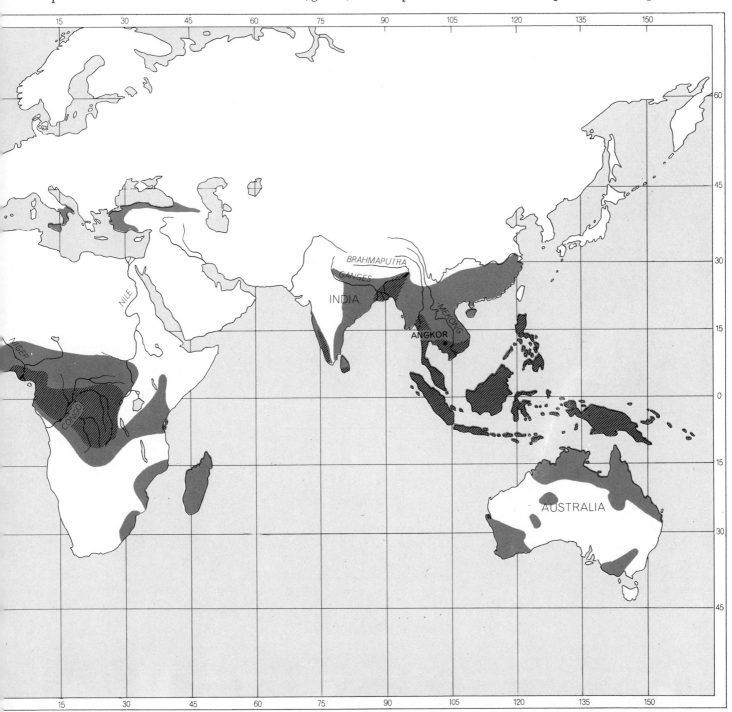

cal climate. Removal of forests, as in efforts to expand agricultural production, tends to quicken laterization, which in turn impairs agriculture. Such an evolution has occurred at Iata in Brazil, where the government undertook to establish an agricultural colony.

the rapid rise in population have given too little consideration to the laterization problem and the measures that will have to be undertaken to overcome it.

In the past nature has provided a measure of control over the process by virtue of forest and jungle growth, which tends to insulate the soil from the eroding effects of the tropical climate and thereby slows down the soil's degeneration. Some of the development plans in underdeveloped areas now call for removing that protective cover to clear the land for agriculture. A recent venture in Brazil vividly illustrates the possible results.

At Iata, an equatorial wonderland in the heart of the Amazon basin, the Brazilian government set up an agricultural colony. Earthmoving machinery wrenched a clearing from the forest and crops were planted. From the beginning there were ominous signs of the presence of laterite. Blocks of ironstone stood out on the surface in some places; in others nodules of the laterite lay just below a thin layer of soil. What had appeared to be a rich soil, with a promising cover of humus, disintegrated after the first or second planting. Under the

equatorial sun the iron-rich soil began to bake into brick. In less than five years the cleared fields became virtually pavements of rock. Today Iata is a drab, despairing colony that testifies eloquently to the formidable problem laterite presents throughout the Tropics.

The small country of Dahomey, adjoining Nigeria in tropical West Africa, had a similar experience on a wholesale scale. There the replacement of forests by plantations resulted in deep leaching of the soil and converted large areas into brick in about 60 years. In an equatorial rain forest there is little growth of vegetation on the dark forest floor, and humus fails to accumulate in the soil. Small clearings in such a forest for the "milpa" type of farming, common among forest people all over the world, will exhaust the soil within a year or two. After the clearing is abandoned a jungle-type growth of shrubs, vines and low trees may take over. This happens, however, only where the clearings are comparatively small. Large areas that have been cleared for plantation cultivation are often permanently lost to agriculture after a few crop cycles have worn out the soil.

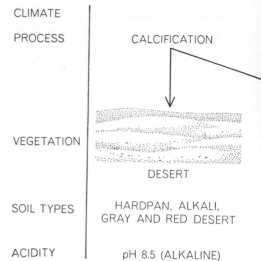

CLIMATE	
PROCESS	CALCIFICATION
VEGETATION	
	DESERT
SOIL TYPES	HARDPAN, ALKALI, GRAY AND RED DESERT
ACIDITY	pH 8.5 (ALKALINE)

SPECTRUM OF SOILS ranges from dry to wet and alkaline to acidic. The method of formation is indicated by the arrows; the

Going back further in history one can see how important a part laterite must have played in the economies of ancient civilizations in the Tropics. The Khmer civilization in Cambodia may well have perished primarily because of the poverty of the lateritic soil. In

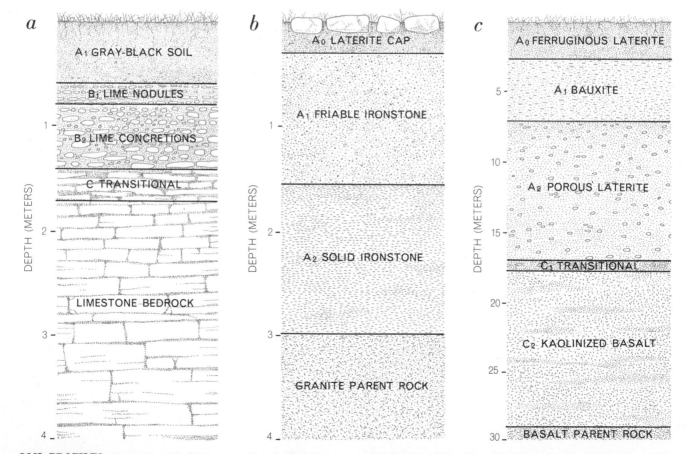

a A₁ GRAY-BLACK SOIL — B₁ LIME NODULES — B₂ LIME CONCRETIONS — C TRANSITIONAL — LIMESTONE BEDROCK — DEPTH (METERS) 1, 2, 3, 4

b A₀ LATERITE CAP — A₁ FRIABLE IRONSTONE — A₂ SOLID IRONSTONE — GRANITE PARENT ROCK — DEPTH (METERS) 1, 2, 3, 4

c A₀ FERRUGINOUS LATERITE — A₁ BAUXITE — A₂ POROUS LATERITE — C₁ TRANSITIONAL — C₂ KAOLINIZED BASALT — BASALT PARENT ROCK — DEPTH (METERS) 5, 10, 15, 20, 25, 30

SOIL PROFILES contrast a typical Temperate Zone soil (*a*), as found in northern India, with lateritic soils as found in Dahomey, West Africa (*b*), and in southern India (*c*). Letters at left identify the principal soil horizons, or layers, as usually classified by soil experts. Temperate Zone soils normally have organic material in horizon *A*; in lateritic soils that material has been leached away.

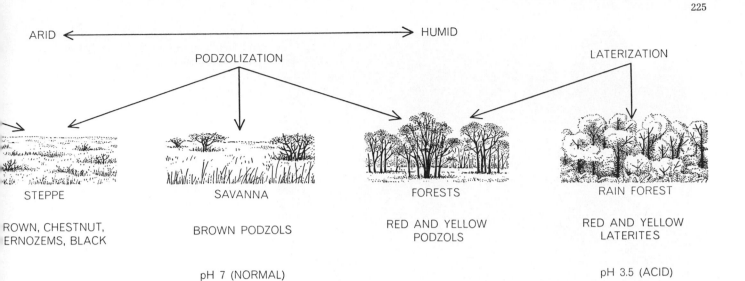

ARID ⟷ HUMID

PODZOLIZATION   LATERIZATION

STEPPE   SAVANNA   FORESTS   RAIN FOREST

ROWN, CHESTNUT,   BROWN PODZOLS   RED AND YELLOW   RED AND YELLOW
ERNOZEMS, BLACK   PODZOLS   LATERITES

pH 7 (NORMAL)   pH 3.5 (ACID)

characteristic soil of steppes, for example, is produced through a combination of calcification, which is an accumulation of hardened carbonates, and podzolization, which is a leaching of the upper layers. Lateritic soils usually lack a hardened layer of accumulated carbonates. Weathering under certain conditions of soil climate, which are found in the Tropics, produces lateritic soil.

Central America the Mayas, contemporaries of the Khmers, who depended greatly on the milpa system of agriculture, were forced to abandon their cities and move north into Mexico; perhaps the reason was the low productivity of the lateritic soil in their old kingdom.

In more modern times the British, French and Dutch colonial empires managed to maintain plantation economies in the Tropics by careful attention to the needs of the soil on an empirical basis. From experience their colonial experts learned to provide the necessary fertilizers, to rotate crops and to move the plantations to new sites every few years. Unfortunately, when they gave up their colonies, the experts went home and the newly independent nations were left with few trained people who knew how to deal with the soil. Cuba is a classic example of the necessity for careful and knowledgeable soil management. Long dependent on its great sugar plantations, the island must contend with a lateritic soil of essentially low fertility that will not produce much more than two successive stands of cane on a given tract. The yields of its plantations will steadily decline unless it finds ways to conserve the island's soil.

The advanced nations now concerned with helping the underdeveloped nations of the Tropics must give serious thought to the laterite problem. If Peace Corps workers attempt to apply the agricultural methods of the U.S. corn belt to Nigeria, or Soviet agriculturists transport the methods of the Ukraine to Cuba, they may well precipitate disas-

ter. Deep plowing of the lateritic soil would probably accelerate leaching and strip the soil of all productivity in short order. The opening up of vast tracts to cultivation, in order to make efficient use of tractors and other modern farming machinery, might lead quickly to the baking of these large expanses into brick by exposing the soil to the action of the sun and wind. And the large river-valley plans now projected for many acres—the Mekong River system of southeast Asia, the Amazon basin of Brazil, the Ganges and Brahmaputra valleys of India, the Niger River of Nigeria, the Volta River of Ghana—might lead to devastation of a more subtle kind.

The Mekong Valley has managed, in spite of conditions strongly favoring laterization of the soil, to sustain a productive agriculture: it is part of Asia's famous "rice bowl." The situation must be credited to an act of God, however, rather than to man's efforts. Each year the growing of crops depletes the highly leached soil of this rainy valley. But then in the monsoon season the overflowing rivers of the system flood the land and replenish it with a new layer of silt. Thus nature continues to renew the soil year after year and keep it reasonably fertile. The designers of the Mekong River Plan must now consider what will happen when they stop the annual floods. Plainly they will have to find a substitute for nature's annual replenishment of the soil. The same is true of most of the other tropical river-control projects, in India, Africa, South America and elsewhere.

Nigeria is a particularly good example of the kind of challenge laterite pre-

sents. Northern Nigeria, the most densely populated part of the country, is a high plateau; many geologists believe it is a remnant of an uplifted plain that covered most of Africa in Tertiary or early Quaternary times. Much of the Nigerian plateau now contains a cap of laterite, which is covered with a thin veneer of soil in some places and shows outcrops of ironstone in others. Only in the valleys of streams and other eroded beds are there deposits of soil sufficiently fertile to support intensive agriculture. The southern part of Nigeria likewise does not look very promising for the large-scale growing of crops. That area consists mainly of a tropical rain forest that could not be turned into agricultural land without destroying its productivity.

Yet Nigeria will have to depend basically on agricultural exports to raise the capital for the development of industry and a better standard of living. Its main hope seems to lie in control of the Niger River and its delta in such a way that a bed of alluvium will be built up. With careful management and the addition of necessary minerals, the reclaimed land could become a fertile agricultural bowl.

Ghana is even more handicapped agriculturally than Nigeria. Thousands of square miles of its area are covered with an almost continuous sheet of laterite, much of it in the form of bauxite. Like Nigeria, Ghana has a large river, the Volta, which by careful management might produce an arable basin for the growing of food.

Aside from efforts to develop a more arable soil, the tropical forests them-

LATERITE CONSTRUCTION appears in the temple of Angkor Wat near Angkor Thom, a major city of the ancient Khmer civilization in what is now Cambodia. Laterite on exposure to air turns into a bricklike form of rock still widely used for construction.

selves could be exploited more effectively than they have been, in Africa and elsewhere. Several countries have shown that cacao trees (which yield cocoa and chocolate) can be a most profitable crop; the planting of teak in the monsoon forests of Asia has been extraordinarily successful; Fordlandia in Brazil and the Firestone Rubber Company in Liberia are now carrying on experiments on the possibility of developing profitable rubber-tree plantations in the equatorial forests. It has also been urged that the savannas of the Tropics, whose lateritic soils would quickly deteriorate if plowed up for cash crops, could be turned by careful management and fertilization into large ranches and pastures for meat animals.

Laterites are not, of course, an unmitigated evil. In the form of bauxite and other economically exploitable metal ores they are a valuable natural resource. In a world that is increasingly concerned about the problem of feeding the multiplying human population, however, it is time to give intensive consideration to how to prevent the laterization of the soil from becoming a major liability.

A generation ago an American geologist, T. H. Holland, remarked that "laterization might be added to the long list of tropical diseases from which not even the rocks are safe." It is no longer a disease of minor proportions. Nor will this disease be easy to cure. The campaign against it will have to include the mapping of the world's laterized areas, research and experiments in the reclamation of lateritic soils and application of the knowledge that is already available, with the United Nations taking the lead in extending this information and help to the tropical countries. There is no single, simple formula for handling the problem of lateritic soil; each situation has to be studied and treated with an individual prescription. The encouraging fact is that the strategy and technology of agriculture have attained a high level of capability in dealing with difficult problems. What can be accomplished in one such area—where the problem is aridity rather than lateritic soil—is beautifully demonstrated by the flowering of the Negev desert in Israel.

20

Mars from *Mariner 9*

INTRODUCTION

Exciting as the moon has been for geological exploration in the past few years, many scientists have been waiting more eagerly for Mars, the planet that is closest and in some ways most similar to earth of all the terrestrial planets. The moon's history is read dominantly in terms of igneous rock petrology and meteorite impacts, much of our knowledge coming from astronauts' explorations and the sophisticated analysis of rock samples returned to earth. Mars's evolution so far has had to be read from the results of unmanned *Mariner* flights, especially from the detailed photographs of *Mariner 9*. So, for this planet, our knowledge is based on deductions from surface morphology and spectroscopy of the atmosphere, and reflected light from surface materials. The evolution of the surface is to be read in terms of weathering and erosion, and transportation and sedimentation by fluids.

Some elements of the surface morphology of Mars, such as the impact craters and associated features, are similar to those of the moon. Others—and that was the startling news sent by *Mariner 9*—looked like those of earth: volcanoes, river valleys, and dune fields. Except for size. One volcano (called Nix Olympica at the time this article was written but now renamed Olympus Mons by the International Astronomical Union) is simply enormous, rising at least fifteen kilometers above its surroundings and 500 kilometers wide at its base. The great valley of Coprates dwarfs any canyon on earth. The volcanoes are so similar to those on earth in all essential respects that their origin seems in no doubt. Dune fields are reasonable identifications of the evenly spaced ridges found in a number of regions, for we know that winds blow strongly on Mars. But the size and detailed morphology of the large canyons and many smaller valleys raise many doubts about their origin as river valleys, especially as there is now no running water on Mars—it's too cold.

These and many other perplexing questions about Mars are explored in this article, "Mars from *Mariner 9*," by a geologist who has been part of the Mars exploration team from the beginning. Bruce Murray has coupled his training and experience as an earth geologist with an ability to think imaginatively about how differently familiar geological processes may operate on other planets. To be a planetologist, one must, like Murray, have an understanding of many fields in geology, ranging from volcanism and igneous processes, through geomorphology and sedimentology, to structural geology and geophysics, not to mention a knowledge of atmospheric composition and dynamics as well as solar system astronomy. All of this knowledge is brought to bear on the puzzling questions of Mars's history—and especially whether there was ever any liquid water flowing in streams on the Martian surface. For if there were, there also may have been (or may be) life in some form. That is one of the many questions that the planned future explorations of Mars will try to answer.

Mars from Mariner 9

by Bruce C. Murray
January 1973

*The first spacecraft to go into orbit around another
planet provides evidence that Mars is just beginning to
heat up internally. Systems of channels and gullies
suggest erosion by water or some other agent*

A little more than a year ago, in November, 1971, the complex robot spacecraft *Mariner 9* fired its braking rocket and was captured in orbit around Mars, thus becoming the first man-made satellite of another planet. From its orbital station, ranging between 1,650 and 17,100 kilometers (1,025 and 10,610 miles) above the surface, *Mariner 9* started sending back to the earth a steady stream of pictures and scientific information that was to continue for nearly a year. By the time its instruments had been turned off *Mariner 9* had provided about 100 times the amount of information accumulated by all previous flights to Mars. It had also decisively changed man's view of the planet that generations of astronomers and fiction writers had thought most closely resembled the earth. As a result of the *Mariner 9* mission it is now possible to make plausible conjectures about the geology of Mars, conjectures comparable, say, to those made about the moon in the early 1960's.

It will be recalled that the first close-up pictures of Mars, made in 1965 by *Mariner 4*, revealed a planetary surface whose principal features were large craters reminiscent of the bleak surface of the moon. Four years later the pictures sent back by *Mariner 6* and *Mariner 7* showed that the Martian surface was not uniformly cratered but had large areas of chaotic terrain unlike anything ever seen on either the earth or the moon. In addition a vast bowl-shaped area, long known to earthbound astronomers as the "desert" Hellas, turned out to be nearly devoid of features down to the resolving power of the Mariners' cameras. None of the pictures returned by the first three Mariners showed any evidence of volcanic activity, leading to the view that Mars was tectonically inactive.

This view has had to be drastically revised in the light of the photographs sent back by *Mariner 9*. The new evidence emerged slowly as the clouds of dust that had shrouded the planet for weeks settled. It revealed, among other things, four large volcanic mountains larger than any such volcanic features on the earth. The *Mariner 9* pictures also show a vast system of canyons, tributary gullies and sinuous channels that look at first glance as if they had been created by flowing water. Elsewhere on the planet's surface there is no suggestion of water erosion. That is probably the major mystery presented by the highly successful mission of *Mariner 9*.

Designed and built by the Jet Propulsion Laboratory of the California Institute of Technology, as were the earlier Mariners, *Mariner 9* was crammed with instruments and electronic gear. After burning 900 pounds of retro-rocket fuel, which it had transported 287 million miles in 167 days, *Mariner 9* weighed 1,350 pounds when it finally went into orbit around Mars. The cameras and instruments it carried were designed by several groups of investigators from Government laboratories and more than a score of universities. The television team, to which I belonged, was headed by Harold Masursky of the U.S. Geological Survey and had nearly 30 members. Somewhat smaller groups were responsible for designing and analyzing data from the ultraviolet spectrometer, the infrared radiometer and the infrared interferometric spectrometer. Other groups had the task of analyzing the trajectory data (which have provided information about the gravitational anomalies of Mars) and the data provided by nearly 100 occultations of the spacecraft's radio signals (which have yielded new knowledge of the planet's atmosphere and surface).

The Old Mars

There were strong reasons for the traditional belief in the resemblance between Mars and the earth. Mars rotates once every 24½ hours and its axis is tipped from the plane of its orbit by almost exactly the same amount as the axis of the earth, thus providing the same basis for the seasonal changes in the amount of solar radiation received by the planet's two hemispheres. Mars has white polar caps, originally thought

CHANGES ON SURFACE OF MARS are depicted in the photograph on the opposite page by a color-difference technique developed by J. A. Cutts of the Jet Propulsion Laboratory and J. J. Rennilson of the California Institute of Technology. It consists of two pictures of the same region in Depressio Hellespontica taken 37 days apart by the wide-angle camera on *Mariner 9*. Both pictures were originally taken through the same filter but at different angles and distances. The two images were rectified by a computer program at J.P.L.'s Image Processing Laboratory. To bring out changes that had taken place on the surface of the planet the earlier picture was printed in green and the later picture was printed in red. When the two pictures are superposed, surface areas that had become darker in the interval between the taking of the two pictures show up with a greenish cast whereas areas that had become lighter appear more reddish. The entire area in the composite pictures corresponds roughly to the smallest area resolvable with earth-based telescopes. Astronomers have observed seasonal changes in the markings of Depressio Hellespontica. Evidently the changes represent an integration of the detailed changes visible in the two *Mariner 9* photographs.

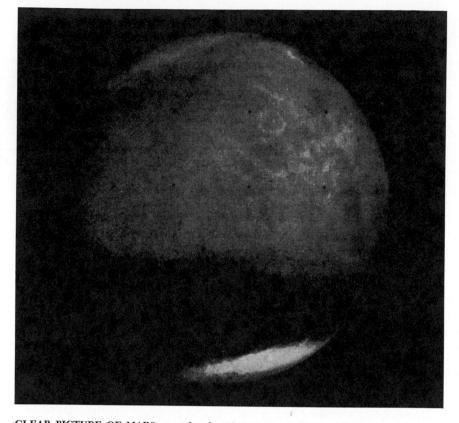

CLEAR PICTURE OF MARS was taken by *Mariner 7* at a distance of 395,249 kilometers as it approached the planet in August, 1969. Since it was then winter in Mars's southern hemisphere the south polar cap was at its maximum size. The north pole was shrouded in haze. The bull's-eye-shaped feature in the upper right quadrant, known as Nix Olympica ("Snows of Olympus"), was thought to be a huge crater. Subsequent closeup photographs taken by *Mariner 9* revealed it to be a gigantic volcanic mountain (*see illustration on page 238*).

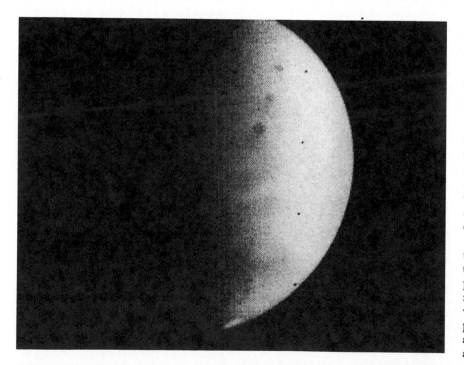

MARS SHROUDED BY DUST was photographed by *Mariner 9* at a distance of 400,000 kilometers, a day and a half before reaching the planet on November 13, 1971. The greatest dust storm in more than a century had obliterated all surface features except for the south polar cap, rapidly shrinking with the approach of spring in the southern hemisphere, and four dark spots in the upper right quadrant. The spot nearest the shadow line is the top of the volcanic mountain Nix Olympica. The other three spots are also volcanic peaks.

to be composed of water, that alternate from one hemisphere to the other once every Martian year (687 earth days). The planet also exhibits dark and light markings that change on a seasonal basis.

Early astronomers speculated that the dark markings might be vegetation. Later and more cautious workers still found it plausible that Mars had had an early history similar to the earth's, which implied the existence of oceans and an atmosphere with enough water vapor to precipitate and erode the surface. Because of Mars's small mass (a tenth the mass of the earth) and low gravity, such an aqueous atmosphere was assumed ultimately to have escaped, leaving the planet in its present arid state. This view of an earthlike Mars strongly influenced proposals for the biological exploration of the planet at the beginning of the space age. It seemed reasonable to suppose that life could have originated on Mars much as it had on the earth, presumably as the result of high concentrations of suitable precursor molecules in primitive oceans. Once life had appeared on Mars, microorganisms, at least, could very well have been able to adapt to changing environmental conditions and so could have survived for discovery and analysis by robot devices launched from the earth.

Such expectations were dampened by the findings of *Mariner 4*. Not only did Mars appear bleak and moonlike but also it was found to lack a magnetic field, which could have shielded its surface against energetic charged particles from the sun. Moreover, Mars's atmospheric pressure was found to be less than 1 percent of the earth's, lower by a factor of at least 10 than had previously been estimated. Since the force of gravity at the surface of Mars is more than a third the force of gravity at the surface of the earth, Mars should have easily been able to hold an atmosphere whose pressure at the surface was a tenth the pressure of the earth's atmosphere at the surface.

Mariner 6 and *Mariner 7* extended these observations. They confirmed that the polar caps are composed of very pure solid carbon dioxide—"dry ice" rather than water ice. The pictures revealing a chaotic terrain suggested that parts of Mars's surface had collapsed and that there had been a certain amount of internal activity. As a result some investigators speculated that the planet might just now be heating up, a circumstance suggested independently by thermal models of the interior. The preponderant view of the Mariner experimenters, however, was still that

Mars was basically more like the moon than like the earth. By then the light and dark markings on Mars seen through telescopes were generally attributed to some kind of atmospheric interaction with dust. Indications that the interaction was controlled by local topography were seen in the second set of Mariner photographs, but no general explanation was deduced. Even so some investigators held to the belief that the markings might instead reflect variations in soil moisture.

The 1971 mission to Mars was origi- nally designed to employ two spacecraft, *Mariner 8* and *Mariner 9,* both of which were to be placed in orbit around the planet. The purpose of the two orbiters was to map most, if not all, of the planet's surface at a resolution high enough to reveal both external and in-

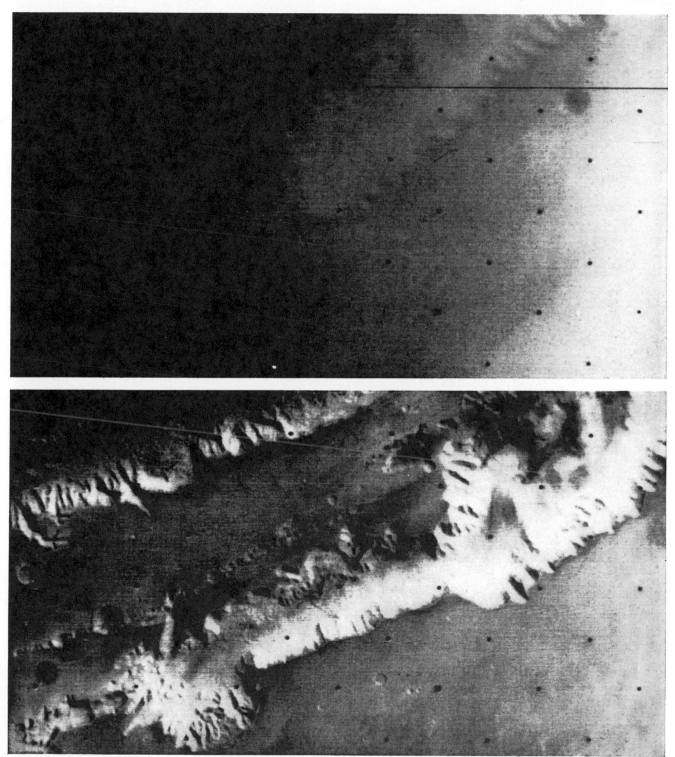

CLEARING OF DUST STORM is shown in these two views of a region in Coprates near Mars's equator. The picture at top, taken when *Mariner 9* had been in orbit 41 days, gives little or no hint of the rugged canyon that was revealed in the bottom picture, taken on *Mariner 9*'s 80th day in orbit. Coprates had long been recog- nized by astronomers as a feature that changes unpredictably in brightness. It now appears that the canyon looks brighter than the surrounding region toward the end of a dust storm, when the can- yon atmosphere is still filled with light-reflecting particles and at- mosphere above surrounding plateau has become largely dust-free.

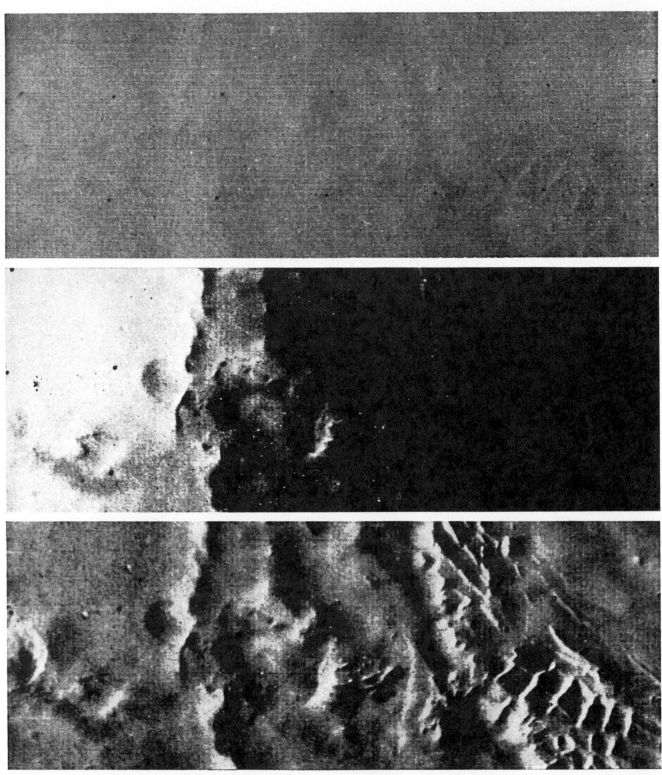

RESULTS OF COMPUTER PROCESSING are demonstrated in these three displays of the same picture showing a region near Mars's south pole. The processing, done within the space of five minutes by J.P.L.'s "mission test computer," was carried out on every picture as it was received from *Mariner 9*. Each image was transmitted as a coded radio signal in which the brightness of each point in the picture was represented by a sequence of nine binary digits rather than as an intensity level (AM signal) or as a frequency tone (FM signal). Each picture frame consisted of 700 lines made up of 832 picture elements per line. The received signal was recorded on magnetic tape and simultaneously displayed on a cathode ray tube. In the initial presentation (*top*) the picture looks rather gray and featureless because the actual contrast on the surface of Mars is quite low. The first stage of computer processing yields the shading-corrected version (*middle*), which indicates the relative brightness in the scene along with some enhancement. The right-hand portion of the picture is dark because it is beyond the terminator and hidden in shadow. A second level of processing (*bottom*) suppresses large-scale differences in brightness in order to enhance preferentially the topographic detail in both the light and the dark areas. The picture is now seen to reveal a complex of transecting ridges two to five kilometers apart that the Mariner television group nicknamed Inca City. The ridges seem to be composed of a resistant material that has filled cross-cutting fractures and that later has been exposed by erosion of the surrounding material. *Mariner 9* took the picture at a distance of 2,937 kilometers.

ternal processes, to study transient phenomena on the surface and in the atmosphere and to provide reconnaissance over a long enough period (from nine months to a year) to observe seasonal changes in surface markings in the hope of clarifying their origins. When *Mariner 8* was lost during launching, the complementary missions of the two spacecraft had to be combined.

When *Mariner 9* reached Mars on November 13, 1971, the greatest dust storm in more than a century was raging on the planet, almost totally obscuring its surface. The first views from a distance of several hundred thousand miles revealed essentially no detail except a glimpse of the south polar cap [*see bottom illustration on page 230*]. The dust storm delighted the investigators who wanted to study the planet's atmosphere, since it promised to reveal how particles were transported by such a thin medium, but it was a disappointment to the investigators concerned with surface features. For example, there had been plans to take a sequence of far-encounter pictures, ultimately to be printed in color, showing the planet getting larger and larger, thereby providing a visual bridge between the level of detail seen through telescopes from the earth and the detail eventually visible in pictures taken from orbit around Mars. A limited effort to produce far-encounter pictures showing Mars in natural color had been made with images taken by *Mariner 7* through separate red, green and blue filters.

The Great Volcanoes

The dust storm delayed the systematic mapping of the Martian surface for nearly three months. Even during the storm, however, four dark spots in the equatorial area were repeatedly seen in the early pictures taken from orbit. The spots clearly represented permanent surface features high enough to stick up through the dust. Presumably they looked dark simply because their surface reflectivity was lower than that of the bright, dusty atmosphere.

One of the four spots corresponded to the location of Nix Olympica ("Snows of Olympus"), so named because it was normally visible from the earth as a bright feature and also a variable one. When this dark spot was observed with the high-resolution, or narrow-angle, camera on *Mariner 9*, the image that emerged was breathtaking. What one saw was the characteristic pattern of coalesced craters that constitute a volcanic caldera. Such calderas are not uncommon on the earth, for example in

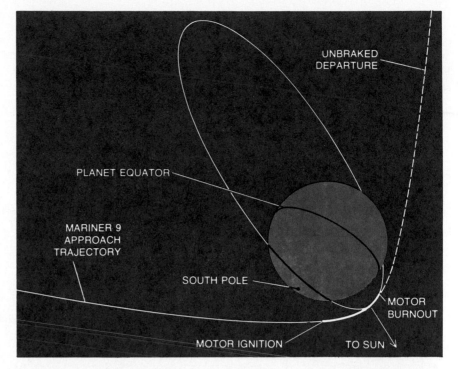

CAPTURE OF *MARINER 9* BY MARS took place on November 13, 1971, after the spacecraft's retro-rocket had generated a decelerating thrust of about 300 pounds for 15 minutes 15.6 seconds. The 287-million-mile voyage from the earth had taken 167 days. *Mariner 9* approached Mars from below and went into an orbit inclined at an angle of about 64 degrees to the planet's equator. The initial orbit ranged from 1,385 kilometers to 17,300 above the planet's surface. A subsequent correction reduced the orbit's high point to 17,100 kilometers and raised the low point to 1,650, achieving the desired orbital period of 11.98 hours.

the Hawaiian Islands. The Martian caldera, however, was 30 times larger in diameter than any in the Hawaiian chain. When the dust had settled, Nix Olympica was seen in full to be an enormous volcanic mountain more than 500 kilometers in diameter at the base,

much larger than any similar feature on the earth; the caldera was only the summit [*see illustration on page 238*]. Atmospheric-pressure maps made later with the aid of the ultraviolet spectrometer and other techniques show that Nix Olympica is at least 15 kilometers high

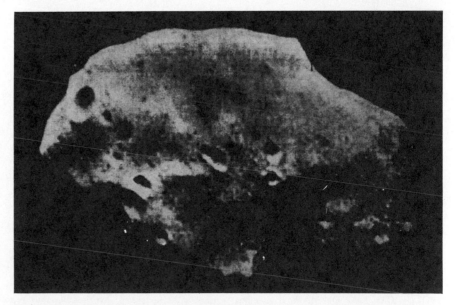

HEAVILY CRATERED MARTIAN MOON, Phobos, was photographed by *Mariner 9* from a distance of 5,500 kilometers. The inner of the two Martian satellites, Phobos is about 25 kilometers long and 20 kilometers wide. Deimos, the outer satellite, is about half the size of Phobos. The picture of Phobos has been greatly improved by computer processing.

and possibly 30. For purposes of comparison, Mauna Loa, the tallest volcanic cone in the Hawaiian Islands, rises less than 10 kilometers from the floor of the Pacific. High-resolution photography revealed that the other three dark spots were also volcanoes, somewhat smaller than Nix Olympica, strung together to form a long volcanic ridge. Following the

traditional name for that area, it is now called Tharsis Ridge [see illustration on these two pages].

The first recognizable features photographed by Mariner 9 presented a fascinating question: How can one explain why one entire hemisphere of the planet, the hemisphere observed by three earlier Mariners, shows scarcely any evidence

of internal activity, whereas the first area to be investigated in detail on the opposite side of the planet has four enormous volcanoes? The explanation apparently is that Mars is just beginning to "boil" inside and produce surface igneous activity. Presumably this process is now well advanced in the Nix Olympica–Tharsis Ridge area but has

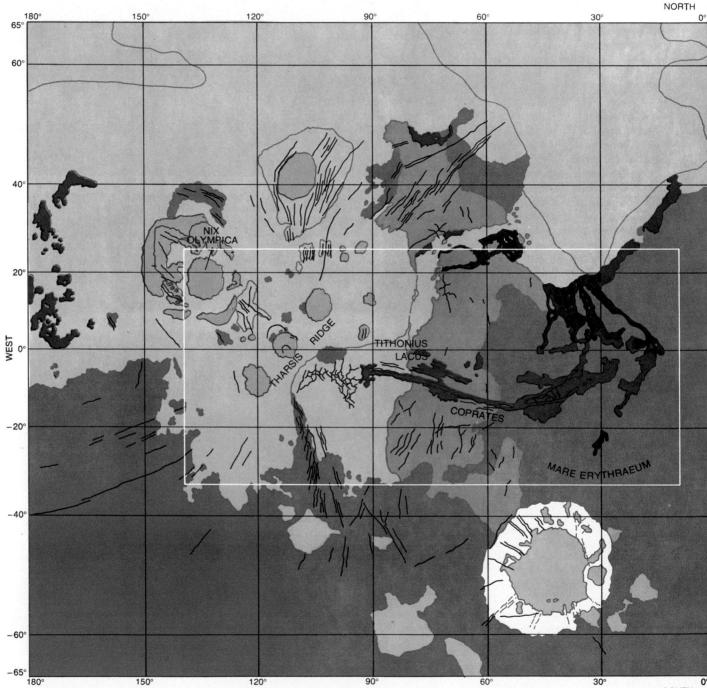

GEOLOGICAL MAP OF MARS is an effort to classify the features that compose the surface terrain. The white rectangle encloses the region depicted in the mosaic on the next two pages. The areas in pale color are smooth plains; areas in medium color are cratered plains; dark-colored areas are old cratered terrain. White areas are mountainous terrain. Areas in light gray are volcanic, embracing Nix Olympica and the three volcanoes that form Tharsis Ridge. Areas in dark gray identify terrain that has been modified in some fashion. Areas in black are channel deposits. Short hatch lines are inferred faults. A number of features referred to in

not yet spread to the planet as a whole. We may be witnessing on Mars a phase similar to one the earth probably went through early in its history, a phase whose record has been totally erased by subsequent igneous and sedimentary processes.

The rate at which a planet's interior heats up depends on a number of factors, chiefly the amount of radioactive material in its original accreted mass and the total mass, which determines the pressure in the interior and the degree of insulation. In very general terms, if Mars had the same original composition as the earth, one would expect it to heat up more slowly because it has only a tenth the mass of the earth. The sheer size of Nix Olympica suggests the possibility that deep convection currents are churning, a process that conceivably could lead some hundreds of millions of years hence to the kind of plate-tectonic phenomena responsible for the slow drift of continents on the earth.

Immediately to the east of the volcanic province is a highly fractured area,

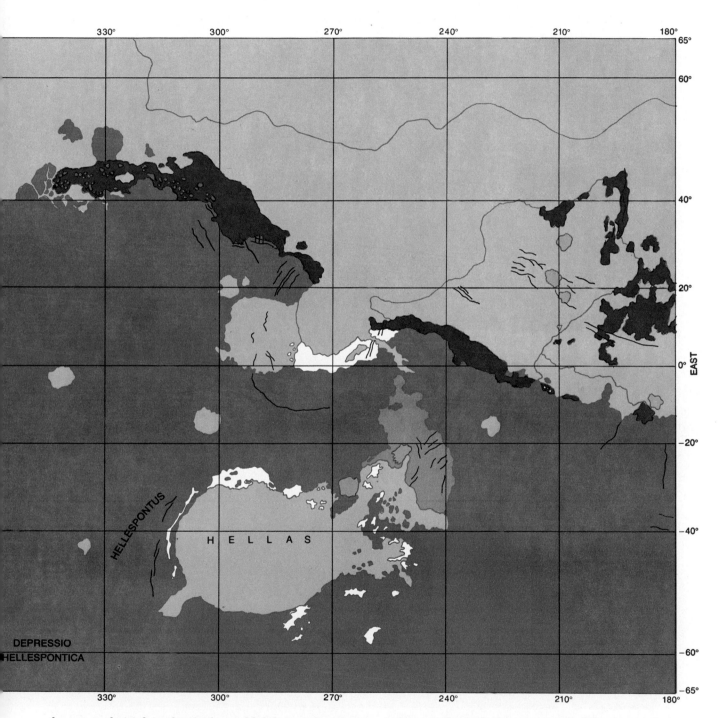

the text or depicted in photographs are labeled. A portion of Depressio Hellespontica is shown in the color photograph on page 228. A canyon in Coprates appears on page 231. A canyon in Tithonius Lacus appears at the top of page 239 and a crater in Hellespontus at the bottom of the same page. A sinuous valley in Mare Erythraeum is shown in the photograph on page 241. Hellas is a nearly featureless bowl more than 1,600 kilometers across. The map was prepared by Michael Carr, John F. McCauley, Daniel Milton and Don Wilhelms of the U.S. Geological Survey in cooperation with several members of television-experiment team of *Mariner 9* project.

and beyond that another extraordinary topographical feature was discovered: a series of huge canyons stretching east and west along the equator [*see illustration on these two pages*]. These canyons, 80 to 120 kilometers wide and five to six and a half kilometers deep, are much larger than any found on the earth. Again we must assume that their origin is due to fairly recent internal activity. Presumably large-scale east-west faulting has exposed underlying layers of the planet whose composition could con- ceivably trigger an erosion process of some kind.

One speculation is that deep permafrost is involved, associated perhaps with the arrival near the surface of juvenile water preceding and accompanying the rise of molten rock near the surface of the planet during the volcanic episode apparent to the west. Mars is everywhere below freezing just a short distance below the surface. Once permafrost was exposed to the atmosphere, its water content would sublimate, making avail- able a loose, friable material sufficiently mobile to serve as an eroding agent in a mass-transport process. We must then ask where the material went. One possibility is that the winds of Mars have transported it as dust to other localities. (Although the Martian atmosphere is thin, its winds may blow at several hundred miles per hour.) Alternatively, the missing material may yet be discovered somewhere to the east of the canyons. Still a third possibility is that it may even have disappeared into the planet's

MOSAIC PANORAMA was made from several dozen *Mariner 9* wide-angle photographs specially computer-processed and matched by Raymond Batson of the U.S. Geological Survey at Flagstaff, Ariz. The mosaic depicts a region extending some 7,000 kilometers along the Martian equator. The area covered by the panorama is indicated on the geological map of Mars on the preceding two pages. The superposed outline map of the U.S. shows vividly the dimensions of the great canyons that parallel the equator for 4,000 kilo-

interior in a complex exchange process.

The largest of the canyons corresponds to a feature long known as Coprates, whose appearance sometimes changes with the seasons. By observing this canyon as the dust storm ended we were able to gain insight into its variable appearance. The canyon is so deep that considerable dust persists in the atmosphere between the canyon walls after the atmosphere above the surrounding region is comparatively dust-free [*see illustration on page 231*]. The dust-filled atmosphere makes the canyon look brighter than the surrounding landscape. Once the canyon atmosphere clears up, there is little contrast between the interior of the canyon and the surrounding area. Hence the variable "surface" markings associated with Coprates probably have nothing to do with the surface at all.

Similar atmospheric processes may well explain some of the other variable markings formerly attributed to seasonal changes on the surface. Other kinds of variation are not so simply explained, but evidently they always involve the interaction of dust, topography and atmosphere.

Like the volcanoes, the great canyons of Mars suggest a fairly recent episode in the history of the planet characterized by large-scale events. On the earth one often finds a reasonably steady state between processes of erosion and processes of restoration; thus one sees a range of surface morphologies from youthful to mature. In the case of the Martian can-

meters. Much larger than anything like them on the earth, the canyons average 100 kilometers in width and reach a depth of more than six kilometers. At the extreme left of the mosaic one can see the giant volcanic mountain Nix Olympica and immediately to the right the series of somewhat smaller volcanoes that form Tharsis Ridge. The sun was generally shining from the lower left when the pictures were taken, so that the shadows fall toward the upper right. For some, relief will be stronger if mosaic is turned upside down.

yons erosion does not seem to be balanced by a corresponding restoration; we do not see old degraded canyons with a mature form.

The Channels

The eastern extremity of the canyons joins a large area of chaotic terrain, a small portion of which was glimpsed by *Mariner 6*. The appearance of the chaotic terrain strongly suggests that it is the result of some kind of collapse and that the collapse is genetically related to the canyons to the west. Extending out from the chaotic terrain in a north-westerly direction are some extraordinary channels, which are also found in a number of other localities on the planet. It is hard to look at these channels without considering the possibility that they were cut by flowing water. Indeed, some of my colleagues think that is the only reasonable explanation.

CLOSE-UP OF NIX OLYMPICA is shown in a mosaic of *Mariner 9* photographs that have been specially computer-processed and matched. The picture is printed with north at the right, so that the sunlight seems to come from the top of the page; the feature is thus seen immediately as a cone and not as a depression. The volcanic mountain, rising from a great plain, is 500 kilometers across at the base, or much larger than similar volcanic mountains on the earth. Pressure mapping of the depth of the atmosphere indicates that Nix Olympica is 25 kilometers high. The main crater, a complex volcanic vent known as a caldera, is 65 kilometers in diameter.

One can estimate the age of the channels by noting the size-to-frequency relation of impact craters on their floors. The channels are clearly younger than the crater-pocked terrain seen over much of the planet, yet they are by no means the youngest features of the Martian landscape.

The discovery of the channels has revived speculation that there may have been an earthlike epoch in the history of Mars. According to this view Mars may once have had a much denser atmosphere and water vapor in such abundance that rain could fall. Given rainfall, the channels could be easily explained. Less easily explained is why channels have survived in only a few areas and why older topography shows no evidence of water erosion. It would seem difficult to explain how the primitive Martian atmosphere, probably dry and reducing (in the chemical sense), could have evolved into a dense, wet one and then have been transformed again into the present thin, dry atmosphere consisting almost entirely of carbon dioxide. Moreover, the present atmosphere is strongly stabilized by the large amounts of solid carbon dioxide in the polar regions. If the channels were created by rainfall, it would seem that one must postulate two miracles in series: one to create the earthlike atmosphere for a relatively brief epoch and another to destroy it.

An alternative hypothesis presents at least as many difficulties. It is suggested that liquid water accumulated in underground reservoirs following entrapment and melting of permafrost. Hypothetically the reservoirs were abruptly breached, allowing the released water to create the channels. The observed channels are so large and deep, however, that a great volume of water must have been involved in their formation. Therefore it would seem even more difficult to ascribe the channels to a "one shot" open-cycle process than to a closed-cycle process such as rainfall.

The origin of the canyons and channels is one of the primary enigmas that has emerged from the *Mariner 9* mission. Because of the importance of liquid water to life as we know it, the possible role of water in creating the canyons and channels has attracted particular interest.

Finally, there are a few areas on Mars to which the term "basins" seems appropriate. The most prominent is the large circular feature Hellas, more than 1,600 kilometers in diameter. Hellas has been observed from the earth for more than two centuries. Sometimes it rivals

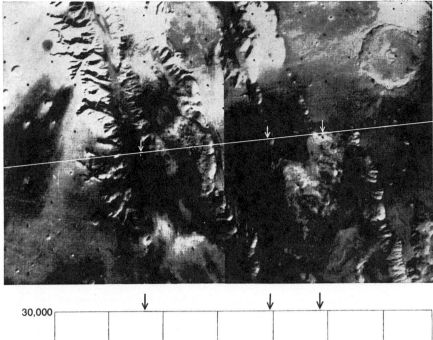

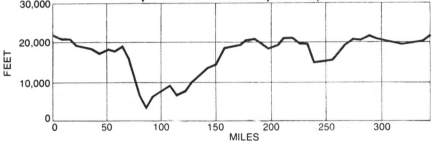

CHASM IN TITHONIUS LACUS is in the extreme western part of the canyon system that runs east and west near Mars's equator. An ultraviolet spectrometer on *Mariner 9* measured the atmospheric pressure at the planet's surface along the track indicated by the white line. The pressure reading was then converted into the jagged depth profile that appears below the pair of photographs. The difference between the highest and lowest points is more than 6,000 meters, making the chasm four times as deep as the Grand Canyon in Arizona. Tithonius Lacus is immediately south of the Martian equator at about 85 degrees west longitude.

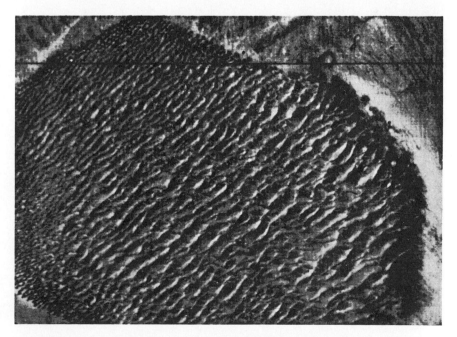

DUNE FIELD some 70 kilometers across was discovered by *Mariner 9*'s narrow-angle camera inside a crater 150 kilometers wide in the region known as Hellespontus. Ridges are one and a half kilometers apart. Dune fields as spectacular as this one appear rare on Mars.

the polar caps in brightness. *Mariner 7* demonstrated that Hellas is indeed a low-lying basin virtually devoid of features. It has been deduced from closeup photographs that the surface of Hellas has probably been smoothed by the influx of large amounts of dust carried into the basin by wind. *Mariner 9*, however, revealed that Hellas exhibited a few faint topographical features just as the planet-wide dust storm was ending. This suggests that variations in the brightness of Hellas may be due to frequent dust storms of a more local nature, a view originally adduced on meteorological grounds by Carl Sagan of Cornell University and his co-workers. Thus Hellas probably acts as a long-term collection basin for dust but may also serve as a source of dust when the Martian winds blow particularly hard. The observation by *Mariner 9* of small-scale dust storms and the recognition that they can alter the brightness of local areas give us further insight into some of the variable features that have been observed over the years from the earth.

One of the crowning achievements of *Mariner 7* was the study at high resolu-

tions of the very large polar cap present during the southern winter of 1969. Measurements of reflectivity and temperature provided by an infrared spectrometer and an infrared radiometer on the spacecraft proved conclusively that the south polar cap was composed of very pure solid carbon dioxide, as had been predicted some years earlier. The photographs showed that the frost cover was thin (probably less than a few meters on the average) and that a variety of unusual surface features were also present in the vicinity of the south pole.

When *Mariner 9* reached Mars, it was late spring in the southern hemisphere, an ideal time for monitoring the wasting of the dry-ice cap and for examining in detail the unusual surface features that should have been further revealed. The disappearance of the south polar cap started out as expected but then clearly showed anomalous behavior. Curiously, the general outline of the shrunken cap persisted throughout the late summer, when the sublimation of the carbon dioxide should have been at a maximum [*see top illustration on page 243*]. This suggested to me that after the

large annual cap of carbon dioxide has sublimated, it exposes a residual cap of ordinary water ice. Ordinary ice, of course, has a much lower evaporation rate than carbon dioxide, and traces of water vapor are present in the Martian atmosphere.

Mariner 9's pictures also disclosed a most peculiar terrain in the south polar area, which we named laminated terrain. Although its outline is not symmetrical, it covers much of the south polar region up to about 70 degrees south latitude. The laminated terrain is composed of very thin layers, alternately light and dark, whose gently sloping faces exhibit a certain amount of texture, or relief [*see bottom illustration on page 243*].

The thin laminas appear to be collected in units of 20 or 30 or more to constitute plates perhaps half a kilometer or more in thickness and up to 200 kilometers across. The plates have outward-facing slopes in which a banded structure can be seen. The laminar deposits have been found only in the polar regions, where carbon dioxide forms an annual deposit of frost. This suggests

BRAIDED CHANNEL adjacent to an impact crater 20 kilometers in diameter is an example of the type of feature that suggests there has been fluid erosion of some kind on Mars. If the agent was actu- ally water, it is hard to understand why eroded terrain is confined to only a tiny fraction of the planet's total surface. The region shown is at six degrees south latitude 150 degrees west longitude.

that the laminations are associated in some way with the coming and going of volatile substances and that they may even retain some solid carbon dioxide or water ice. Since the laminations are marked by very few impact craters, one can deduce that they are a recent development in the history of Mars.

The North Pole

The north polar region of Mars finally became available for observation by *Mariner 9* rather late in the mission as a result of gradual changes in lighting, associated with the change in season and with the lifting of the haze that characteristically develops in the fall over each pole. The quasicircular structures characteristic of laminated terrain were found to be even more abundant around the north pole than around the south pole. One can see 20 or 30 individual plates arranged in a pattern reminiscent of fallen stacks of poker chips. The existence of the laminated terrain and circular-plate structures in the north polar regions as well as in the south polar ones indicates beyond any reasonable

doubt that their formation must be associated in some way with the periodic deposition and evaporation of volatile material.

Michael C. Malin, a graduate student at the California Institute of Technology, and I have speculated that the distribution of the circular plates and their overlapping arrangement can be explained by changes in the tilt of Mars's axis. We posit that the rotational axis of the planet has been displaced over the past tens of millions of years as a result of convection currents deep in the mantle, currents that are probably associated with the production of volcanoes in the equatorial areas. As the spin axis has shifted, the laminated plates have formed concentrically around each successive position of the poles.

This speculation is at least consistent with information about Mars's gravity distribution deduced from changes in *Mariner 9's* orbit. The planet exhibits gravitational anomalies suggestive of deep density differences of the kind that could be associated with deep convection. Moreover, there is a strong correlation between the gravitational anomalies

and the location of the equatorial volcanoes.

The regular appearance of the laminas and of the plates themselves suggests that they are also associated in some way with periodic alternations of the climate of Mars. In collaboration with two other graduate students, William Ward and Sze Yeung, I have investigated the theoretical variations in the orbit of Mars over a period of time. We find that perturbations in the orbit caused by other planets, analyzed a number of years ago by Dirk Brouwer and G. M. Clemence, alter the orbit's eccentricity in a way that turns out to be quite favorable for our hypothesis. The eccentricity of Mars's orbit varies from .004, or nearly circular, to .141. Its present value is .09 [*see top illustration on page 246*].

The consequence of this variation in eccentricity is a variation in the yearly average amount of sunlight reaching the poles of the planet, together with a much stronger variation in the maximum solar flux when the planet is closest to the sun [*see bottom illustration on page 246*]. Although the variation in average radiant

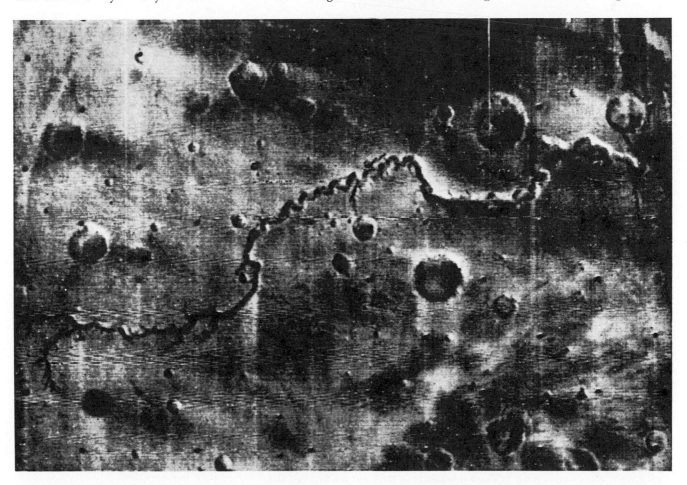

SINUOUS VALLEY 400 kilometers long and up to six kilometers wide is located at 29 degrees south latitude 40 degrees west longitude in the region known as Mare Erythraeum. The feature resembles the outline of a meandering river or one of the lunar rilles, which may have been created by flowing lava. The origin of such features on Mars has not been satisfactorily explained.

input at the poles is only a few percent, it is sufficient under some circumstances to cause a cyclical variation in the growth and sublimation of permanent carbon dioxide frost caps. Assuming that dust storms regularly deposit dust during the sublimation phases, thin laminas of the type observed could be produced. The plates themselves would therefore correspond to a periodicity of the order of two million years. Hence the laminated terrain seems to closely reflect the fluctuations in average radiant flux reaching the planet both in

the short run (roughly 90,000 years) and in the long run (two million years). Inasmuch as a total of 20 or 30 plates are visible in the northern hemisphere the laminated terrain constitutes a record reaching back something like 100 million years. An alternative view regards the origin of the laminated terrain as being primarily erosional rather than constructional.

Evolution of the Atmosphere

If all the polar laminations accumu-

lated in only a few hundred million years at most, representing no more than the past 5 percent of the history of Mars, what happened earlier? One encounters a basic difficulty in understanding Mars if one tries to apply the famous dictum of the 18th-century geologist James Hutton: "The present is the key to the past." Whether one looks at the volcanic terrain, the canyon lands, the channels or the polar laminations, all seem to record a remarkable degree of activity and change during the most recent part of Mars's geological past. I was led by these considerations to wonder if it is possible that the atmosphere of Mars as we know it may be a fairly recent acquisition. Malin and I are presenting this view as a "contentious speculation." It may be that Mars had no atmosphere at all, or only a very thin, unimportant atmosphere throughout the middle period of its history, lasting perhaps several billion years. Presumably there was an initial primitive atmosphere associated with the accretion of the planet, but this atmosphere may have been lost quite early, particularly if it consisted chiefly of hydrogen and methane.

We think a significant fraction of the mass of the present Martian atmosphere was released during the formation of Nix Olympica and the other three volcanoes in the Tharsis Ridge area. The existence of widespread blankets of rock material and other evidence of somewhat earlier and more extensive volcanism and sedimentation suggest that rather large volumes of volatile substances have issued from the planet's interior in the later geological episodes. Thus it may be that as Mars matured enough to boil it simultaneously began to produce an enduring atmosphere. The atmosphere in turn has produced the laminated terrain and provided the wind-transport and erosion mechanisms to form the channels and the great canyons. On this hypothesis Mars is still far from reaching a steady state in which erosion would be balanced by modification, with a resulting development of a spectrum of morphological features. As part of our contentious speculation one might even imagine that in the early stages of the development of the present Martian atmosphere, before the polar cold traps for carbon dioxide were well established, enough water might have been brought to the surface to have flowed down the channels under peculiar, nonrecurring conditions. At least this possibility avoids the problem of positing two miracles in series and lets us settle for just one if ultimately liquid

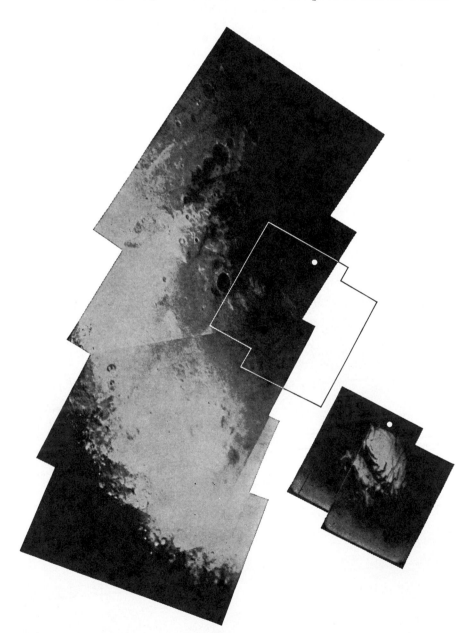

SOUTH POLE IN DIFFERENT SEASONS can be compared in these views taken by *Mariner 7* in August, 1969 (*mosaic at left*), and by *Mariner 9* in November, 1971 (*right*). The area covered by the two *Mariner 9* pictures is indicated by the shape superposed on the *Mariner 7* mosaic. The geometric south pole is shown by a white dot in both views. In August, 1969, it was winter in the southern hemisphere of Mars and the frost cap of dry ice (solid carbon dioxide) was close to its maximum size. The right portion of the mosaic is dark because it is in shadow. When the *Mariner 9* pictures were taken 27 months later, the frost cap was shrinking rapidly with the approach of spring in Mars's southern hemisphere.

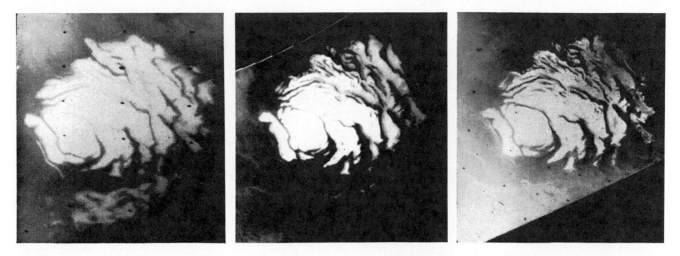

SHRINKING OF SOUTH POLAR CAP is depicted in these three views taken by *Mariner 9*. The original images have been enhanced, stereographically rectified and printed at the same scale. Viewed in the usual order, the pictures show the appearance of the frost cap after the spacecraft had been in orbit 14, 36 and 94 days. The first picture is still somewhat hazy because of the planet-wide dust storm. It can be seen, nevertheless, that the outline of the surface frost changed significantly in the three weeks between the first two views but then changed surprisingly little in the next eight weeks, when sublimation of the frost cap should have been at a maximum.

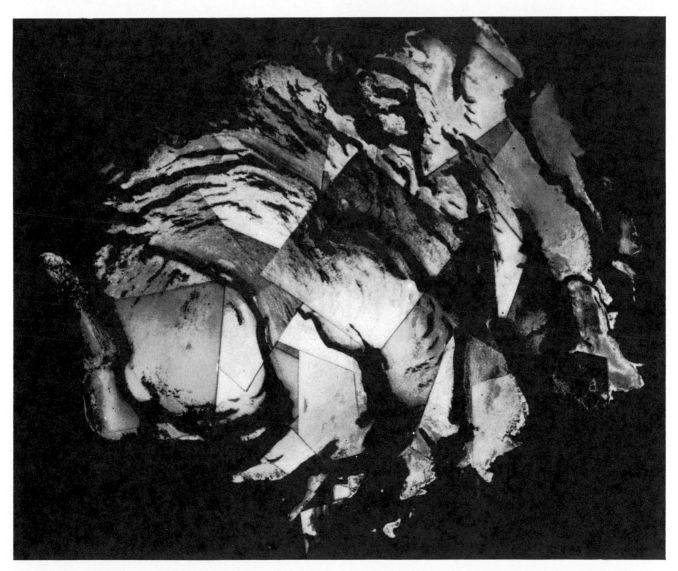

MOSAIC OF RESIDUAL SOUTH POLAR CAP was assembled (by Laurence A. Soderblom of the U.S. Geological Survey Center of Astrogeology in Flagstaff, Ariz.) from closeup pictures taken between the 58th and 94th day of orbital flight. The residual cap, which measures about 300 by 350 kilometers, is centered about 200 kilometers from true south pole (*white dot*) at about 45 degrees west longitude. Shades of gray have been optimized to bring out detail in frost-covered areas, hence frost-free surface appears black.

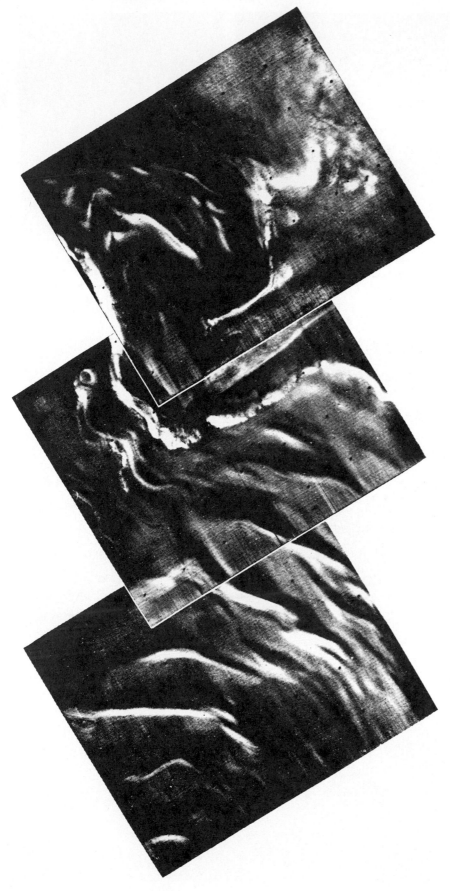

LAMINATED TERRAIN NEAR NORTH POLE resembles that in the south polar region. The author speculates that the distribution of circular plates and their overlapping arrangement may be evidence that the tilt of the axis of Mars has shifted over the past 100 million years. Such a shift could have resulted from convection currents deep in mantle of planet.

water is really required to explain the genesis of the channels.

The young-atmosphere hypothesis may also help to explain why "permanent dark areas" (for example the two-pronged feature Sinus Meridiani) should survive in the face of the frequent planet-wide dust storms. Again Mars somehow does not seem to us to be in a steady state, although others do not share our viewpoint. According to our hypothesis the dark markings may be the site of older surface materials not yet affected by the chemical weathering associated with the new atmosphere. In fact, there seems to be some correlation between the permanent dark markings and the terrain populated by the oldest craters.

Other *Mariner* 9 investigators, such as Sagan and W. K. Hartmann, have developed a quite different view of Mars's history. The nature of the old cratered terrains suggests to them that a long period of atmospheric erosion preceded the spectacular events of the more recent past. Thus the concept of an earthlike Mars is not by any means dead. Nonetheless, concepts of the geological history of Mars are changing rapidly in the light of *Mariner* 9's highly successful mission. Perhaps ultimately some intermediate interpretation will fit the observations best.

Is There Life on Mars?

The present *Mariner* 9 results suggest to me, however, a view very different from that of the early astronomers who thought that Mars was once earthlike and is now a dried-up fossil. I would argue that Mars is probably just now starting to become earthlike with the development of a durable atmosphere. "Just now" is hard to pin down quantitatively because the dates assigned to the craters on the basis of meteor flux rates are still highly uncertain. My guess now would be that the atmospheric "event," if it really happened, took place within the past quarter of Mars's history and certainly within the past half. If this contentious speculation should become widely accepted, it would necessarily imply pessimism about the possibility that past conditions were favorable for the appearance of simple forms of life on Mars. If Mars indeed was like the moon and lacked a significant atmosphere for much of its history, and if the maximum amount of water on the surface has been at most enough to create a few channels, it seems highly unlikely that there has ever been a sufficient accumulation of liquid water in the surface layers of Mars to allow the

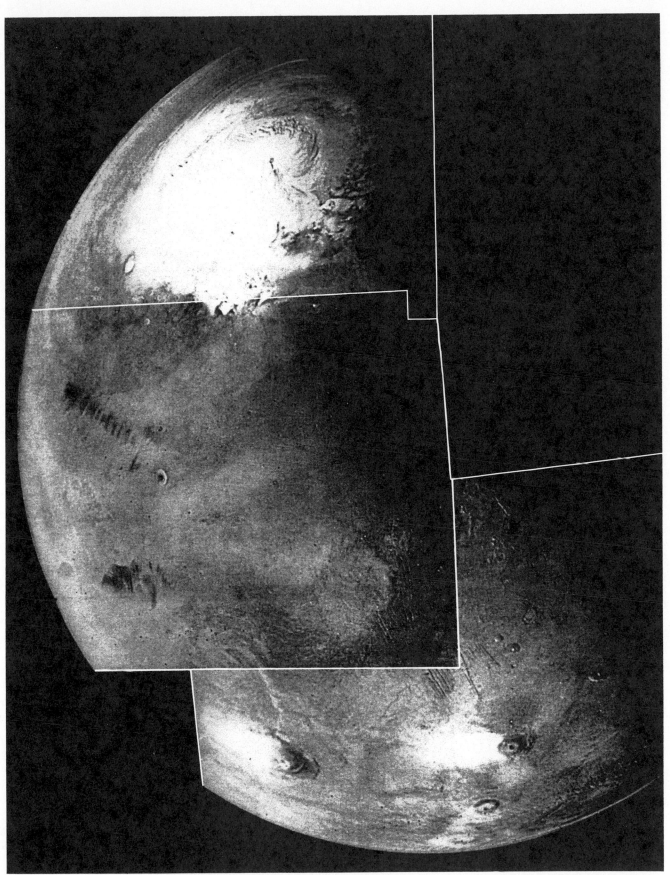

NORTHERN HEMISPHERE OF MARS was photographed in three frames taken only 84 seconds apart by swiveling the wide-angle camera aboard *Mariner 9*. The bottom frame clearly shows Nix Olympica, the volcanoes of Tharsis Ridge and at the lower right the huge canyon that lies just below the equator. The pictures, taken on August 7, 1972, at an altitude of 13,700 kilometers, were among the last of the 7,273 produced by *Mariner 9*'s two cameras. Clouds of water ice or of carbon dioxide crystals obscured the planet north of the 50th parallel until the final weeks of the mission. The north polar cap is shrinking during the late Martian spring.

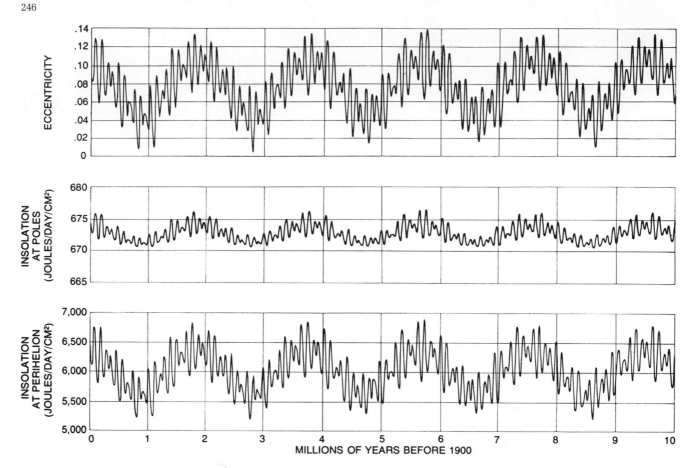

CHANGES IN ORBIT OF MARS over the past 10 million years may account in part for the peculiar circular plates and laminations observed in the planet's polar regions, according to a hypothesis developed by the author and his students. The eccentricity of Mars's orbit (*top curve*) has varied from .004, or nearly circular, to .141. This would lead to changes in the average amount of solar energy reaching the poles (*middle curve*) and in the peak energy reaching the planet when it was closest to the sun (*bottom curve*).

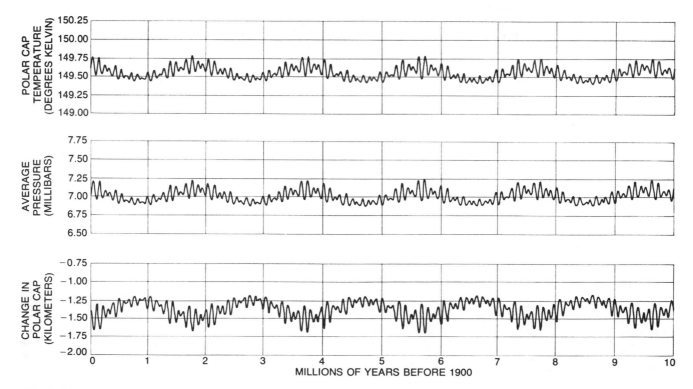

CHANGES IN SOLAR INPUT TO MARS, due to the changing eccentricity of the planet's orbit, could produce cyclical variations in the growth and sublimation of the carbon dioxide frost caps. The top curve shows the long-term changes in the temperature of solid carbon dioxide at the poles. The middle curve shows variations in average atmospheric pressure at the planet's surface that might have resulted from the variations in insolation. The bottom curve shows changes in the height of the permanent frost cap.

accidental development of life from prebiological organic materials. On the other hand, life-on-Mars enthusiasts argue otherwise and emphasize that if water has been available at all in surface layers, it would provide a favorable environment for the development of life. Obviously such a debate cannot be settled by the kinds of information collected by *Mariner 9*. The answer must wait for sophisticated chemical and mineralogical analysis of the surface soil itself.

Simultaneous with the flight of *Mariner 9* the U.S.S.R. undertook an ambitious mission whose objective was to land a capsule on Mars and conduct some analyses of the surface. Unfortunately the Russian lander *Mars 3* failed shortly after reaching the surface and apparently transmitted no useful information. I expect the U.S.S.R. to repeat this kind of mission late in 1973, and I look forward to seeing pictures from the surface and probably the results of some simple chemical analyses.

One hopes that by 1976 we shall be getting information back from a complex U.S. lander being developed in the Viking program and possibly from a second-generation Russian lander. The Viking capsule is being designed not only to look for organic compounds directly but also to perform some simple but important determinations of the basic inorganic composition of surface minerals. Such measurements will provide an important clue to the past chemical evolution of the surface minerals, including whether or not they have reacted chemically with water.

Will the Viking lander finally tell us if life now exists or ever did exist on Mars? I personally rather doubt it. I think the difficulty of obtaining an unambiguous yes or no is so great that it is probably beyond the grasp of even an investigation as ambitious as the Viking mission. My own view is that the final answer to the great search for life on Mars may have to await the return, probably by unmanned means, of a sample of Martian soil for sophisticated analysis in terrestrial laboratories. My guess is that the U.S.S.R., having demonstrated the ability to return unmanned samples from the moon, should be in a position to repeat this feat for Mars around 1980 (assuming that it continues to give unmanned space exploration the same priority that it has had in the past). The U.S. has no ambitious plans for exploring Mars beyond Viking. Thus *Mariner 9* will long be remembered as one of the high points in the American exploration of Mars.

V

GEOPHYSICS AND PLANETARY RESEARCH

The main emphasis in this volume is on topics in the geosciences that represent new directions, such as plate tectonics, or that are of special importance in the modern world, such as energy or the effect of natural cycles and pollutants on the biosphere. However, a book dealing with the geosciences would be incomplete without some reference to modern research in geophysics.

To this end we have included three articles as examples of the methods used in the study of the earth's evolution and the structure of its interior. Walter Elsasser's classic discussion deals with the origin of the magnetic field by dynamo action deep in the earth. Earth magnetism provides an important clue about the physical state of the fluid iron core as well as a tool for studying sea-floor spreading and continental drift. Peter Goldreich reviews what is known of the tidal interaction of sun and moon with earth. Tidal friction affects the moon's orbit as well as slowing earth's rotation, thus increasing the length of the day.

Seismology is the chief tool for probing the earth's deep interior. As the earth evolved, it became chemically differentiated, that is, divided into zones with distinct compositions. Bruce Bolt's article shows how seismic waves can be used to discover the major divisions in the interior of the planet corresponding to changes in chemical composition and physical state.

For a concluding article we go back to the very beginnings of the system of planets. J. S. Lewis describes one of the newest theories about the condensation of the planets from the primitive solar nebula. This approach, which speculates on the chemical reactions that would be likely in a cooling nebula, explains much of the information developed from recent spacecraft missions.

21

The Earth as a Dynamo

INTRODUCTION

Although earth's magnetic field has been used for more than five hundred years as an aid in navigation, to this day it has defied explanation. To be sure, many theories have been advanced, including the dynamo theory, the subject of Walter M. Elsasser's article, "The Earth as a Dynamo." However, even for this, the most generally accepted explanation, the mathematics are so formidable that proponents of the theory are forced to resort to simplified models and arguments by analogy.

Despite the lack of a confirmed theory, earth magnetism plays an important scientific role. Because of its importance in navigation, governments have subsidized worldwide expeditions to map the magnetic field and keep track of its variations. Thus for the last few centuries, its properties have been well known. Geophysicists have learned how to study earth's ancient magnetic field going back hundreds of millions of years, using the new techniques of fossil magnetism. Rocks often record the direction of the local magnetic field at the time they were formed and, from readings of the rock record, ancient wanderings of the magnetic pole have been charted and reversals in polarity of the magnetic field have been discovered. Polarity changes have become a major tool for determining rates of sea-floor spreading, as explained in earlier articles in this book.

Elsasser's article is an elegant piece of writing, a clear and simple exposition of an extremely complicated theory. Although written sixteen years ago, it is surprisingly up-to-date in its explanation of dynamo action in the earth's fluid iron core.

A problem for the theorists, posed by recent results of space exploration, is that Jupiter has a magnetic field but Venus and Mars do not. The moon had a magnetic field some three billion years ago (evidenced by fossil magnetism) but lacks one now. Why?

The Earth as a Dynamo

by Walter M. Elsasser
May 1958

*Why is the earth a magnet? The best answer at present
is that the slow flow of matter in its fluid core generates
electric currents, which in turn set up a magnetic field*

The problem of explaining the earth's magnetism has been with us ever since William Gilbert, the Elizabethan scientific genius, first showed that the earth as a whole acts as a gigantic magnet. It may continue to be with us for a long time to come, because the main source of the magnetic field undoubtedly lies deep in the interior of the earth, forever beyond reach of our direct investigation. But as we learn more about physics, and about the properties of our planet, we are coming closer and closer to a reasonable theory which rests on established physical principles and fits the known facts. I want to tell of some recent work on the problem that makes us confident we are now on the right track.

Gilbert thought that the field might be produced by a large body of permanently magnetic material inside the earth. This idea had to be given up a long time ago, when it became plain that the temperature of the earth's interior must be much too high to allow any material to retain magnetism. Physicists were left completely at a loss to explain the earth's field. It became fashionable to look for some new cosmic process for producing magnetism which eluded detection in the laboratory but applied to very large rotating bodies such as planets and stars. The main proponent of this line of thought in modern times has been the famous British physicist P. M. S. Blackett. But Blackett and his school have now abandoned such ideas because astrophysical evidence shows very clearly that the magnetism of stars does not conform to any reasonable scheme involving a new fundamental law.

We come back, then, to the simple and familiar process by which we generate magnetic fields every day—namely, by the use of electric currents. An electric current flowing around a bar of iron induces magnetism in the metal. But the magnetized material need not be a solid bar. We know that electricity can produce magnetic fields in a gas or other fluid: such phenomena have lately been much studied under the name of magnetohydrodynamics. Let us see whether magnetohydrodynamics can explain the existence of the earth's magnetic field.

We must assume first of all that there are electric currents in the core of the

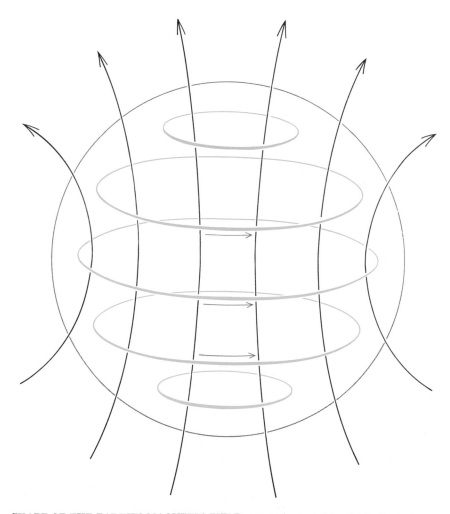

SHAPE OF THE EARTH'S MAGNETIC FIELD (*arrows*) strongly suggests that it is produced by circular electric currents (*ellipses*) flowing at right angles to its lines of force.

earth. Possibly they were started originally by chemical irregularities which separated charges and thereby initiated a battery action that generated weak currents. But how the currents originated is less important than the question of how such currents could be built up and perpetuated to maintain the earth's continuing magnetic field. We do not have to look far for a process that could accomplish this: it is the familiar principle of the dynamo.

A dynamo converts the energy of mechanical motion into electric current. The motion may be relative motion between a coil and a ferromagnetic material or between two coils (or other conductors of electricity). The simplest illustration of such a machine is the disk dynamo invented by Michael Faraday. He put a disk of copper on a spindle and spun the disk over a bar magnet set up near the outer edge of the disk [*see diagram at the right*]. The motion of the conductor through the magnetic field of the magnet induced a small current in the disk, as shown in the diagram. Now we can replace the bar magnet with a coil, and this will induce a current in the disk in exactly the same way, provided we start with a current in the coil. And if we feed the current induced in the disk back into the coil, we have a self-contained system for generating current simply by spinning the disk. This is exactly the principle on which power-station generators operate.

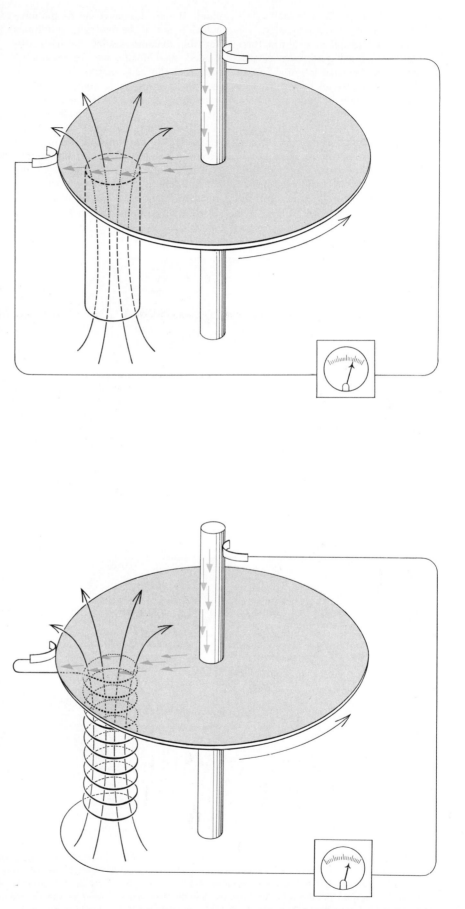

The Faraday disk itself could not maintain a current for very long, because the current induced in the disk is so weak that it would soon be dissipated by the resistance of the conductor. If we had a material that could conduct electricity a thousand times better than copper, the system would indeed yield a self-sustained current. We could also make it work by spinning the disk very fast, but here again the solution is only theoretical, because the necessary speed is far beyond feasible limits. There is, however, a third way we could make such a dynamo self-sustaining, and this has a direct bearing on our problem of explaining the earth's magnetism. All we need to do is to increase the size of the system: theory says that the bigger we make such a dynamo, the better it will function. If we could build a coil-and-disk apparatus of this kind on a scale of many miles, we would have no difficulty in making the currents self-sustaining. And the larger we made it, the less speed of rotation we would need. If the disk had the same diameter as the earth, its

SIMPLE DYNAMO generates electric current (*colored arrows*) when a copper disk is turned through the magnetic lines of force of a bar magnet (*top drawing*) or a coil (*bottom*).

outer edge could rotate at a snail's pace and we would still have a dynamo.

We have a plausible mechanism, then, for maintaining self-sustaining electric currents in the earth's interior. Let us now look at the interior to see whether currents can actually be generated there.

The study of the travel of earthquake waves through the earth has told us some very definite facts about its internal structure and composition. First of all, we know that the earth has a distinct core 4,316 miles in diameter—a little more than half the total diameter of the globe. We know also that this core is

fluid. It may therefore be in motion—which is one of the necessary conditions for the dynamo model we have discussed. And finally, we are reasonably sure that the main material of the core is iron: it is probably composed of an iron-nickel alloy with some dissolved impurities. Thus we have a molten metal or alloy filling the inner part of the earth. It is at once a good conductor of electricity and a fluid in which motions can easily take place. In other words, the core is exactly the kind of medium we need for the Faraday type of dynamo. It allows both mechanical motion and

the flow of current, and the interaction of these can generate self-sustaining currents and magnetism.

So far, so good, but the most important evidence is yet to come. The nature of the earth's magnetic field itself is the strongest argument for the dynamo theory. For many years geophysicists have known that the magnetic field is irregular and constantly changing [*see maps below*]. The compass needle does not point exactly to the north, and it deviates from true north by different amounts in various parts of the world.

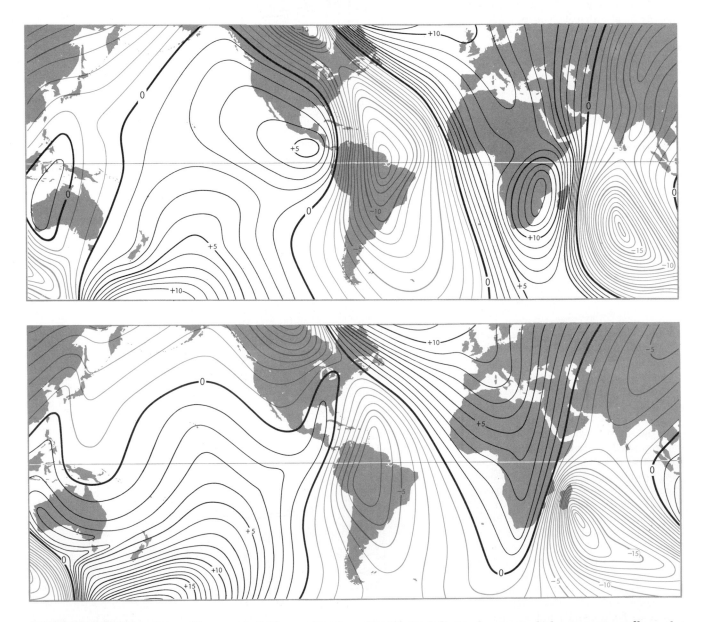

CHANGING EDDIES in the earth's magnetic field are reflected by the differences in the contours on these two maps of the world. The contours plot the rate of change in minutes of arc per year of magnetic declination: the deviation of a compass needle from true north. The map at top shows the contours in 1912; the map at bottom, the contours in 1942. A contour labeled with a positive number (*black*) indicates the rate at which a compass needle on the contour moved away from true north; a contour labeled with a negative number (*color*), the rate at which a needle on the contour moved toward true north. On the North Island of New Zealand in 1942, for example, a compass needle would have moved away from true north at a rate of five minutes of arc (+5) per year.

There are many eddies, as it were, in the field. They are eddies in a literal sense, because they change as time goes on. Seamen and fliers who use magnetic charts for navigation know only too well that the maps must be brought up to date every few years. Not only does the field fluctuate but the rate of change itself varies from time to time.

Moreover, even the over-all strength of the earth's field as a whole is not constant. In the century since accurate measurements of the field began to be made, its strength has declined about 5 per cent. We need not be alarmed into thinking that by the year 4000, say, people will no longer be able to use magnetic compasses for navigation. There is evidence that in the last few years the decline in strength has begun to slow down, and the trend may turn upward in the near future. Finally, we know that the north magnetic pole has been wandering about the Arctic in the course of geological time; it is now just north of Canada near the 70th latitude.

All this is precisely what we should expect on the dynamo theory. It represents overwhelming evidence that the core of the earth is in motion; the variations and changes in the field must reflect these motions. So the pieces of the puzzle fit together. Seismology tells us that the earth's core is fluid; the study of the magnetic field confirms that there are motions in this same core. Indeed, from the measured changes in the field we can compute the speed of these motions. It turns out that matter in the core is moving at the rate of about a hundredth of an inch per second, which means that in the course of a few centuries a particle may travel through a large part of the core. In a very true sense the variations in the magnetic field that we observe on the surface of the earth tell us something about what is going on in the deep interior.

The question arises at once: What is the source of energy for these motions? We do not, as yet, have a conclusive answer. Some scientists believe that heat flowing outward from the center of the earth gives rise to convection currents in the core, as happens when a pot of water is heated at the bottom. Small differences in the chemical constitution of parts of the core also would produce slow movements of the fluid, as differences in salt content do in the oceans. Or it may be, as Harold Urey of the University of Chicago has suggested, that the earth's interior is changing and moving because it has not yet reached a state of equilibrium. In any event, mo-

CIRCULAR ELECTRIC CURRENT (*gray arrows*) which gives rise to the earth's magnetic field may be driven by smaller eddies of current (*closed arrows*). This schematic diagram is a section through the Equator, *i.e.*, the poles are perpendicular to the plane of the page.

tion there is, and there is one more thing we can say safely. Calculations show that the energy required to maintain motion in the earth's core is surprisingly small.

The dynamo theory of the earth's magnetic field takes off from the two basic facts that seem well established: the existence of a core composed of molten metal and the presence of motions in this core. It requires no additional assumptions. From these facts we can derive the magnetic field merely by applying laws of classical physics which have been known and uncontested for well over 100 years. But the reasoning is not altogether simple and straightforward: we have some tricky questions to answer.

Suppose we take a fishbowl, say a foot in diameter, fill it with mercury and heat it from below. We will certainly get thermal convection in the mercury. But by no stretch of the imagination can we expect to detect any electric currents or magnetism in the bowl. Now our bowl of mercury is very similar to the earth's liquid core, and we often use scale mod-

els satisfactorily to investigate fluid motions: *e.g.*, a ship model in a tank or a model of an airplane in a wind tunnel. What, then, is wrong with our model of the earth's core? Mathematical analysis answers this question. The model fails because electrical processes and mechanical motions do not scale down (or up) in the same way.

If we suddenly shut off a current in a coil in the laboratory and instantly hook up the coil to an oscilloscope which displays what happens immediately thereafter, we find that the current does not stop at once but takes a small fraction of a second to die out. This is known as the decay time. A circular current in the mercury of our fishbowl would have a decay time of about a hundredth of a second. But now as we increase the size of the vessel we discover the striking and surprising fact that the decay time increases out of all proportion: in fact, it goes up as the square of the increase in the vessel's diameter. In a vessel 10 times larger than our first, the current will take 100 times longer to decay. Calculations show that an electric current circulating in the earth's core would require about 10,000 years to decay. Ten thousand years would be ample time for this current and its associated magnetic field to be "pushed around" considerably by motions in the fluid, however sluggish they might be.

We can visualize how this pushing-around process might generate new electric currents and might amplify magnetic fields. Let us consider just the magnetic fields. We can think of the field lines as so many ribbons which are pulled and stretched by the motion of particles in the fluid: this is not merely a fanciful way of speaking but actually a literal description of the way they behave. Let us assume that a group of particles in the fluid, moving at different speeds, pulls laterally on some lines of force which are straight to begin with. The lower part of these ribbons, where the particles are moving faster, will be pulled farther than the upper part, and therefore the ribbons will be stretched. In the process of stretching they gain energy—energy which is imparted by the motion of the particles. This basic process of the conversion of mechanical energy into electrical and magnetic energy —not essentially different from the operation of a dynamo—can be shown mathematically to account satisfactorily for the electric currents and the magnetism of the earth's core.

If the core contained no electric currents at the beginning of the earth's history, it could soon have acquired small currents by some chemical battery action. These tiny currents would rapidly be amplified by being pushed around by the fluid motion, and the amplification would go on until an equilibrium state was reached. Thereafter the core would contain self-replenishing electric currents and the earth would have a permanent magnetic field.

One major question remains: Why does the earth have a unified general field? The process I have described would produce only localized eddies, not a general field. But we have a reasonable answer to the question. The stage director that lines up the eddies in some sort of order must be the earth's rotation, for the general field is aligned approximately along the axis of rotation.

The greatest difficulty in all attempts to explain the earth's magnetic field has been the problem of introducing the driving force that produces the general symmetry of the over-all field. We have to assume that the field is generated by circular electric currents closed upon themselves. In such a setup there is no apparent place where we can insert a driving force—either a battery or any other. But the dynamo theory allows the earth's rotation to act as a driving force. The rotation causes the closed currents of the eddies to flow in the same direction [see diagram on preceding page]. If we were in the earth's core we would observe only the individual ed-

dies. But at our observation posts outside the core their aggregate effect is nearly equivalent to a single current flowing in a large circle around the whole core, as the diagram indicates. Over-all the field is fairly stable, but, since eddies start, grow and decay in more or less random fashion, the magnetic field is unpredictable in its details.

Of late, strong support for the dynamo theory has come from the camp of the astronomers. They have found that not only our sun but also many other stars possess strong magnetic fields, and we can easily picture dynamo processes operating in the stars. The sun has extremely strong magnetic fields in the sunspots, and its archlike prominences must be formed by magnetic lines of force [see photograph below]. So far some 60 other strongly magnetic stars have been identified; no doubt a large proportion of the whole star population, if not all of this population, is more or less magnetic.

The stars rotate; their gaseous matter conducts electricity well; their magnetic fields change rapidly—all this points to a dynamic rather than static explanation of their fields, such as we postulate for the core of the earth. Although our fishbowl of mercury is an inadequate model of what goes on inside the earth, the earth's core seems to be a quite tolerable model for the magnetohydrodynamic processes that fill the universe on the vastest possible scale.

PROMINENCE on the sun was photographed with a coronagraph at the High Altitude Observatory of the University of Colorado. The artificially eclipsed disk of the sun is at bottom. Shape of the prominence suggests that it travels along magnetic lines of force.

22

Tides and the Earth-Moon System

INTRODUCTION

Among the forces that shape the earth, the tides are perhaps the least obvious, as Peter Goldreich points out in his article, "Tides and the Earth-Moon System." Nevertheless, they have had a profound effect. Tidal currents on the continental shelves can produce erosion and transport sediments. The frictional dissipation of energy in these currents slows down the rotation of the earth, thus increasing the length of the day and also increasing the distance of the moon from the earth. When the moon was much closer to the earth, huge catastrophic tides in the ocean and solid earth may have occurred. Tides are known to trigger moonquakes, though it is uncertain whether a similar connection with earthquakes occurs. Man may someday derive energy on a large scale from tides to meet his growing needs. Billions of years from now, when the moon once again approaches the earth as a result of tidal friction, catastrophic tides may once again wreak destruction on the earth.

The tidal attraction of the moon and the sun toward the earth's equatorial bulge forces the earth's axis to precess with a period of twenty-seven thousand years. This precession may well affect climatic variations, ice ages, and sea level changes. The tides themselves are important in navigation, and in flushing pollutants out of estuaries and bays.

Tides and the Earth-Moon System

by Peter Goldreich
April 1972

*Tidal friction in shallow waters has controlled
the evolution of the earth-moon system for aeons.
It is causing the length of the day to increase
and the moon to recede from the earth*

Anyone who has spent any time at the seashore is aware of the tides, the periodic advance and retreat of the sea from the land. It is well known that the tides are driven by the gravitational pull of the moon and the sun on the earth. Not so well known is the important role the tides have played in the evolution of the earth-moon system. As the tidal currents flow over the bottom, a turbulent boundary layer is created in which the mechanical energy of the water's movement is dissipated as heat of friction. This mechanical energy is derived from the rotation of the earth and the motion of the moon in its orbit. The degradation of mechanical energy into heat results in an inexorable increase in both the length of the earth day and the distance of the moon from the earth. The best modern estimates of the rate at which this evolution is proceeding imply that the earth-moon system has been fundamentally altered by the action of tidal friction. Currently the length of the day is increasing by approximately two milliseconds per century and the moon is spiraling away from the earth at about three centimeters per year.

The moon raises tides both in the oceans and in the solid body of the earth. (It also raises tides in the atmosphere, but the effect of these tides on the earth-moon system is so small that they will not be discussed here.) For the moment consider the idealized case of the tides raised by a satellite traveling in a circular orbit around the equator of a solid, elastic planet (a planet that might be made out of rubber). The gravitational force of the satellite would distort the planet into an ellipsoid with its long axis through the equator pointing toward the satellite. The distortion arises because the satellite's gravitational attraction is greater on the near side of the planet than on the far side. Although the gravitational force diminishes as the inverse square of the distance, the height of the tidal bulges varies as the inverse third power of the separation between the satellite and the planet, reflecting the fact that the tide is due to the *difference* in the satellite's attraction at opposite sides of the planet. An observer standing at one point on the surface of the planet experiences two equal tidal maximums during each revolution that the planet makes under the satellite. The planet and the satellite orbit around their mutual center of mass. The centrifugal acceleration that results from this whirling varies with distance from the center of mass across the planet. This effect, together with the differential force of the satellite's gravity across the planet, is responsible for the raising of two tidal bulges.

The restoring forces that limit the height of the tides are due to the planet's elasticity and self-gravity. In its response to the periodic tidal stretching force (periodic as it is felt at a point fixed to the planet) the planet behaves rather like a weight that is hung from a spring and subjected to a periodic force. If the weight is set in motion and then left alone, it will bob up and down at its natural frequency, which depends only on the mass of the weight and the stiffness of the spring. If the weight is not allowed to oscillate freely but is driven up and down in a periodic fashion, the response of the system depends on whether the frequency of the applied force is higher or lower than the natural frequency of the system. If the forcing frequency is lower than the system's natural frequency, the displacement of the mass is in phase with the force, that is, the maximums and minimums of the displacement occur simultaneously with the maximums and minimums of the force. If the forcing frequency is higher than the natural frequency, the displacement is half a cycle out of phase with the applied force.

The solid body of the earth is more complex than such a simple harmonic oscillator and is capable of simultaneous oscillation in modes of many different frequencies. The longest oscillation period, which is determined by the time it takes seismic waves to travel through the earth, is about 55 minutes. In recent years detailed studies of the earth's free oscillations have been made following their excitation by great earthquakes [see "Resonant Vibrations of the Earth," by Frank Press; SCIENTIFIC AMERICAN, November, 1965]. Because the tidal period of slightly less than 12 hours is longer than the free-oscillation periods, high tides in the solid earth are located along an extension of the earth-moon center line.

The range of free-oscillation periods of the oceans is set by the time it takes tidal waves to propagate across them. (These are not the sometimes disastrous "tidal waves" generated by earthquakes.) The speed of tidal waves can be computed by multiplying the earth's surface gravitational acceleration by the ocean depth and taking the square root of this quantity. For a typical ocean depth of four kilometers the speed is .2 kilometer per second. Thus a tidal wave can cross 10,000 kilometers of ocean in about half a day, and the oscillation periods of the oceans are comparable to the tidal forcing period. This coincidence, and the irregular shape of the ocean basins, conspire to make the ocean tides exceedingly complicated. Accordingly we shall continue to restrict our attention to solid-body tides whenever possible.

In a real planet the dissipation of energy by friction modifies the periodic tidal distortion. Once again we shall find it helpful to exploit the analogy between a planet and a weight-spring system. The presence of friction causes the displacement of a spring to lag in phase behind the applied force. By the same

token high tides at points on opposite sides of the earth occur some time after these points are aligned along the direction of the center of mass of the planet-satellite system. If the planet's rotation period is shorter than the satellite's orbital period (as is the case with the earth and the moon), the delay will result in the displacement of the tidal maximums in the direction of the planet's rotation. The asymmetrical position of the tidal bulges with respect to the line of centers gives rise to a net torque, or twisting force, on the satellite. This torque increases the satellite's orbital angular momentum. For every action there is a re-

action; a torque of equal magnitude but opposite direction acts on the planet and brakes its spin. A net torque results because the satellite is attracted more strongly by the closer tidal bulge that is leading it in longitude. The tidal torque varies as the inverse sixth power of the planet-satellite separation.

The tidal torque transfers angular momentum and energy from the planet's spin into the satellite's orbit. It requires less energy to give the moon a certain amount of angular momentum than it takes to give the earth the same amount. Therefore, although the total angular

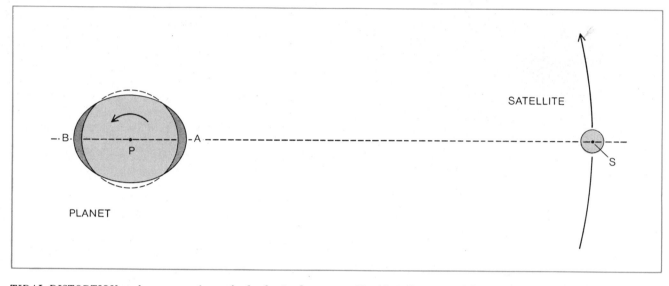

TIDAL DISTORTION at the equator of a perfectly elastic planet is caused by a satellite traveling in a circular orbit. The gravitational force of the satellite would deform the planet into an ellip-soid with its long axis *A–B* pointing toward the satellite. View is from the top of the system; *P* indicates the pole of the planet. The planet rotates in the same direction as the satellite orbits.

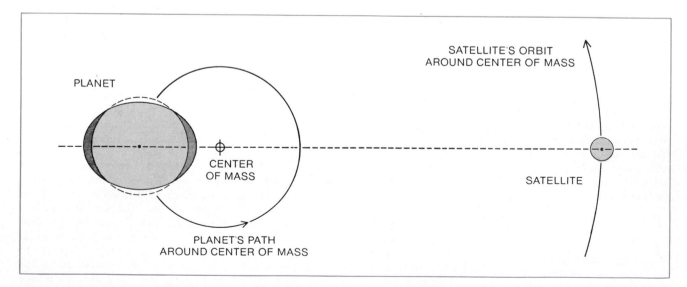

TWO EQUAL TIDAL BULGES are felt by an observer standing at one point on the planet's surface. The bulges are raised by the differential force of the satellite's gravity across the planet coupled with the centrifugal acceleration that results from the system's whirling around its center of mass. The centrifugal acceleration increases with distance from the center of mass across the planet. The center of mass of the earth-moon system is actually about 1,000 miles below the surface of the earth, not outside as shown.

momentum of the planet-satellite system remains invariant, its total mechanical energy decreases as the result of frictional losses in the planet. This loss of energy through friction plus the transfer of angular momentum means that the moon is gaining angular momentum at the expense of the earth's spin. The direction of flow of angular momentum is entirely determined by the relative lengths of the month and the day.

This picture of tidal evolution is not significantly changed by the inclusion of the ocean tides. It is true that simple schematic drawings do not accurately depict the shape of tidally distorted oceans.. Nevertheless, the qualitative evolution I have outlined of the transfer of angular momentum from the planet's spin into the satellite's orbit still applies.

The eccentricity, or oval shape, of a satellite's orbit is affected by tidal friction. Imagine a satellite moving on an elliptical orbit around a planet and that the transfer of angular momentum from the planet's rotation into the satellite's orbit takes place in just two impulses, a large one at pericenter (nearest approach) and a smaller one at apocenter (farthest separation). The impulsive addition of angular momentum at pericenter does not alter the pericenter position but does increase the distance to apocenter. Similarly, the impulsive addition of angular momentum at apocenter does not affect the position of apocenter but does act to increase the pericenter distance. Because the tidal torque varies as the inverse sixth power of the planet-satellite separation, more angular momentum is added near pericenter than near apocenter, and the orbital eccentricity increases.

The tides raised by the planet in the satellite tend to decrease the orbital eccentricity. In all cases where this effect is significant these same tides will have "despun" the satellite to synchronous rotation, that is, the satellite's period of rotation around its own axis equals its period of revolution around the planet and it always presents one face to the planet. In this case the tidal bulges on the satellite are purely radial, along the planet-satellite line of centers, and do not produce a tidal torque. Tidal energy is still dissipated by friction as the height of the tides in the body of the satellite oscillates in response to the variation of the planet-satellite separation. The energy dissipated as heat in the satellite is derived from the satellite's orbital energy, and since for a fixed angular momentum the lowest-energy orbit is a circle, these tides will tend to make the satellite's orbit rounder.

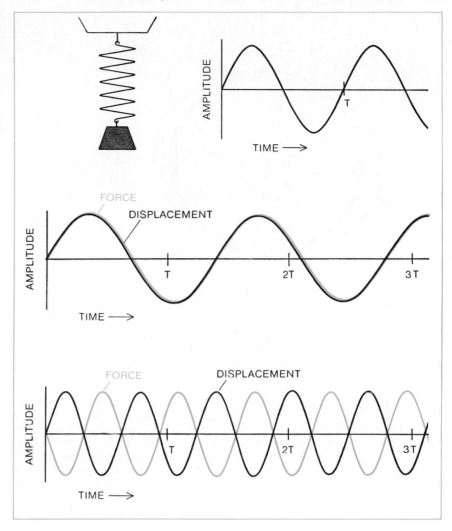

WEIGHT-SPRING OSCILLATOR is a model of the periodic tidal stretching force the planet feels as it is pulled by the satellite. If the weight-spring system is set in motion and then left alone, it will bob up and down at its natural frequency (*top*). If the weight is periodically driven up and down by a forcing frequency (*color*) lower than the system's natural frequency, the displacement (*black*) of the weight is in phase with the force (*middle*). If the forcing frequency is higher than the oscillator's natural frequency, then the displacement of the weight is a half-cycle out of phase with the applied force (*bottom*).

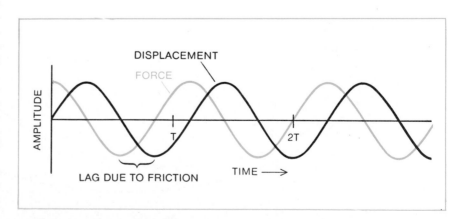

FRICTION CAUSES A DELAY in the response of the weight-spring system forced below its natural frequency. This delay means that the displacement of the weight (*black*) lags in phase behind the forcing frequency (*color*), in analogy to the way that the tidal bulges on opposite sides of the earth occur sometimes after these points align with the moon.

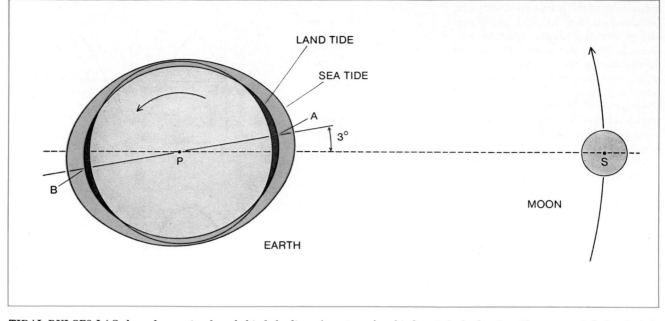

TIDAL BULGES LAG three degrees in phase behind the line of centers *P–S* between the earth and the moon because of friction. In the case of the earth, whose rotation period is shorter than the moon's orbital period, the lagging tides are carried ahead of the moon in longitude by the rotation of the earth. Land tides on the earth are black and sea tides color. View is from the pole.

Because tidal torque varies as the inverse sixth power of distance, its effect can be felt only over rather small separations, thus limiting its importance in the dynamics of the solar system. No planetary orbit has been appreciably altered by the tides the planet raises on the sun.

The tides the sun raises on the planets have significantly despun the two innermost planets, Mercury and Venus. They have had a more modest effect on the spin of the earth and no appreciable effect on the rest of the planets. Tidal friction is more important in satellite systems than in the planetary system because the distances involved are smaller. Thus the tides raised by the earth on the moon, and by Jupiter and Saturn on their large inner satellites, have brought these satellites into synchronous rotation. The evolution by tidal friction of

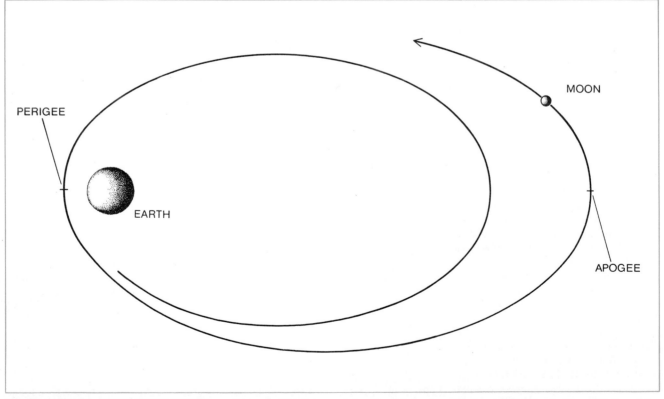

MOON SPIRALS AWAY FROM EARTH as the earth transfers some of its angular momentum to the moon. The moon's orbit is an ellipse. The impulsive addition of angular momentum at perigee (moon's closest approach to earth) changes the orbit to a larger, more eccentric ellipse. The moon will return to the original perigee if no further changes of angular momentum take place.

satellite orbits, other than the moon's, has not been established observationally, although there are several theoretical arguments that suggest the possibility of some tidally induced changes.

The synchronous rotation of the moon is the most easily observable consequence of tidal friction in the solar system. In 1754 Immanuel Kant first proposed tidal friction as the explanation of the moon's rotation. At its present distance from the earth the despinning of the moon from a much faster initial rotation rate would have taken no longer than 10 million years, a mere instant of geologic time. The stability of the moon's synchronous rotation is provided by its irregular figure, which is slightly elongated in the direction of the earth. This stability prevents the synchronous lock from being broken by the collision of comets and asteroids with the moon.

Direct measurement of the rate at which the lunar orbit is expanding is not yet possible, although lunar laser ranging provides some hope for the future [see "The Lunar Laser Reflector," by James E. Faller and E. Joseph Wampler; SCIENTIFIC AMERICAN, March, 1970]. Instead two less direct methods based on the relation between the radius and the period of the lunar orbit have been used.

The first of these methods relies on "modern" (over the past 250 years) observations of the longitudes of the sun, Mercury and the moon as seen against the background of the distant stars. The observed longitudes of the sun and Mercury are compared with the longitudes that are predicted based on a constant rotation rate for the earth. Discrepancies between the two sets of longitudes allow a determination of the irregularities in the earth's rotation rate, some of which are the result of nontidal causes such as changes in the earth's moment of inertia and the transfer of angular momentum between the core and the mantle by the redistribution of mass inside the earth. When the effects of variations in the earth's rotation rate are subtracted from the discrepancy in the longitude of the moon, a unique determination of the rate of increase of the lunar orbital period is obtained.

The second method of measuring the rate at which the lunar orbit is expanding involves studying the records of ancient solar eclipses. Such records were exploited for the first time by Edmund Halley in 1693. The method has the advantage of employing a longer time base than the one founded solely on modern observations, but it suffers from ambi-

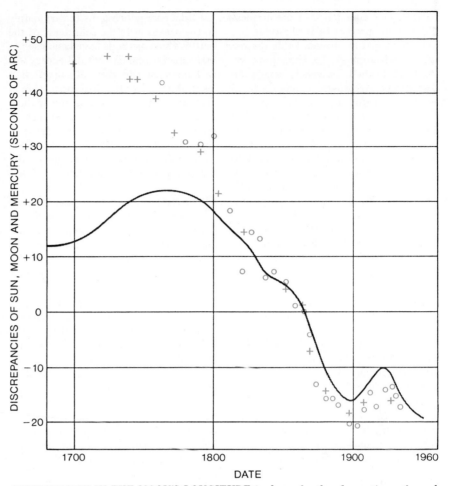

DISCREPANCY IN THE MOON'S LONGITUDE is shown by the observations of occultations of stars by the moon (gray). The moon has a tendency to be behind its position as predicted assuming a constant rotation rate for the earth, that is, the moon's orbital period is lengthening. Discrepancies in the positions of the sun (O) and Mercury (+) occur between their predicted longitudes and their observed longitudes. The measurements allow a determination of the irregularities in the earth's rotation rate, some of which are due to nontidal causes. When these irregularities are subtracted from the discrepancy in the longitude of the moon, the rate of increase of the lunar orbital period is determined.

guities and inaccuracies in the ancient records. In a recent study Robert R. Newton of Johns Hopkins University concluded that some records of ancient eclipses were biased by their authors' desire to enhance their historical or literary impact. We shall rely on the results obtained from the "modern" observations, although they are currently subject to some dispute.

As I have mentioned, these observations imply an expansion of the moon's orbit by about three centimeters per year and an increase of approximately 20 microseconds per year in the length of the earth day. The associated value of the angle of the tidal lag with respect to the earth-moon line of centers is roughly three degrees. The rate of tidal-energy dissipation is about 2.6×10^{12} watts, which is close to the world's power consumption and about 15 percent of the total heat flow from the earth's interior.

If we assume that the present value of the tidal-lag angle is characteristic of its value in the past, we can extrapolate the lunar orbit back in time and deduce when, if ever, the moon was close to the earth. This procedure yields the startling result that the moon was near the earth less than two billion years ago. It is generally believed that the planets, including the earth and the moon, are 4.6 billion years old. One would think that, if the moon was ever near the earth, it would have been so early in the history of the earth-moon system. Is there any way out of the discrepancy in the two dates? The most plausible way would be to discard the assumption that the tidal phase lag has remained constant over geologic time. Certainly it would seem foolhardy to extrapolate back two billion years relying on a parameter that has been determined from data extending back no more than 4,000 years, the time scale covered by the ancient solar-eclipse records. Our confidence in the applicability of the present tidal phase lag has been bolstered by a remarkable discovery in paleontology. In 1963 John W. Wells of Cornell University suggested that the yearly growth bands on fossil corals are made up of daily growth ridges. From counts of the number of daily increments per yearly band on fossil corals the number of days per year has been determined all the way back to the mid-Devonian period of some 380 million years ago [see "Corals as Paleontological Clocks," by S. K. Runcorn; SCIENTIFIC AMERICAN Offprint 871]. The data that now exist are consistent with a constant phase lag of three degrees over the past 380 million years. As remarkable as the coral data are, they only take us back to a time when the length of the day was 22 hours and the moon's distance from the earth was 58 earth radii instead of the present 60.25. To assess what is involved in a further backward extrapolation it is necessary to locate the region where the tidal energy is being dissipated.

Although the astronomical data can reveal the total rate at which tidal energy is being lost in the earth, they provide no information on the location of the dissipation. The first question is: Is

the tidal energy being dissipated mainly in the oceans or in the solid body of the earth? Three methods have been used to estimate the total rate of tidal-energy loss in the oceans and they all suggest that most, if not all, of the tidal dissipation in the earth is taking place in the shallow seas. The crudest method consists in consulting tidal charts that show the tidal distortion of the oceans. From the shape of the tidal distortions the torque on the moon can be directly calculated.

Two more accurate methods were suggested by Sir Geoffrey Taylor in 1919. Both methods rely on the fact that the tidal currents in the deep oceans are so slow that hardly any energy should be dissipated there. Thus the dissipation of tidal energy should be confined to the shallow seas where the tidal currents are amplified to much greater velocities. Taylor's first method is to estimate the tidal currents in the shallow seas and then directly calculate the frictional dissipation of energy in the turbulent boundary layers along the ocean bottom. This method is very sensitive to errors in the tidal velocities because the rate of energy dissipation is proportional to the cube of the velocity. At one time this method was used to predict that more

than half of the total dissipation took place in the Bering Sea. Later estimates revised the adopted tidal currents downward. It is now maintained that the Bering Sea is responsible for only 10 percent of the total tidal-energy loss. Taylor's second method is currently the most favored one. It requires knowledge of both the tidal height and the flux of water at the entrance to a shallow sea. From the tidal height the pressure at any depth can be calculated, and the net work done on the water entering and leaving the shallow sea can be estimated. A compilation of calculations of this type by Gaylord R. Miller, then at the University of California at San Diego, shows that the shallow seas account for roughly two-thirds of the astronomically determined rate of energy dissipation.

If we accept that tidal energy is dissipated mainly in the shallow seas, then there is little basis for assuming that the tidal phase lag was constant during the past two billion years. Evidence of continental drift and sea-floor spreading points to a very different configuration of oceans and continents as recently as 500 million years ago. This might well account for the factor of two or three by which the present phase lag must exceed its past average in order to reconcile the rate of tidal evolution with the age of the earth-moon system.

The dynamics of the earth-moon system is complicated by the influence of the sun. At present the plane of the moon's orbit is inclined to the ecliptic (the plane of the earth's orbit) by five degrees; the earth's Equator is inclined to the ecliptic by 23.5 degrees. The sun forces the moon's orbital plane to rotate in space while maintaining a constant inclination to the ecliptic; this precession of the lunar orbit has a period of 18.6 years. The combined attraction of the moon and the sun for the earth's rotational bulge forces the earth's axis to precess with the much longer period of 27,000 years. The attraction of the moon by the earth's rotational oblateness is insignificant, and the lunar orbit is controlled by the sun. Thus the inclination of the moon's orbit to the earth's Equator varies between 18.5 and 28.5 degrees with a period of 18.6 years.

In the past the moon was closer to the earth, and the earth's equatorial bulge was greater. Both factors enhanced the strength of the interaction of the earth's oblateness and the lunar orbit. Simple calculations show that if the moon were ever closer than 10 earth radii, this interaction would have dominated the interactions of the lunar orbit and the

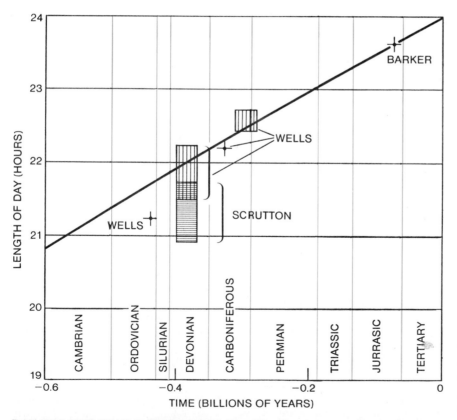

DAY WAS LESS THAN 21 HOURS LONG 600 million years ago according to the counts of daily growth bands on corals. Names of investigators appear next to their results. Length of the day is extrapolated back through the ages on the assumption that the present tidal phase lag of three degrees has remained constant throughout the past half-billion years.

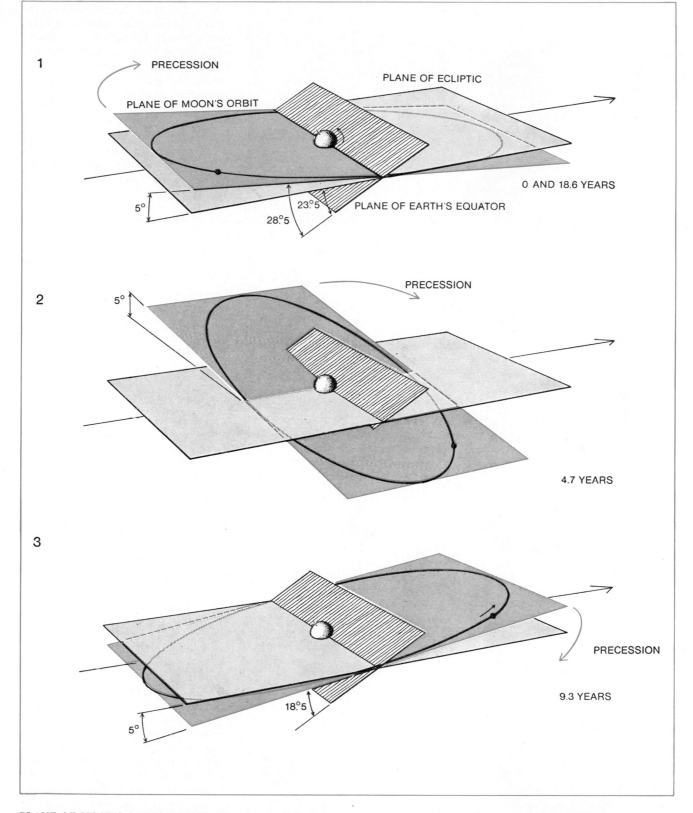

1

PRECESSION

PLANE OF ECLIPTIC

PLANE OF MOON'S ORBIT

0 AND 18.6 YEARS

5°

23°5

28°5

PLANE OF EARTH'S EQUATOR

2

5°

PRECESSION

4.7 YEARS

3

18°5

5°

PRECESSION

9.3 YEARS

PLANE OF MOON'S ORBIT PRECESSES with a period of 18.6 years because of the gravitational force of the sun. The moon's orbital plane (*dark color*) is inclined to the ecliptic (plane of the earth's orbit) by five degrees; the plane of the earth's Equator (*hatched area*) is inclined to the ecliptic by 23.5 degrees. The sun forces the moon's orbital plane to rotate in space while maintaining a constant inclination (five degrees) to the ecliptic. Once every 18.6 years the plane of the earth's Equator will tilt above the ecliptic in the same direction (*long arrow*) that the plane of the moon's orbit will tilt below (*top*). The angle between them is 28.5 degrees, the most it can ever be. In 4.7 years the plane of the moon's orbit will have precessed a quarter of the way around and will be at right angles to its former position (*middle*). In 9.3 years the plane of the moon's orbit will have gone halfway around the precessional cycle (*bottom*). The plane of the earth's Equator and the plane of the moon's orbit will both be tipped above the ecliptic in the same direction, and the angle between the two planes will be at its minimum amount of 18.5 degrees. In 18.6 years the moon's orbital plane will have continued the rest of the way around and returned to its initial position, where it will repeat the precessional cycle.

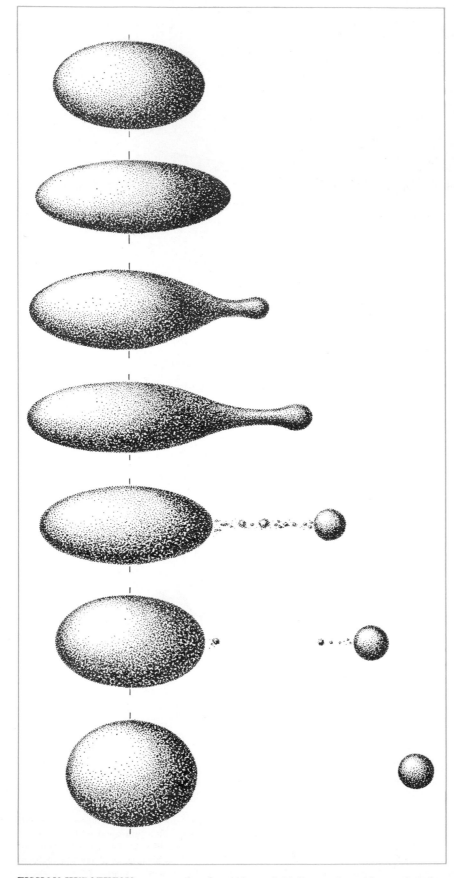

FISSION HYPOTHESIS supposes that the earth was initially rotating with a period close to 2.6 hours. As iron was drawn from the mantle into the core the earth's moment of inertia decreased, increasing its rotation rate. The body became unstable, changing from an oblate spheroid (*top*) to a pear-shaped figure with its long axis in the plane of its rotation (*middle*). Eventually the neck of the pear split off and formed the moon (*bottom*).

sun. At less than 10 earth radii the inclination of the moon's orbit to the equatorial plane of the earth would have remained constant. As the lunar orbit and the earth's Equator precessed, the angle the Equator makes with the ecliptic and the inclination of the lunar orbit to the plane of the ecliptic would have varied periodically.

Armed with an understanding of tidal friction and earth-moon-sun rigid-body dynamics, it is possible to appreciate the calculated history of the earth-moon system. Calculations similar to the ones I carried out several years ago have been performed by many others, starting with the pioneering work of George Darwin (the son of Charles Darwin) in the late 19th century. More recent investigators include Horst Gerstenkorn, William M. Kaula, Gordon J. F. MacDonald, S. Fred Singer and Louis B. Slichter. Such calculations show that the moon never moved in an equatorial orbit. Its orbital inclination to the earth's Equator would have exceeded 10 degrees if the moon had ever been closer than 10 earth radii in the past. At the closest approach of the moon the length of the day would have been about five hours.

We are on less certain ground when we consider the variation of the eccentricity of the lunar orbit in the past. The value of the tidal phase lag in the moon is not yet known. (Recall that the tides raised by the earth in the moon act to decrease the eccentricity.) Even if tidal dissipation in the moon is about the same as the dissipation in the earth, the dominant effect on the eccentricity of the lunar orbit would be the one due to the tides raised in the earth. In fact, the absence of oceans on the moon suggests that its tidal phase lag is smaller than the earth's. Thus it seems highly likely that the moon's orbital eccentricity is increasing at the present time, and that it was therefore even smaller in the recent past.

The results of the calculations I have been discussing are not in themselves very controversial. Speculation on the origin of the moon is largely concerned with deciding where along the calculated evolutionary path the moon actually originated. If there were some readily discernible landmark to which the evolutionary path pointed, there would undoubtedly be more agreement on the origin of the moon. In the absence of any landmarks a bewildering variety of hypotheses have been put forward. No one article can do justice to all of them, but we can examine the general classes of proposals to see how well they satisfy the constraints imposed by the evolutionary calculations. Theories of lunar origin divide naturally into three categories: fission, capture and binary formation.

The idea that the moon was split off from the earth was originally put forward by George Darwin. The modern version of this hypothesis relies on the idea of rotational instability of the primordial earth; this instability caused the earth to split apart into the present earth-moon system. This class of theories is faced with enormous difficulties. Foremost among them is the fact that the earth's day would have been about five hours when the moon had just split off. An initial rotation period close to 2.6 hours is required to induce rotational instability. Proponents of the fission hypothesis must account for the disappearance of about half of the initial angular momentum of the earth-moon system. This has not proved to be an easy task. The fission hypothesis also cannot readily explain the current five-degree inclination of the lunar orbit to the ecliptic plane. The evolutionary calculations imply that the lunar orbit would have been inclined by at least 10 degrees to the earth's Equator when the two bodies were close together. Fission would result in an initially equatorial orbit. Computations show that an initially equatorial orbit would have evolved into an orbit that lay in the ecliptic plane at the present earth-moon separation.

Capture theories of lunar origin also have their problems. These theories propose that the moon was once an independent planet that was caught in the earth's gravitational field and swung into a lunar orbit. Problems with capture theories are in large part due to the implausibility of the capture process, which requires the velocity of an initially unbound moon to be reduced by some kind of dissipation process. Tidal friction and collisions with material already in earth orbit have been proposed as dissipative sources leading to capture. Tidal-friction losses during a single close encounter between the earth and the moon would be very small, and energy losses due to collisions with orbiting material would require a significant fraction of a lunar mass to be in orbit around the earth.

This brings us to the one really new hypothesis of lunar origin that has appeared during the past half-century. It was conceived by Gerstenkorn, who found by tracing the orbit of the moon back past the time of minimum earth-moon separation that still earlier the moon would have moved in a highly inclined eccentric orbit. Gerstenkorn reasoned that the moon had initially been captured in such an orbit, and that tidal

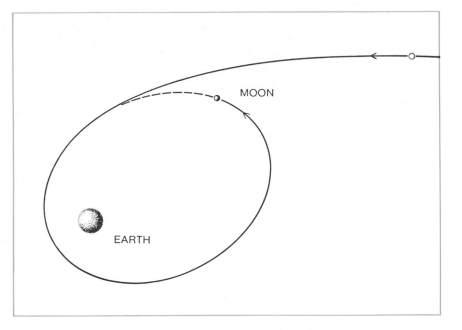

CAPTURE HYPOTHESIS proposes that the moon was once drifting free in space and was caught in the earth's gravitational field and swung into lunar orbit. Its excess energy would have been dissipated by tidal friction and collisions with debris orbiting the earth.

friction had subsequently decreased the orbital eccentricity and inclination and then increased the radius to its present value. As ingenious as Gerstenkorn's idea is, it has not met with wide acceptance. One of the obstacles it faces is the improbability of the capture process, which Gerstenkorn attributes to tidal dissipation in the body of the moon. Capture by this process requires that the moon approached the earth with a relative velocity of less than one kilometer per second aimed at a distance of no more than two earth radii from the surface of the earth. The source of concern is that at closest approach the moon would have been within the Roche limit of 2.86 earth radii. Within this distance the earth's tidal gravitational force exceeds the lunar self-gravity and would tear the moon apart. It is not clear whether or not the moon's elastic strength would have been sufficient to prevent its destruction during the short time it would have spent within the Roche distance.

The hypothesis that the moon formed from material orbiting the earth faces no real dynamical problem. There is little doubt, however, that the bulk composition of the moon, which has a density of 3.3 grams per cubic centimeter, is significantly different from the bulk composition of the earth, whose mean density is 5.5 grams per cubic centimeter. Advocates of the fission theory have appealed to the similarity between the density of the earth's mantle and that of the moon as support for their hypothesis. Capture enthusiasts have used the den-

sity difference to argue that the moon did not form from the same batch of material as the earth. In spite of the difficulties associated with the chemical differences between the earth and the moon (which may largely involve the much lower abundance of iron in the moon), I favor the hypothesis that the moon formed from material in orbit around the earth. As yet there is no compelling reason to prefer any particular distance from the earth as the site of formation.

After all this discussion of the history of the earth-moon system it is appropriate to describe briefly what the future holds in store. It is worth mentioning that all future predictions are based on the assumption that the sun does not suffer a catastrophe such as a supernova explosion before these changes have taken place.

Calculations show that the moon will continue to recede from the earth until it reaches a distance of about 75 earth radii. At that stage the length of the earth day and the lunar month will be equal. The most likely possibility is that, once the day and the month are equal, the synchronous rotation of the earth with respect to the moon's revolution would be stabilized for all future time by the shape of the earth's figure, which would be slightly elongated in the direction of the moon. The earth would point its axis of least inertia toward the moon, just as the moon now keeps its axis of least inertia toward the earth. This relation would be maintained as the solar tides continued to attempt to brake the earth's rotation. Most of the angular momentum of the earth-moon system would be held in the moon's orbit. The earth feels a torque from the moon as well as from the sun. As the sun drew angular momentum from the earth-moon system by braking the earth's spin, the moon would impart some of its angular momentum to the earth, enough to compensate for the sun's braking plus a little more. The result is that the earth's rotation would speed up again and the moon would approach the earth, thus decreasing its moment of inertia around the center of mass of the earth-moon system.

The possibility that the moon may once have been as close as a few earth radii deserves some additional comment. We can estimate that tides several kilometers in height would have sloshed about as the earth rotated with a five-hour day under a moon whose orbital period would have been only slightly longer. The energy dissipated by tidal friction would have exceeded the solar heating of the earth. Much of the ocean water would have evaporated, and there is the additional possibility that the earth's mantle would have melted to a considerable degree. All this activity would have been confined to the short time that

the moon remained near the earth—perhaps a few thousand years at distances of less than six earth radii. If this event was sufficiently early in the earth's history, subsequent evolution of the earth's surface could have erased all traces of it from the geological record. It might be expected that the moon, by virtue of its more stable surface, would be the place to look for the evidence of such a catastrophic event.

The returns from the two lunar landing missions *Apollo 11* and *Apollo 12* have already enormously increased our understanding of the origin of the moon. It is fair to say, however, that we cannot yet distinguish between the three rival theories of lunar origin. Nevertheless, there are clues in the data obtained so far that bear rather directly on all these hypotheses.

Analysis of the rocks returned by *Apollo 11* indicates that they crystallized 3.6 billion years ago; similar analysis of the rocks from *Apollo 12* shows that they crystallized only 3.25 billion years ago. These crystallization ages are substantially different, ruling out the simultaneous filling of maria ("seas") all over the moon. The filling of the maria must not have been associated with any unique event in lunar history. It had been specu-

lated that the flooding of the lunar maria resulted from melting induced by tidal friction in the moon following its capture by the earth. This idea can now be laid to rest.

Before the Apollo missions relative ages had been assigned to various areas on the moon by counting craters within those areas. It was reasoned that the more craters there were in an area, the older that area was, since it must have been exposed for a longer period of time to accumulate larger numbers of craters. Because the flux of impacting bodies in the past was entirely unknown, absolute ages could not be deduced. The radioactive dating of the returned samples has for the first time provided absolute ages for portions of the lunar surface. These age determinations lead to the conclusion that the flux of impacting bodies diminished by about three orders of magnitude (1,000 times) during the first billion years after the formation of the moon. This conclusion is forced on us by the fact that the flux of impacts on the highlands is about three orders of magnitude greater than the flux responsible for the craters on the lower Mare Tranquillitatis (the *Apollo 11* site). Since the age of the highlands must be less than 4.6 billion years (the probable age of the moon), they are at most one billion years older than Mare Tranquillitatis. The high initial bombardment rate is almost certainly connected with the process by which the moon was assembled out of smaller bodies.

Analysis of the returned samples indicated that the mare material was chemically fractionated at about the time the moon formed. We do not yet know to what extent this fractionation took place in space before the moon assembled and how much of it was due to subsequent internal fractionation in the body of the moon. The problem is probably connected to the difference in composition between the earth and the moon. Its solution would be a significant step forward. We have also learned that the composition of the mare rocks and soil cannot be representative of the bulk lunar composition [see "The Lunar Soil," by John A. Wood; SCIENTIFIC AMERICAN, August, 1970].

The return of lunar-highland samples is eagerly awaited. Perhaps these, the oldest rocks on the moon, will reveal the crucial evidence we need to distinguish between the rival theories of lunar origin. It is more likely that we shall still be speculating about the origin of the moon a century hence.

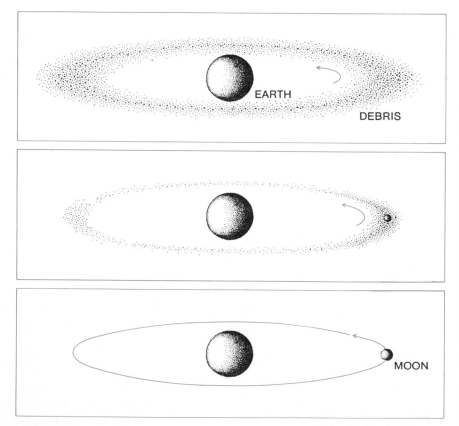

ACCRETION HYPOTHESIS suggests that the moon was formed after the earth from material of somewhat different composition in earth orbit. As yet there is no compelling reason to prefer any particular distance from the earth as the site of the moon's formation.

23

The Fine Structure
of the Earth's Interior

INTRODUCTION

Seismology is the principal tool for exploring earth's deep interior and no review of geoscience is complete without the inclusion of the subject. Bruce A. Bolt's article, "The Fine Structure of the Earth's Interior," is an appropriate choice because it summarizes the earlier methods and results that led to the discovery of the major internal divisions of our planet between the surface and very center. It then takes up the newest discoveries of the finer details of earth structure, made possible by the use of arrays of seismographs and seismic waves from nuclear explosions and conventional earthquakes.

It has long been known that the earth is divided into an outermost crust between ten and seventy kilometers thick, a rocky mantle extending to a depth of about twenty-nine hundred kilometers, and a central core made up mostly of molten iron on the outside, with solid iron below. What the new results tell us is something of the roughness of the boundaries, of the sharpness of the discontinuities between the divisions, and of smaller changes within the major divisions that may be due to partial melting, or closer packing of atoms due to the high pressures.

The earth is zoned into major divisions because, early in its history when its interior was much hotter, light compounds floated up and heavy materials sank into the interior. In a sense, then, we can study the earth's internal chemistry with seismology by using seismic waves like radar to detect chemical and physical boundaries. From the speed of seismic waves within the zones we can make inferences about the composition and physical state of material in regions that we will never be able to reach directly.

The Fine Structure of the Earth's Interior

by Bruce A. Bolt
March 1973

The waves sent out by earthquakes and nuclear explosions have been studied in detail with new seismometer arrays. They show, among other things, that the core of the earth has a solid kernel

The echoes of an earthquake ring deep into the earth. Depending on the medium through which they pass, the seismic waves are bent, speeded up, slowed down and in the case of some vibrations stopped altogether. When the faint echoes of the earthquake emerge at the earth's surface, they actuate the sensitive pendulum recorders known as seismographs. By correlating the records of seismographs at different locations it is usually a straightforward matter to establish the time and location of the original event, provided that it exceeded a certain magnitude.

During the past decade the worldwide interest in man-made seismic events (nuclear-test explosions) has led to a new generation of sensitive seismographs and the development of seismic arrays that have provided investigators with remarkably effective probes for studying the earth's interior. The delicate traces of the new instruments have enabled seismologists to confirm many of the older hypotheses of the earth's internal structure and have led to the discovery of fascinating new features. Just as radio telescopes have revealed celestial objects that were once invisible, the new generation of seismic instruments has detected fine details of our globe that were once unobservable.

As an example, geophysicists have speculated for years on the nature of the earth's center. Is it a solid or a liquid, and what is its density? The answers have been put on a much firmer basis within the past year. At the earth's center is a solid inner core with a density about 13.5 times the density of water. The radius of this core is some 1,216 kilometers, which makes it a little larger than the moon. The solid inner core is surrounded by a transitional region a little more than 500 kilometers thick. The transitional region in turn is surrounded by a liquid outer core about 1,700 kilometers thick.

Outside the liquid outer core is a mantle of solid rock some 2,900 kilometers thick, which approaches to within 40 kilometers of the earth's surface under the continents and to within 10 kilometers under the oceans. The thin rocky skin surrounding the mantle is the earth's crust. No drill has penetrated the earth's crust deeper than a few kilometers.

The basic technique in deep-earth "prospecting" is to measure the time of travel of a seismic wave from the instant of its generation by an earthquake or explosion to its arrival at a seismographic station. The distance between the seismic event and the station is expressed as the angle (designated delta) formed at the earth's center between radii drawn to these two points. Thus if the earthquake were anywhere on the Equator, it would be 90 degrees "distant" from a recording station at the South Pole. Similarly, if the earthquake were on the Equator in the Andes at 80 degrees west longitude, it would be 90 degrees distant from a recording station on the Equator in Gabon on the west coast of Africa at 10 degrees east longitude.

Earthquakes and explosions generate two types of wave that travel through the interior of the elastic earth. In seismology they are referred to as P (compressional) waves and S (shear) waves. The speed of the waves depends on the density and elastic properties of the rocks through which they pass. S waves are slower than P waves and do not pass through regions that are liquid. A wave that arrives at the seismograph without reflection is designated by a single P or S. If the wave is reflected once at the earth's surface, it is labeled PP or SS. In addition various letters and numbers can be inserted between the two (or more)

P's or S's to indicate more specifically the region through which the wave has passed. Thus a compressional wave that passes through the earth's central core is called PKP, or P' for short. The wave that is reflected from the opposite side of the earth is labeled $P'P'$.

During the first half of this century seismologists painstakingly built up the curves of travel time against angular distance for P and S waves using hundreds of specially selected earthquakes from around the world. In 1906 the British seismologist R. D. Oldham first proposed that earthquake waves show that the interior of the earth is not featureless. He explained the pattern of arrival of the dominant P and S earthquake waves by postulating that the earth has a large central kernel. Although the details of his argument proved to be incorrect, Oldham's general conclusion was later verified in many ways. Somewhat stronger evidence of discontinuous features in the earth's structure was discovered in 1909 by Andrija Mohorovičić of Yugoslavia. He found that when he plotted the time-distance curve for P waves from Balkan earthquakes, there was a sharp bend near a distance of about 200 kilometers (an angular distance of about two degrees). Mohorovičić explained the bend by supposing that at a depth of about 50 kilometers there was an abrupt change in the properties of the earth's interior. That discontinuity, which separates the superficial crust from the mantle, was later found to be worldwide.

The detection of this discontinuity below the surface was soon followed by the firm location of the much greater discontinuity, nearly halfway to the earth's center, that had been proposed by Oldham. Seismologists had noted that at distances of up to 100 degrees P waves from earthquakes were recorded with more or less uniform amplitudes but that at distances beyond 100 degrees the amplitudes decayed dramatically. The short-period *P* waves appeared strongly and consistently again only at distances greater than 140 degrees, but at these larger distances they arrived some two minutes later than one would expect on the basis of simply extrapolating their velocity for distances less than 100 degrees.

An explanation that fitted the observed pattern of travel times was worked out numerically in 1913 by Beno Gutenberg of the University of Göttingen. He calculated that at a depth of about 2,900 kilometers the velocity of *P* waves falls precipitously by about 40 percent. This discontinuity marks the structural boundary between the mantle and the core. Seismograms show that *S* waves can travel anywhere above this major boundary, indicating that the mantle material must be rigid. For at least 2,000 kilometers below the core boundary, however, no *S* waves have been observed to propagate. For this reason and others the outer 2,000 kilometers of the core are regarded as being molten.

One might expect that a sharp boundary between the mantle and the core would reflect seismic waves. Indeed, such echoes are clearly observed on seismograms. A direct reflection of *P* waves from the boundary between the mantle and the core is designated *PcP* [*see illustration below*]. Echoes of the *PcP* type provide a direct means of determining the depth of discontinuities in the deep regions of the earth.

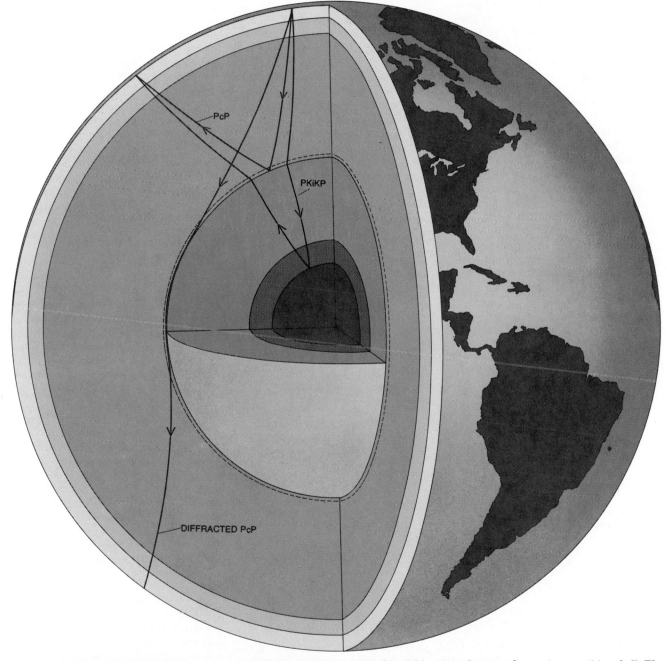

CROSS SECTION OF THE EARTH is based on the most recent seismological evidence. The outer shell consists of a rocky mantle that has structural discontinuities in its upper part and at its lower boundary that are capable of reflecting or modifying earthquake waves. Below the mantle an outer fluid core surrounds a solid ker- **nel at the earth's center; between the two is a transition shell. The paths taken by the three major kinds of earthquake wave are depicted. The waves reflected from the outer liquid core are designated *PcP*; the waves reflected from inner solid core are *PKiKP*; the waves that creep around the liquid core are diffracted *PcP*.**

In the 1920's, with the aid of somewhat improved seismographs and more refined earthquake surveillance, it became possible to detect the onset of delayed waves of the P type at distances between 110 degrees and 140 degrees. At such distances the waves evidently passed through the core and were of the PKP type. Inge Lehmann of Denmark suggested in 1936 that the pattern of travel times for these PKP waves could be explained if the core consisted of two regions, an outer one and an inner one. This notion was endorsed by Gutenberg, who now was working with Charles F. Richter at the California Institute of Technology, and independently by Harold Jeffreys of the University of Cambridge. In one of the most impressive sustained studies ever carried through in the physical sciences the two Cal Tech seismologists and Jeffreys (who worked in the early stages with K. E. Bullen) independently computed average velocity distributions for the whole of the earth's interior based on analyses of thousands of seismographic records of P and S waves. In general these independent solutions agreed as well as the measurements of that time allowed.

With the discovery of the inner core all the major boundaries within the earth had apparently been located. Other discontinuities were suggested from time to time to account for observed discrepancies in seismograms, but by and large

measurements from the available instruments did not have the resolving power to clinch the case for further significant worldwide deep structures. In the period immediately following World War II, however, many seismologists became convinced (particularly from studies of seismic waves of long wavelength traveling around the earth's surface) that the structure of the upper few hundred kilometers of the earth was complex and also that it varied from place to place. The velocities of P and S waves in many geographical regions inexplicably decreased in a layer below the crust.

Around 1960 observational seismology took a major leap forward. Largely as a consequence of the attempt by several countries to find ways to discriminate between underground nuclear explosions and natural earthquakes, seismology was transformed from a neglected orphan of the physical sciences into a family favorite. A global network of more than 100 standardized sensitive seismographic units was established with U.S. support, and many other earthquake observatories were modernized throughout the world. Arrays of seismographs, comparable to giant radio antennas, were constructed by the U.S., Britain and other countries. In these seismic arrays the seismographs are connected in such a way that microseisms—the random small quivers of the earth—

can be filtered out. One such giant antenna is the Large Aperture Seismic Array (LASA) located near Billings, Mont. It consists of 525 linked seismographs distributed over an area 200 kilometers in diameter [see illustration on this page].

In the more than three decades following Lehmann's suggestion that the earth's core might itself have a core, no unequivocal evidence on the nature of the inner core's boundary had ever come to light. The evidence was finally supplied by LASA. In 1970 E. R. Engdahl of the National Oceanographic and Atmospheric Agency, Edward A. Flinn of Teledyne Incorporated and Carl F. Romney of the Air Force Technical Applications Center announced that the Montana array had detected the echoes, designated $PKiKP$, that had bounced steeply back from the boundary of the inner core. The source of the echoes was underground nuclear test explosions in Nevada as well as earthquakes [see top illustration on opposite page]. There were two immediate conclusions. First, the inner core has a sharp surface. Second, its radius is within a few kilometers of 1,216 kilometers (a value, incidentally, that I had predicted eight years earlier on the basis of other seismological evidence).

Even more can be done with these remarkable observations. By comparing the relative strengths of the PcP and $PKiKP$ pulses one can calculate the density of the rocks at the top of the earth's inner core. There are a number of uncertainties in such calculations. The equations assume that the speeds of P waves are well determined throughout the earth and that the densities of the materials on each side of the boundary between the mantle and the core are known. Some recent calculations I have made in collaboration with Anthony I. Qamar indicate that the density at the center of the earth cannot be much greater than 14 times the density of water. Earlier estimates had ranged as high as 18 times the density of water. Our value agrees quite well with the density that iron is estimated to have when it is subjected to the pressure that exists at the earth's center.

Let us now turn our attention from the core to the fine structure of the upper part of the earth's mantle, the region just under our feet, so to speak. The presence of great mountain ranges, mid-ocean ridges and deep ocean troughs, together with a considerable amount of geological and geophysical evidence, indicates that large portions of the upper

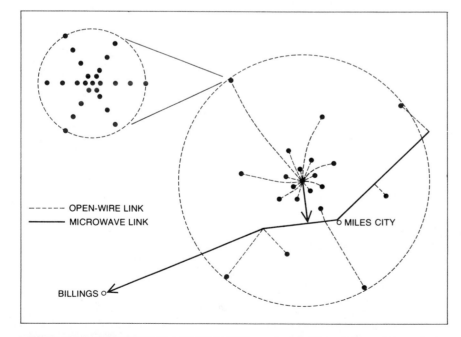

LARGE-APERTURE SEISMIC ARRAY (LASA) was installed near Billings, Mont., in the mid-1960's by the Department of Defense. One of the largest of more than 100 seismographic stations established throughout the world with U.S. support to detect underground nuclear explosions, it has provided much new knowledge of the earth's interior. LASA consists of 525 linked seismometers grouped in 21 clusters. In each cluster 25 seismometers are arranged as shown at the upper left. The array covers an area 200 kilometers in diameter.

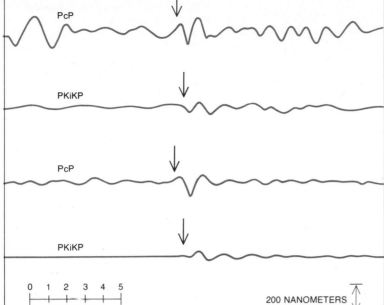

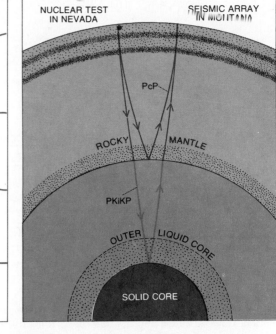

UNDERGROUND NUCLEAR EXPLOSION in Nevada on January 19, 1968 (code-named "Faultless"), produced the traces shown at the left on LASA seismometers. Time proceeds from left to right. The vertical scale shows the magnitude of ground movement involved; 200 nanometers is only half the wavelength of violet light. The *PcP* echoes from the outer core (*diagram at right*) are the closest to a straight-down-and-back path yet reported. Angular distance between explosion and recording instruments was only 11 degrees; this was the angle between lines from two points to the earth's center. *PKiKP* pulses represent echoes from solid inner core.

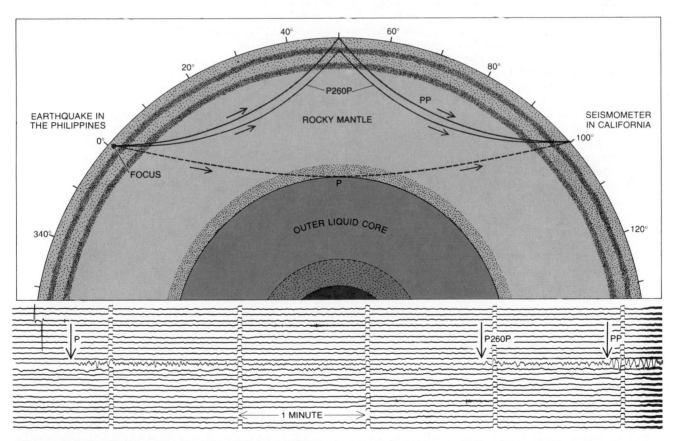

PHILIPPINE EARTHQUAKE generated waves that followed three distinct paths (*top*) before arriving at a high-gain seismograph located in an abandoned mine in the Sierra Nevada. The installation, the Jamestown Station, is operated by the University of California at Berkeley. In the seismograph a spot of lights is focused on a rotating drum, producing a series of parallel lines (*bottom*). The line in the Jamestown seismogram that contains three distinct echoes from the Philippine earthquake is shown in color. The *P* wave arrived first, followed some three minutes later by the echo *P260P* that probably bounced off a reflecting surface 260 kilometers deep in the earth under the Pacific Ocean. The reflection from the underside of the ocean floor, *PP*, arrived another minute later.

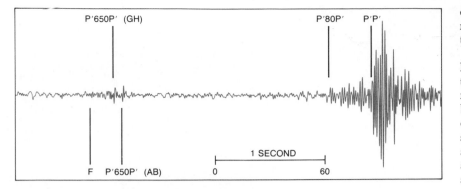

UNDERGROUND NUCLEAR TEST on October 14, 1970, on the Russian island of Novaya Zemlya produced this seismic trace at the Jamestown Station. The large phase *P'P'* was produced by a compressional wave reflected from the other side of the globe under Antarctica, as depicted in illustration on the opposite page. It was preceded by echo *P'80P'*, evidently reflected from a structure 80 kilometers below surface of Antarctica. Two minutes earlier still record shows doublet *P'650P'* (*GH*) and *P'650P'* (*AB*), evidently reflected from a layer 650 kilometers below surface of Antarctica. Origin of wave train starting at *F* is unknown.

part of the earth are in continuous slow motion. Indeed, it now appears that the surface of the earth is divided into six to eight large platelike regions that move with respect to one another. Africa is on one such plate; North America is on another. Dynamic processes operating particularly at the edges of plates would seem to account for much of the topographic relief on the surface of the globe. Most volcanoes are located and most earthquakes occur near the plate margins. One would expect large-scale movements of crustal masses to be reflected in the architecture of at least the upper few hundred kilometers of the earth. One would also expect this variable architecture to give rise to varied patterns on seismograms, and indeed different patterns are observed.

A few years ago, working with Qamar and Mary O'Neill at the University of California at Berkeley, I found a series of unexpected waves on seismograms. Measurements indicated that the waves might have been reflected from the underside of layers located perhaps hundreds of kilometers below the Mohorovičić discontinuity. The waves arrived at seismometers in the Berkeley network as much as 150 seconds before the corresponding waves reflected from the underside of the earth's surface.

We refer to these echoes generically as *PdP* waves. When we are able to calculate the depth below the surface of the layer from which the waves are reflected, we insert the figure in place of *d*. Thus a wave reflected from a layer 260 kilometers below the surface is designated *P260P*. A clear example of such an echo is one that was produced by an earthquake on May 22, 1972, near the Philippines and recorded at Jamestown,

Calif., near the Nevada border [*see bottom illustration on preceding page*]. The distance between the focus of the earthquake and California is about 100 degrees.

Such reflections do not always show up on seismograms, and for this reason and others some seismologists have suggested that the wave paths are not as symmetrical as we have proposed. Notwithstanding the reservations of our colleagues, we are satisfied that many observations of *PdP* waves can be explained in terms of a roughness of the rocky material in the upper part of the earth's mantle that is capable of producing reflections. It is noteworthy that a significant discontinuity in the upper mantle was suggested as far back as 1926 by Perry Byerly of the University of California at Berkeley on the basis of travel times for waves produced by an earthquake in Montana. Byerly's plot of observed travel times against distance showed peculiarities that could be explained if the *P* waves had encountered some kind of surface at a depth of about 400 kilometers.

Our interpretation has been strengthened by slightly different types of analyses. In 1968 R. D. Adams of the Seismological Observatory in New Zealand and a year later Engdahl and Flinn in the U.S. independently observed small waves that arrived slightly earlier than the usual *P'P'* echoes. Ordinary waves of the *P'P'* type make the long journey from the focus of an earthquake to the other side of the earth and are reflected back to a station in the same hemisphere as the earthquake, having passed through the core twice. Adams and Engdahl and Flinn interpreted the precursor waves as *P'P'* waves that did not quite reach the

opposite surface of the earth but were reflected back from a discontinuity in the upper mantle.

Waves of the *P'P'* type are particularly useful for probing the earth's structure. Their path is so long that they arrive some 39 minutes after they have begun to be generated by an earthquake. Therefore when they reach the seismograph most of the other waves sent out by the earthquake have already arrived at the observatory and the instrument is quiescent.

A particularly striking example of multiple long-distance reflections was provided by an underground nuclear explosion at the Russian test site in Novaya Zemlya on October 14, 1970. The *P'P'* waves passed through the earth's core, were reflected under Antarctica and returned to the Northern Hemisphere. In a recording made at Jamestown the main echo *P'P'* is the most prominent feature on the seismogram [*see illustration on this page*]. About 20 seconds before the onset of the large *P'P'* reflections a train of much smaller waves begins that can be explained as reflections from the underside of layers located in the 80 kilometers of rock below the surface of Antarctica. These forerunner waves are thus designated *P'80P'*.

As the eye scans the seismogram further from right to left only inconsequential waves are seen for more than a minute and a half; they are minor jiggles continuously produced by the background microseismic noise of the earth. All at once, almost precisely two minutes before the first *P'80P'* waves, one can see a beautiful doublet: two sharp peaks, separated by a few seconds, that stand out clearly above the background shaking. These sharp pulses agree nicely with the expected travel time of rays reflected by a layer located some 650 kilometers below the surface of Antarctica; hence they are designated *P'650P'*. The presence of a doublet means that there was only a slight variation in the paths of the two rays reflected from the 650-kilometer layer. Presumably one of the rays entered the transition layer, or *F* layer, that is thought to lie between the inner solid core and the outer liquid core and the other ray did not [*see illustration on opposite page*].

Many other examples have now been reported, particularly by James H. Whitcomb of the California Institute of Technology, of *P'P'* waves arriving earlier than one would expect if the upper mantle of the earth were uniformly smooth. Most seismologists now agree that these precursor waves at least indicate the existence of a rather sharp boundary at a depth of 650 kilometers and that the boundary is probably a worldwide feature.

It should be noted, however, that the particularly clear seismographic record of the Novaya Zemlya test explosion shows no spikes between the $P'650P'$ waves and the $P'80P'$ waves, as one would expect if there were a sharp reflecting surface at some intermediate depth, such as the 260-kilometer depth I have mentioned in connection with the record produced by the earthquake near

the Philippines. The absence of intermediate spikes indicates either that such shallower discontinuities are not present under Antarctica or that they are not so easily detected by short-period P waves that pass through the earth's outer core because the shallower discontinuities are less sharp than the one at 650 kilometers.

Although the geophysical extent and the precise depth of some of the abrupt changes of structure in the upper mantle remain indefinite, the newly studied class of PdP and $P'dP'$ reflections strongly indicates that sharp variations in rock

properties do exist in the top few hundred kilometers of the earth. That is just the region where, according to the new plate-tectonic view of the earth's crust, there must be a driving mechanism to account for the movements of continents and crustal plates. The mechanism most frequently proposed is the slow flow of viscous hot rock in the form of convection cells immediately below the plates in the upper mantle. It is not obvious how such a flowing region could be reconciled with the observations of sharp structural discontinuities. Geophysicists

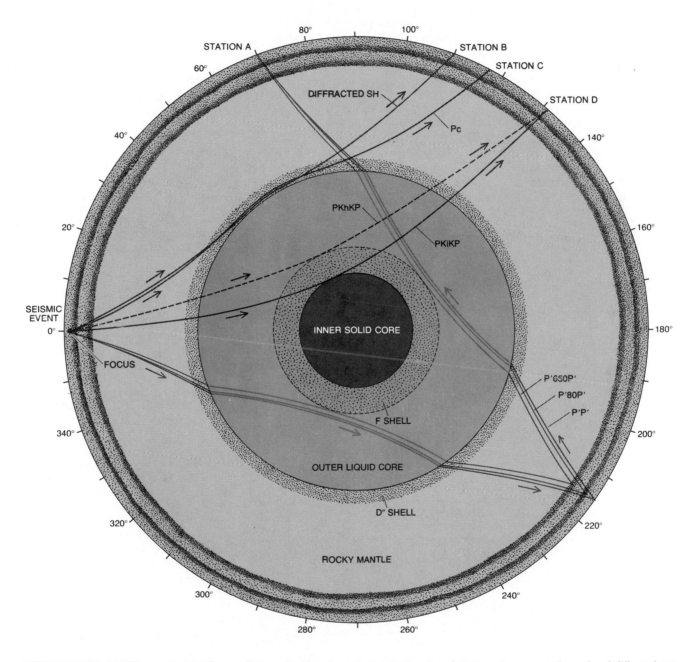

SEISMIC-WAVE PATHS are superposed on a diagram showing the shells that make up the earth's interior. The three nearly parallel rays arriving at Station A represent the paths taken by seismic waves from the Russian underground nuclear test whose seismographic record is shown on the opposite page. The two rays arriv-

ing at Station B and Station C represent the paths of diffracted SH waves and Pc waves depicted in the illustration on the next page and in the top illustration on page 277. Studies of such waves clarify the nature of D'' shell. Deeper rays arriving at Station D represent the paths of waves shown in the bottom illustration on page 277.

are now busy examining whether or not convecting rocks can maintain sharp boundaries at certain depths.

Seismological evidence has long existed that suggests some kind of transition region or thin shell at the base of the rocky mantle just above the liquid outer core at a depth of about 2,900 kilometers. This inferred shell was given the label D'' even before much was known about it. Recently fresh light has been thrown on the velocities of P and S waves in the D'' shell, allowing some inferences about its properties.

In the first place modern instruments, unlike early seismographs, do detect clear waves of the short-period P type at distances well beyond 100 degrees. Such observations are surprising, because if the wave speeds remained constant or increased at the bottom of the mantle, the core would act as an opaque screen that would quickly cut off the direct, short-period seismic waves at about 105 degrees. Beyond this distance the seismographic stations would be in the shadow of the earth's core.

The actual situation can be compared to what happens when a shadow is cast on a wall by an opaque object illuminated by the sun. The shadow is not quite sharp because the light waves are diffracted, or bent, by the edge of the object. In the earth some seismic waves get diffracted into the shadow produced by the earth's core. Regardless of the actual velocity of P waves at the base of the mantle, therefore, the shadow region is always dimly illuminated with earthquake waves. It turns out, from work done by Lehmann and by Robert A. Phinney of Princeton University, as well as from the studies of others, that the strength of the short-period waves actually observed in the shadow of the core is greater than it should be if diffraction in a zone of constant velocity were the only mechanism operating.

Recently I have completed a study of waves designated Pc that have traveled out to more than 118 degrees and have arrived at seismographic observatories in the core shadow zone with substantial strength [see illustration below]. The time of travel and the amplitudes of such waves indicate that they have passed through a shell where the speed of propagation is perceptibly less than the speed that prevails only 100 kilometers or so above the core boundary. Putting the evidence together, the best explanation seems to be that the P-wave velocity drops by a few percent in the D'' shell above the earth's core.

The evidence from the Pc waves does not stand alone. Because the outer core is liquid, certain kinds of shear (S) waves are inhibited from creeping around the core boundary; their energy leaks off into the core in the form of P waves. There is, however, one type of shear wave, called the SH wave, that cannot produce P waves in the liquid core. Its energy gets trapped at the base of the mantle, and thus it can travel great distances. Very large SH pulses have been recorded at distances ranging from 90 degrees to more than 115 degrees from an earthquake [see top illustration on opposite page]. From the measured travel times one can derive the velocity of the shear wave as it travels around the core boundary. Calculations carried out in recent years by John Cleary of the Australian National University, by Anton L. Hales and J. L. Roberts of the University of Texas at Dallas, by Mansour Niazi, then working at Berkeley, and by others indicate that the S-wave velocity decreases at the base of the mantle in the D'' shell in the same way that the P-wave velocity does.

What kind of earth model can be based on the fine-structure results I have been discussing? The model should also take other facts into account, notably the observed periods of the earth's free oscillations [see "Resonant Vibrations of the Earth," by Frank Press; SCIENTIFIC AMERICAN, November, 1965]. Such a model was computed at Berkeley in 1972. Named CAL 3, it shows the average variation in P-wave and S-wave velocities and in rock density from the surface of the earth to the center [see illustration on page 279].

What does the low velocity in shell D'' mean? One suggestion is that in this narrow shell the mantle rocks become less rigid as they approach the liquid core. Another possibility is that the rock composition changes slightly in the D'' shell. Below the mantle-core boundary the measured physical properties indicate a material that consists mainly of iron. Perhaps there is about 10 percent more iron in the D'' shell than there is in the rocks above it. Such iron enrichment may represent iron that never settled into the core during the formation of the earth or it may represent liquid iron that has diffused outward to form an alloy with the solid mantle.

Currently there is controversy concerning the possible presence of bumps, perhaps 500 kilometers from one side to the other, on the boundary between the mantle and the core. The existence of such bumps has been suggested by Raymond Hide of the British Meteorological Office and by others to explain the variations in the strength of the earth's magnetic and gravity fields as they are measured at different points on the earth's surface. The only way to "X ray" the earth for such fine structure is to use short-period seismic waves that interact with the boundary of the core. It can be calculated, however, that undulations on the mantle-core interface of less than 10 kilometers cannot easily be resolved with waves of the PcP type that return to the surface after a single reflection.

Fortunately, with the advent of sensitive seismographs a method has come into being that has improved resolving power. Some of these seismographs are at sites so quiet that they regularly de-

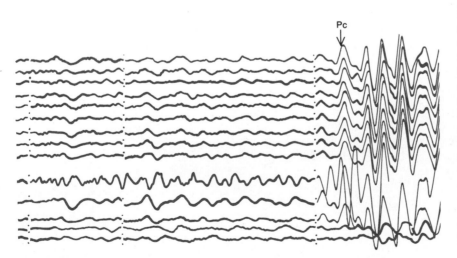

Pc WAVES FROM AN EARTHQUAKE in the South Sandwich Islands were guided with only small loss around the earth's core by the D'' shell. They were recorded at an angular distance of 118 degrees by an array in Uinta Basin of Utah. Here each wavy line is produced by a separate seismometer. Distance between breaks on the bottom line is 10 seconds.

tect waves that have been trapped inside the core and internally reflected four to seven times before emerging and returning to the surface. A wave that has been internally reflected seven times is designated *P7KP* [*see top illustration on next page*]. Such multiple reflections had been predicted but were never seen until the advent of the modern sensitive seismograph and of seismic arrays. It was a great thrill when we scanned along a seismogram made at Jamestown of an explosion on Novaya Zemlya and there —Eureka!—at the travel time predicted for a wave reflected seven times was an unmistakable tiny pulse nestling in the valley of microseismic background noise [*see bottom illustration on next page*].

It is doubtful that we will ever see a natural earthquake generate such a clear example of a *P4KP* or *P7KP* wave. Sharp onsets can be seen when the source is a nuclear test because the energy is released explosively in a way that is simpler than the way energy is released in most earthquakes. In time planned experiments using nuclear explosions, set off in particularly favorable locations and designed to present no hazard to the human environment, should provide still more sensitive probes of the fine detail of the interior of the planet.

What conclusions can be drawn on the nature of the mantle-core boundary from the records of *P4KP* and *P7KP* waves? First, the onset of the waves is quite abrupt. This confirms that the mantle-core boundary is a sharp discontinuity, perhaps extending over no more than two kilometers. Second, the additional distance represented by the extra three legs in the *P7KP* wave reduces its amplitude to about a third of the amplitude of the *P4KP* wave. The small size of the decrease implies that the liquid outer core transmits short-period *P* waves very efficiently indeed.

Finally, if the multiple reflections encountered topographic bumps on the mantle-core boundary more than two kilometers high, the travel times of the waves would be altered enough for the variations to be measurable. By comparing the travel times of many multiply reflected waves at the same seismographic station one should be able to derive the height of the bumps, if any, from the amount of variation in the times. Although studies of this kind are barely a year old, the present indications are that the variation is no greater than it is for waves that do not bounce from the core boundary. Therefore one can say tentatively that if there are topographic undulations, either their height is less than a few kilometers or, if their height is sig-

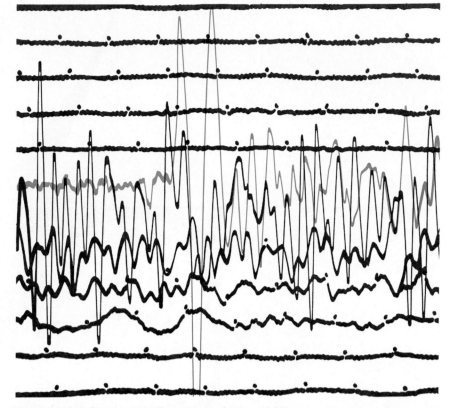

HORIZONTAL SHEAR WAVE of the type designated *SH* was produced by a major earthquake in Iran on August 31, 1968. It was recorded at a distance of nearly 100 degrees by a seismograph at Port Hardy in Canada that responds only to horizontal ground motion. The large *SH* pulse sent pen flying across recording drum, crossing traces made earlier and later.

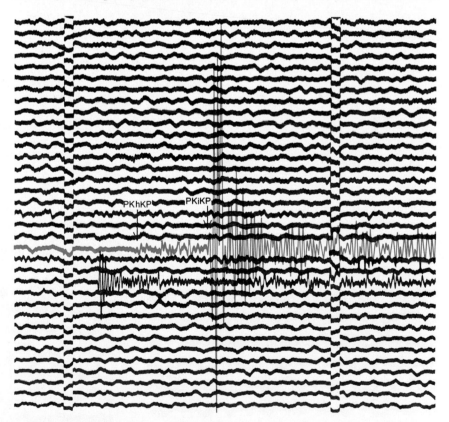

WAVES THAT PENETRATED EARTH'S CORE were created by a deep earthquake near Java and recorded at a distance of 132 degrees by a seismograph at Golden, Colo. The large-amplitude wave *PKiKP* was refracted through the earth's solid inner core. It was preceded 17 seconds earlier by smaller *PKhKP* waves, which were probably reflected from *F* layer.

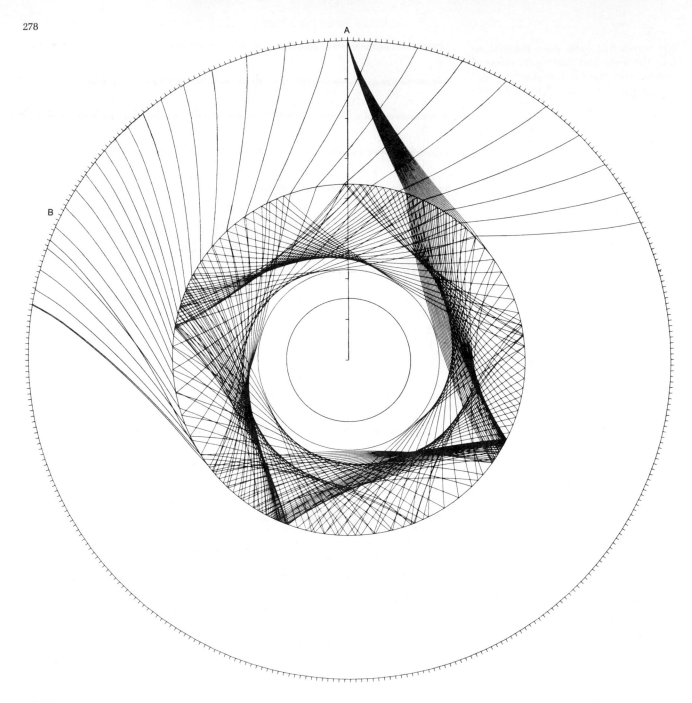

MULTIPLE REFLECTIONS can result from P waves that get trapped inside the earth's liquid outer core. This computer plot depicts the paths of waves, generated by a seismic event at A, that have bounced inside the core seven times before reaching the surface, for example a station at B. Such waves are assigned a K for each bounce, hence the waves shown would be designated *PKKKKKKKP*, or *P7KP*. Computer program that produced ray paths was devised by C. Chapman for seismological average earth model called CAL 3.

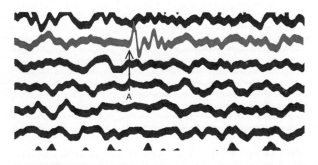

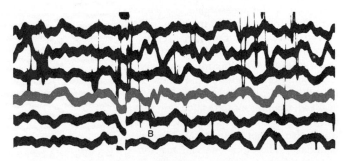

FAINT PULSE OF *P7KP* ECHO can be seen at the right (B) in this seismogram made at Jamestown of an underground explosion on Novaya Zemlya in 1970. The stronger *P4KP* pulse, labeled A at the left, was recorded about 20 minutes earlier; it represents a wave that was reflected only four times inside the core of the earth. Such evidence indicates that the boundary of the core is rather smooth.

nificantly higher than that, they are few in number.

Let us now focus our attention still deeper in the earth. Even before 1940 the better-equipped seismological observatories reported tiny but clear *P*-wave precursors about 10 seconds before the main onset of the core waves (*PKP*) at distances of between 130 and 140 degrees. Quite unmistakable observations of such precursor waves were reported on seismograms from the nuclear explosions at Bikini Atoll in 1954. A recent typical precursor wave was generated by an earthquake near Java and recorded 132 degrees away at a seismic station in Golden, Colo. The precursor waves arrived 17 seconds before the onset of the much stronger *PKP* wave [*see bottom illustration on page 277*].

One straightforward explanation for these precursors occurred to me in 1962. Perhaps the wave velocities around the inner core were significantly different from the values that were accepted at the time. I suggested that between the inner and the outer core there might be a transition shell, designated *F*, that has a small jump in velocity at its upper boundary. After working out many mathematical models of the earth's core I found that, although other explanations are possible, the hypothesis of an *F* shell predicts travel times in reasonable agreement with all the *PKP* observations, including the precursor waves. Independent studies by Adams and M. J. Randall at the Seismological Observatory in New Zealand soon confirmed the general results I had obtained. The precursor waves reflected from the surface of the *F* shell were named *PKhKP*. Just in the past year an alternative explanation of some of the precursor waves has been put forward by R. A. Haddon of the University of Sydney. He suggests that they are *PKP* waves scattered by bumps on the mantle-core boundary.

There were several fruits of my proposal of 1962. Revised travel times for the core waves, based on the *F*-shell model, not only helped other seismologists to find more examples of the small *PKhKP* precursors but also helped in the search for the reflections (*PKiKP*) from the boundary of the inner core. It was these latter reflections that led to the conclusions that the inner core has a sharp surface and that its radius is about 1,216 kilometers.

An important by-product of the 1962 core-velocity estimates concerns the question of whether the inner core is solid or not. The notion that the inner core might be solid was put forward in 1940 by Francis Birch of Harvard University.

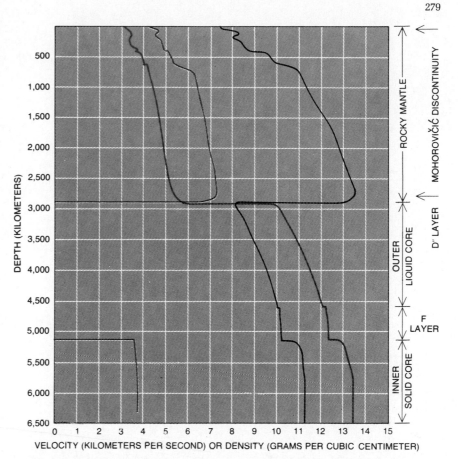

SEISMOLOGICAL AVERAGE EARTH MODEL (CAL 3) was computed last summer in author's laboratory at the University of California at Berkeley. The model, which takes all available seismological information into account, is defined by the three curves. The colored curve shows the variation in the earth's density with depth. The solid black curve shows the average velocity of *P* waves. The gray curve shows the average velocity of *S* waves. Since *S* waves are not propagated through liquids, the gray curve is interrupted by the earth's outer liquid core. The energy in an *S* wave, however, can be transmitted as a *P* wave through the liquid core and then reconverted into an *S* wave for transmission through the solid inner core. Direct evidence for a solid inner core has only been obtained in the past year.

versity. Now, if the inner core is rigid, it should transmit not only compressional waves but also shear waves, giving rise to waves, designated *PKJKP*, observable under specially favorable conditions at the earth's surface.

Working at the University of Sydney with the travel times of *P* waves and the wave velocities then assumed for the core, Bullen made estimates in 1950 of the travel times for *PKJKP* waves. There followed an enthusiastic but unsuccessful search for them. Seismologists at observatories around the world scrutinized seismograms to see if there were unexplained wiggles near the predicted times for *PKJKP* waves. In my own work on the core I estimated in 1964 an S-wave velocity in the inner core of about 3.7 kilometers per second. As a result the search for *PKJKP* waves was shifted to arrival times several minutes later than the ones that had first been tried.

Early last year Bruce Julian, David Davies and Robert Sheppard of the Mas-

sachusetts Institute of Technology provisionally announced that the large-aperture seismic array in Montana had recorded *PKJKP* waves close to the arrival time predicted by the 1962 model. If this announcement is confirmed, a long-sought key to the nature of the earth's kernel will have been found.

What further consequences follow from our new knowledge of the planet's fine structure? Many could be mentioned. For example, since wave speeds depend on rock densities, the revised *P* and *S* velocities in the earth make it possible to estimate the density variation with greater assurance, as incorporated in the CAL 3 model. Geochemists who try to describe the composition of the materials in the various shells will be aided by the more precise determinations of structural boundaries and density. Ultimately the entire body of geophysical and geochemical knowledge must be incorporated into a satisfying account of the evolution of the earth.

24

The Chemistry of the Solar System

INTRODUCTION

What motivates scientists to speculate on the origin of the system of planets? Aside from natural curiosity, which may be reason enough, they know that a planet's beginnings determine its evolution. The initial size and composition of the earth and its distance from the sun set the course for its evolution to the planet we inhabit—one with continents, oceans, mountains, volcanoes, beneficent atmosphere and climate. The earth is unique in our solar system, but knowing something about how it started, we might discover whether similar planets, hospitable to life, might exist in other solar systems of the universe.

Speculation and controversy have marked the subject of planetary origins for hundreds of years. The theory described by John S. Lewis in "The Chemistry of the Solar System" represents a new approach. It makes use of the known abundances of elements in the cosmos, and of the obviously reasonable assumption that about 5 billion years ago the primitive solar nebula was hot at its center and cool at its edges. Chemistry tells us that the reactions governing condensation from the nebula depend primarily on temperature. Given these starting conditions, one can calculate the composition of matter that condenses from the nebula at different distances (hence temperatures) from the sun. A major uncertainty concerns the rate of cooling of a nebula, for the reactions would proceed differently if the gas cooled rapidly or slowly compared to the time of formation of the planets.

Fortunately lunar and planetary space flights, as well as observations from earth, provide data on the compositions of the planets and their atmospheres, thus providing a check on the various theories of planetary formation. Any theory must explain the division of the planetary system into two groups: the small, inner terrestrial planets and the outer, massive Jovian planets. Within these groups there are major similarities as well as striking differences that must be accounted for. It is interesting that Lewis can use a chemist's approach to predict the compositional zoning and structure of planetary interiors. In the next few years, spacecraft will examine atmospheres and surfaces, measure planetary masses, moments of inertia, and magnetic fields and, on Mars and Venus, land instrumental packages. Before the end of the decade we should be able to verify many of the predictions of this new approach to planetary formation through chemistry.

The Chemistry
of the Solar System

by John S. Lewis
March 1974

*The sun, the planets and other bodies in the system
formed out of a cloud of dust and gas. What processes
in the cloud could account for the present
composition of these objects?*

The era of space exploration, with the landing of spacecraft on the moon and Venus and the flyby missions to Mars, Venus and Jupiter and now to Mercury and Saturn as well, has greatly enlarged our knowledge of the composition of the planets and satellites and the evolution of the solar system. Our modern understanding of the composition of the solar system nonetheless has a substantial history. It began in the 1930's with the work of Rupert Wildt on the physics and chemistry of the Jovian planets: Jupiter, Saturn, Uranus and Neptune. Wildt's investigations revealed that the atmosphere of Jupiter contains large amounts of the gases ammonia (NH_3) and methane (CH_4) and gave strong reason to suspect that all the Jovian planets consisted largely of hydrogen.

Around 1950 there was a second period of vigorous inquiry into the chemical nature of the solar system. At that time Harrison Brown proposed that the objects in the solar system could be divided into three classes according to their density and their chemical composition. The first class was the rocky objects such as the terrestrial (earthlike) planets, their satellites, the asteroids and the meteoroids. The second class was the objects consisting of both rocky and icy materials such as the nuclei of comets and the satellites of the outer planets. The third class was the sun and the Jovian planets, which consist mainly of matter in the gaseous state. At the same time that Brown suggested these classes Gerard P. Kuiper and Harold C. Urey were working on complex theories of the origin of the solar system. Urey stressed the importance of meteorites as clues to the origin of the planets, and he calculated the stability of the chemical equilibrium of numerous meteoritic minerals. Since that time the detailed study of the rocks brought back from the moon, the spectra obtained of comets and the numerous photographs made

from spacecraft of the moon and the planets, together with what we know about the geology of the earth, have all contributed to the current picture of the solar system and its origin.

The Solar Nebula

The solar system is not a homogeneous mixture of chemical elements. Moving outward from the sun, there is a general trend for the abundance of the volatile, or easily evaporated, elements to increase with respect to the abundance of the nonvolatile elements. That general trend appears to be evidence for the hypothesis that the Jovian planets and the other bodies in the outer solar system formed at low temperatures and the terrestrial planets formed at high temperatures. According to this hypothesis, the solar system formed out of the solar nebula: a large cloud of dust and gas. The center of the nebula was hot (several thousand degrees Kelvin) and the edges were cold (a few tens of degrees K.).

The difference in temperature meant that the composition of the dust must have varied radically with distance from the center. Close to the protosun in the center all materials would have been totally evaporated. At a distance of perhaps 20 million miles from the protosun, a fifth of the way to the present orbit of the earth, a very few nonvolatile materials could have condensed into solid particles. From that point out to the present location of the asteroid belt the dust would have been composed primarily of grains of rocky material with only a limited content of volatile substances. Beyond the location of the asteroid belt the nebula would have been cold enough to allow the volatile substances such as water, ammonia and methane to freeze into

THE PLANET JUPITER, seen here from the spacecraft *Pioneer 10*, is composed of much the same elements in the same proportion as the sun. It is a representative of one of the three compositional classes of objects in the solar system. This class includes the sun and Saturn. The other two classes are the rocky objects such as the terrestrial (earthlike) planets and the objects composed of both rocky and icy material, such as the nuclei of comets. In this picture the cloud belts of Jupiter are resolved in a wealth of detail invisible from the earth. The planet appears gibbous because of the position of the spacecraft with respect to the planet and to the sun, a view that can never be seen from the earth. The terminator, the line between the sunlit hemisphere and the dark one, is at the left of the image. The picture was made at a distance of 2,020,000 kilometers from the planet at 20:58 Universal Time on December 2, 1973, about 30 hours before the spacecraft's closest approach of 130,000 kilometers. It was made with the imaging photopolarimeter experiment of the University of Arizona. The photopolarimeter consists of a telescope with an aperture of 2.5 centimeters (one inch) coupled to an optical system that splits the light from the planet into a blue channel and a red one. The device scanned the planet as the spacecraft spun on its axis 4.8 times per minute, building up the image in a raster pattern somewhat like the image on a television set. The photopolarimeter alternated between the blue channel and the red channel every half millisecond and telemetered the raw data for each channel to the earth. At the University of Arizona the image from each channel was rectified to reduce distortion and increase its sharpness; then the two images were combined to produce a color composite.

SUN

MERCURY

VENUS

EARTH

MARS

JUPITER

SATURN

URANUS

NEPTUNE

PLUTO

OBJECT	DISTANCE FROM SUN (MILLIONS OF MILES)	DIAMETER (THOUSANDS OF MILES)	MASS (EARTH = 1)
SUN	–	867	343,000
MERCURY	36	3.0	.1
VENUS	67	7.6	.8
EARTH	93	7.9	1.0
MARS	142	4.2	.1
JUPITER	486	89	317.8
SATURN	892	76	95.2
URANUS	1,790	30	14.5
NEPTUNE	2,810	28	17.2
PLUTO	3,780	3.7	<.2?

NUMBER OF MOONS	DENSITY (WATER = 1)	ROTATION PERIOD (DAYS)	REVOLUTION AROUND SUN (YEARS)	AVERAGE TEMPERATURE (DEGREES K.)	GRAVITY (EARTH = 1)
–	1.4	27	–	5,800	28
0	5.4	55	.24	~ 600	.37
0	5.1	−243	.62	750	.89
1	5.5	1	1.00	180	1.00
2	3.9	1	1.88	140	.38
12	1.3	.4	11.86	128	2.65
10	.7	.4	29.48	105	1.14
5	1.6	.4	84.01	70	.96
2	2.3	.6	164.79	55	1.53
0	?	6.4	248.4	?	?

their ices, thus forming solid particles.

It is believed the sun formed when the center of the heated solar nebula became unstable and collapsed under the influence of its own gravity. The massive Jovian planets could have been formed by the same process. Alternatively they could have begun with a rocky and icy core so large that its gravitational attraction captured large masses of undifferentiated gaseous material. On this basis the similarity of their composition to the composition of the sun is easily understood.

The composition of the terrestrial planets and similar objects varies markedly with distance from the sun. There is a simple conceptual model, the equilibrium-condensation hypothesis, that can account for these differences. Let us assume that the chemical equilibrium between the dust and the gas in the solar nebula governed the composition of the solid materials that eventually accumulated into the planets, their satellites, the asteroids and the comets. On that basis one can calculate how the chemical composition of the condensed matter depends on its temperature at its formation. For this purpose one additional fact is needed: the relative abundances of the various chemical elements of which the primordial nebula was composed.

Our knowledge of the abundances of most elements in the galaxy is derived almost exclusively from the study of bodies in the solar system. The sun, however, is so much like most other stars that the relative amounts of the various elements in the solar system can be assumed to be typical of the large majority of stellar systems. Observations of the sun reveal that 99.996 percent of the mass of any sample of its material would consist of 15 elements. The least abundant element of the 15 (nickel) would still be commoner than all the other 80-odd elements put together.

There is a second model of how the terrestrial planets formed out of the solar nebula. The equilibrium-conden-

DATA ON THE PLANETS, incorporating information from various spacecraft missions as well as from ground-based observations, are displayed next to drawings of the planets that illustrate their relative sizes. Minus sign in front of rotation period of Venus indicates that the planet rotates in a direction opposite to the direction in which all other planets rotate. Tenth moon of Saturn, named Janus, was discovered in 1966.

DE-GREES KELVIN	EQUILIBRIUM-CONDENSATION MODEL	INHOMOGENEOUS-ACCRETION MODEL
1,600	1. Condensation of refractory oxides such as calcium oxide (CaO) and aluminum oxide (Al_2O_3) and also of titanium oxide and the rare-earth oxides	1. Condensation of refractory oxides such as calcium oxide (CaO) and aluminum oxide (Al_2O_3) and also of titanium oxide and the rare-earth oxides
1,300	2. Condensation of metallic iron-nickel alloy	2. Condensation of metallic iron-nickel alloy
1,200	3. Condensation of the mineral enstatite ($MgSiO_3$)	3. Condensation of the mineral enstatite ($MgSiO_3$)
1,000	4. Reaction of sodium (Na) with aluminum oxide and silicates to make feldspar and related minerals, and the deposition of potassium and the other alkali metals	4. Condensation of sodium oxide (Na_2O) and the other alkali-metal oxides at about 800 degrees K.
680	5. Reaction of hydrogen sulfide gas (H_2S) with metallic iron to make the sulfide mineral troilite (FeS)	
1,200 –490	6. Progressive oxidation of the remaining metallic iron to ferrous oxide (FeO), which in turn reacts with enstatite to make olivine (Fe_2SiO_4 and Mg_2SiO_4)	
550	7 Combination of water vapor (H_2O) with calcium-bearing minerals to make tremolite	
425	8. Combination of water vapor with olivine to make serpentine	
175	9. Condensation of water ice	5. Condensation of water ice (H_2O)
150	10. Reaction of ammonia gas (NH_3) with water ice to make the solid hydrate $NH_3 \cdot H_2O$	6. Condensation of ammonium hydrosulfide (NH_4SH)
120	11 Partial reaction of methane gas (CH_4) with water ice to make the solid hydrate CH_4 $7H_2O$	7 Condensation of ammonia ice (NH_3)
65	12. Condensation of argon (Ar) and leftover methane gas into solid argon and methane	8. Condensation of solid argon (Ar) and methane
20	13. Condensation of neon (Ne) and hydrogen, leading to 75-percent-complete condensation of solar materials	9. Condensation of neon (Ne) and hydrogen, leading to 75-percent-complete condensation of solar materials
~1	14. Condensation of helium (He) into liquid	10. Condensation of helium (He) into liquid

MAJOR REACTIONS that would have occurred in the formation of the solar system according to two different hypotheses are shown with respect to temperature in the primordial solar nebula. The equilibrium-condensation model assumes that the nebula cooled very slowly as the planets were forming. The inhomogeneous-accretion model assumes just the opposite: that the nebula cooled quickly with respect to the rate at which the planets were forming. The two assumptions result in radically different predictions for the composition of the planets and satellites. In the author's view the equilibrium-condensation model seems to describe the planets of the solar system more accurately. Color area indicates two steps of each hypothesis that probably did not occur because the temperature never fell that low.

sation model assumes that there were no substantial changes in the temperature of the nebula at the location of planets that were accreting solid material onto their surface. Perhaps the material accreted rapidly in comparison with the rate at which the nebula was cooling. Or perhaps the composition of the solid material was determined by the temperature when it had last reacted with the gases of the nebula; after that time the gases could have dissipated and the solids could have accreted slowly onto the planets, each of which would have been homogeneous in composition. In either case the chemical equilibrium between the gas and the dust would determine the composition of both the gases and the condensates present at any time.

Comparison of the Two Models

The second model, the inhomogeneous-accretion hypothesis, is exactly the opposite of the first. Here the solar nebula cooled rapidly in comparison with the rate at which the planets accreted solid material. Since the composition of the material would have been determined by its temperature its accretion onto a planet could have given rise to onionlike layers of different condensates. It is important to add that the two hypotheses are extreme cases; the truth may lie anywhere in between them. Since we have no a priori grounds for choosing between them, however, we should understand the consequences of both extremes.

The results of the two assumptions are quite different [*see illustration at left*]. If one starts with a temperature of 2,000 degrees K. in the solar nebula, the equilibrium-condensation hypothesis would lead first to the condensation of refractory compounds containing calcium oxide (CaO), aluminum oxide (Al_2O_3), rare-earth oxides and so on. Then when the nebula had cooled to about 1,500 degrees, a metallic iron-nickel alloy similar to the one found in meteorites would condense. That step would be followed by the condensation of enstatite ($MgSiO_3$) and then by the formation of various minerals such as feldspar through the deposition of sodium, potassium and the other alkali metals. As the temperature dropped to 680 degrees metallic iron would be corroded by hydrogen sulfide gas (H_2S) to make the mineral troilite (FeS); the remaining iron would be progressively oxidized to form minerals such as olivine, which would combine with water vapor at a still lower temperature to yield serpentine. Finally, as the temperature of the nebula dropped below 170 degrees, water vapor would condense as ice. The ice would later react with ammonia gas to make a hydrate, written $NH_3 \cdot H_2O$. At 100 degrees water ice would also combine with some of the methane gas in the nebula to yield another hydrate: $CH_4 \cdot 7H_2O$. Eventually argon and the leftover methane gas would freeze out at about 60 degrees. If the temperature continued to drop to as low as 10 degrees, neon and hydrogen would condense. Finally even helium would condense if the temperature ever fell below one degree. There is, however, no evidence that the temperature within the solar nebula was ever that low.

How does this sequence of events differ from the sequence predicted by the inhomogeneous-accretion hypothesis? One important difference is that those minerals that can be formed only by chemical reactions between gases and previously formed minerals cannot be made. This constraint rules out processes such as the corrosion of metallic iron by water vapor and by hydrogen sulfide gas, and the reaction of water ice with ammonia and methane to form the hydrates $NH_3 \cdot H_2O$ and $CH_4 \cdot 7H_2O$. Thus the sequence of chemical reactions in the process of inhomogeneous accretion is far simpler than the one in the process of equilibrium condensation.

At first the condensation of the refractory oxides, of the iron-nickel alloy and of enstatite would proceed in the same way as they would in the equilibrium-condensation process. Thereafter sodium oxide (Na_2O) would condense. Then water, ammonium hydrosulfide (NH_4SH) and ammonia would freeze into their respective ices. At the end methane and argon would freeze. Again the temperature would probably never drop low enough to allow the condensation of neon, hydrogen and helium. It is clear, however, that the dependence of composition and density on temperature for the material formed by inhomogeneous accretion would be quite different from that for material formed by equilibrium condensation.

There are ways to check the predictions of both models against reality. For example, we have quite a lot of information about the composition and the structure of one planet: the earth. Which of the two hypotheses is better able to account for what is known about that planet? Since the two models predict the content of volatile substances in the material that formed the planets, which hypothesis is in better accord with our knowledge of planetary atmospheres? Can we discriminate between the two hypotheses on the basis of the substantial amount of information we have about carbonaceous chondrites, the oldest and most primitive meteorites?

Evidence from Meteorites and Planets

Let us look first at the evidence from the carbonaceous chondrites. That evidence strongly favors the process of equilibrium condensation. Unlike the crust of the earth, chondrites are composed of unheated, unmelted, undifferentiated primordial material. The mineralogy and relative abundances of the elements of this preplanetary stuff are available for direct inspection. A mineral that is nearly universal in meteorites is troilite, or FeS. Troilite is an important feature of the sequence of events in equilibrium condensation, but it is absent from the sequence in inhomogeneous accretion. The next most widely distributed minerals are pyroxene and olivine, most often containing between 5 and 20 percent ferrous oxide (FeO). Ferrous oxide is an essential part of the equilibrium-condensation sequence.

Carbonaceous chondrites are rich in volatile materials, including a large quantity of water bound to other substances such as the mineral serpentine. It is an interesting fact that the spectra of many asteroids are indistinguishable from laboratory spectra of carbonaceous chondrites. That would be expected on the basis of the equilibrium-condensation model. Finally, carbonaceous chondrites are highly homogeneous. There is absolutely no tendency for the material in these meteorites to be strongly sorted into many different pure minerals; instead they tend to be remarkably well mixed.

The available data on the atmospheres of the planets are also much more easily understood on the basis of the equilibrium-condensation process than on that of the inhomogeneous-accretion one. The content of volatile elements in the planets formed by inhomogeneous accretion would be zero until the time when water had condensed out of the solar nebula; thereafter the condensed material would be more than 60 percent water! On the other hand, the equilibrium-condensation model predicts the right amount of water for the earth while leaving Venus extremely arid; it also predicts a water content for Mars that is six times higher than that for the earth.

The amount of carbon in the planets is predicted well by both models. The source of the oxygen to make carbon dioxide on the terrestrial planets would apparently have to be ferrous oxide. The content of ferrous oxide in Venus, the earth and Mars predicted by the equilibrium-condensation model is sufficient in all three cases to make the observed amounts of carbon dioxide. Ferrous oxide, however, is totally absent in the sequence predicted by the inhomogeneous-accretion model.

The basic chemical and physical properties of the earth all favor the equilibrium-condensation model. The chemical composition of the earth as a whole cannot be distinguished from that of the iron-rich carbonaceous chondrites. Unfortunately for students of such problems, however, some 99 percent of the mass of the earth consists not of crustal material but of mantle material and core material, neither of which can be analyzed directly. One must be content with inferring the composition of these materials from remote observations of such properties as density and the velocity of seismic waves.

It is generally agreed that the earth's mantle largely consists of compounds containing magnesium oxide (MgO), silicon dioxide (SiO_2) and ferrous oxide (FeO). Currently it seems probable that the entire mantle is about 10 percent ferrous oxide. The mantle is divided into two distinct layers: the upper layer is mainly silicates of iron and magnesium and the lower one, where the pressure is higher, is largely a mixture of oxides of iron and magnesium.

The core is also divided into two layers. The inner core, which accounts for only 1 percent of the earth's mass, has a density that corresponds to the density of a solid iron-nickel alloy under immense pressure. The outer core is liquid and has a density appreciably less than that of liquid iron alone at the same pressure, indicating that there must be a lighter element mixed in. The lighter component has generally been thought to be either silicon or sulfur. It is a remarkable coincidence that the equilibrium-condensation model alone provides the correct abundance of sulfur to explain the observed density of the outer core. Moreover, the equilibrium-condensation model explains why the earth has a slightly higher density than Venus by predicting that the earth has sulfur in its core and Venus does not.

Whether or not Venus actually is deficient in sulfur compared with the earth is not yet known, but it is certainly true that Venus' atmosphere lacks detectable sulfur compounds. If Venus emits volcanic gases as the earth does, it should have in its atmosphere a readily detectable amount of carbonyl sulfide (COS). Carbonyl sulfide has not been found, however, even by very sensitive spectroscopic methods.

Internal Structure of the Planets

On the basis of all the evidence from meteorites and planets let us accept the equilibrium-condensation model as yielding a good approximation of the primordial diversity in the composition of the solar system. Now let us briefly survey the nature of the physical and chemical processes going on inside the planets and satellites in order to gauge what types of internal structure could have been formed in their parent bodies.

The easiest question to ask is how the present internal structure of differentiated objects such as the earth depends on the temperature at which they formed. (Small bodies such as meteoroids and asteroids that were never warm enough to differentiate would of course remain as homogeneous mixtures of their constituent minerals.) In order to answer this question we must retrace the steps of the equilibrium-condensation process as the temperature within the solar nebula dropped from 2,000 degrees K.

The first minerals formed, the refractory oxide minerals, would have given rise to a class of protoplanets with a high concentration of calcium, aluminum, titanium and the rare-earth elements. The protoplanets would also have been rich in uranium and thorium, and hence they would have been heated rapidly by the radioactive decay of those elements in their interior. Thus in spite of the fact that the refractory oxide minerals have high melting points, the temperature inside the protoplanets would have been high enough for them to melt readily. The present-day interior of a large body of this type would be mostly homogeneous. Sulfides, iron and iron oxides would be totally absent. It has recently been suggested by Don L. Anderson of the California Institute of Technology that the moon is a member of this com-

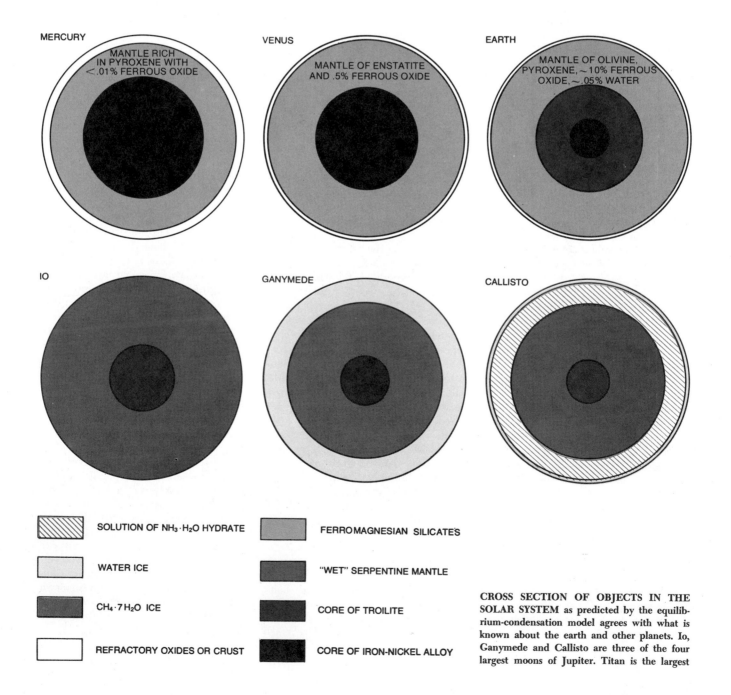

CROSS SECTION OF OBJECTS IN THE SOLAR SYSTEM as predicted by the equilibrium-condensation model agrees with what is known about the earth and other planets. Io, Ganymede and Callisto are three of the four largest moons of Jupiter. Titan is the largest

positional class. The moon would have had to have been extensively contaminated with sulfides and iron oxides, however, in order to account for its observed composition.

The second compositional class would have consisted of protoplanets formed just after the metallic iron-nickel alloy condensed out of the solar nebula. After these bodies had differentiated they would have had a thin crust of oxide minerals resting directly on a massive metallic core. They are not represented in the solar system today.

The protoplanets of the third class would have been formed after the temperature in the nebula had dropped far enough for enstatite to condense. The chemical makeup of a planet that came into being at this point would have been

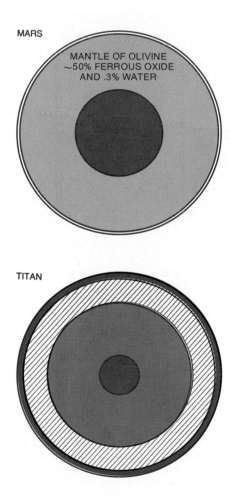

MARS

MANTLE OF OLIVINE ~50% FERROUS OXIDE AND .3% WATER

TITAN

moon of Saturn. "Wet" mantle of serpentine contains much water bound into the rock. The boundary between the inner mantle and the outer mantle of the earth is not shown because it is a change in the mineral structure and not a difference in chemical composition.

dominated by silicates. A crust of oxide minerals would have formed a thin layer on top of a mantle rich in pyroxene containing a negligible amount (less than .01 percent) of ferrous oxide. The planet Mercury is a member of this class.

The protoplanets of the fourth class would have differed from those in the third in that their crust would have retained alkali metals, resulting in a composition chemically similar to the crust of the earth except that the content of volatile elements would have been much lower. Such a body would have had a mantle containing perhaps .5 percent ferrous oxide. The bodies in this class would have corresponded to Venus.

The protoplanets of the fifth class would have retained sulfur as a constituent of troilite. On melting, the troilite would have sunk to form a dense outer core. Most of the mass of the core would have consisted of this sulfide melt rather than solid metal. The mantle, which would have been composed of olivine and pyroxene, would have contained at least 2 percent ferrous oxide. This class is not represented in the present solar system except perhaps in meteorites.

The sixth major class would have been made up of protoplanets that had formed when the temperature in the solar nebula had dropped enough (to about 600 degrees K.) for water to be retained, bound in the crystal structure of minerals such as tremolite. Up to .3 percent of the mass of these bodies could have been water. The earth, with a core rich in sulfur, its mantle containing some 10 percent ferrous oxide and its water content about .05 percent, belongs to this class.

Mars and Beyond

The seventh major class would have consisted of protoplanets that had formed after metallic iron had been completely oxidized. Such a planet would have contained some .3 percent water. Its olivine mantle would have been very rich in ferrous oxide, perhaps consisting of as much as 50 percent ferrous oxide. The core would have been either solid or liquid troilite, devoid of metallic iron. The resulting object would closely resemble Mars.

The protoplanets of the eighth class, which would have formed at a temperature low enough for the olivine in the mantle to have combined with the hydroxyl radical in the solar nebula to yield a "wet" mantle of serpentine, would have had a tremendously high water content: about 14 percent by mass. The

difference in the density between the minerals in the crust and those in the mantle would have essentially disappeared, but it is possible that a core of ferrous sulfide could have existed. A homogeneous parent body of this composition could have differentiated enough to form a core even at such a relatively low temperature as 400 degrees because the melting point of silicates with a large water content is fairly low. The asteroids probably belong to this eighth class. Calculations based on the known melting points of silicates should reveal whether or not such small objects could be warmed enough by the decay of radioactive elements in their interior for the minerals to be differentiated in their feeble gravitational field. If the asteroids are in fact like the carbonaceous chondrites, then perhaps the ordinary drier, less oxidized chondrites originated between the orbits of the earth and Mars instead of in the asteroid belt.

The protoplanets of the ninth class, formed below the condensation temperature of water, would have contained water ice. In such a body the water content, both as ice and as water bound in minerals, would have been about 60 percent by weight. Thus its density would have been quite low: only about 1.7 grams per cubic centimeter and less than a third the average density of the earth. Jupiter's largest satellite, Ganymede, may fall into this class.

The protoplanets of the 10th class would have retained ammonia as the solid hydrate $NH_3 \cdot H_2O$. A mixture of this hydrate with ice begins to melt at the low temperature of 173 degrees K., which is only a few degrees above the daytime surface temperature of Jupiter's second-largest satellite, Callisto. If a protoplanet in this class had ever been heated enough by radioactive decay to differentiate, it would have had a deep mantle of a liquid water-ammonia solution surmounted by a thin crust of water ice. During the differentiation a core of sulfides, oxides and hydrous (water-containing) silicates would have been formed as a sediment. The overall result would have been a stratified ball of mud and slush.

The protoplanets of the 11th class would have been about 4 percent methane by weight. The heat generated by radioactive decay would have melted the solid hydrate $CH_4 \cdot 7H_2O$, liberating methane gas to form an atmosphere. Differentiation would have given rise to a solid crust of the hydrate. A prime candidate for membership in this class is Saturn's largest satellite, Titan, which has a massive atmosphere rich in methane.

The protoplanets of the 12th and last class would have held solid methane as the temperature in the solar nebula dropped to almost 50 degrees K. Since carbon is nearly as abundant as oxygen in the solar system, and since the density of solid or liquid methane is very low (about .6 gram per cubic centimeter), the density of such a protoplanet would have been only one gram per c.c. Several of the smaller satellites of Saturn have been reported to have densities of near one gram per c.c., but the margin of error in such measurements is at least a factor of two. It is entirely possible that some of the satellites of Uranus or Neptune belong to this class, but their masses and radii will have to be determined by spacecraft flyby missions before we can confirm that speculation with any certainty.

I have mentioned the satellites of Jupiter, Saturn, Uranus and Neptune but not the planets themselves. There are reasons for the absence of the Jovian planets from the compositional classes. As we have seen, these classes are based on the temperature at which the protoplanetary material condensed out of the solar nebula according to the equilibrium-condensation model. Now, since the composition of Jupiter and Saturn is essentially the same as the composition of the sun, they could have formed just about anywhere in the solar nebula with no strong dependence on the temperature in that region. All that would be required is an initial protoplanetary core whose gravitational field was strong enough to accrete the undifferentiated solar material. Uranus, Neptune and Pluto are left off the list simply because not enough is known about their composition.

Important Results

One important result of the sequence of reactions I have been discussing is that the densities of the terrestrial planets, like their chemistry, can all be explained as a direct consequence of the variation of the temperature at which

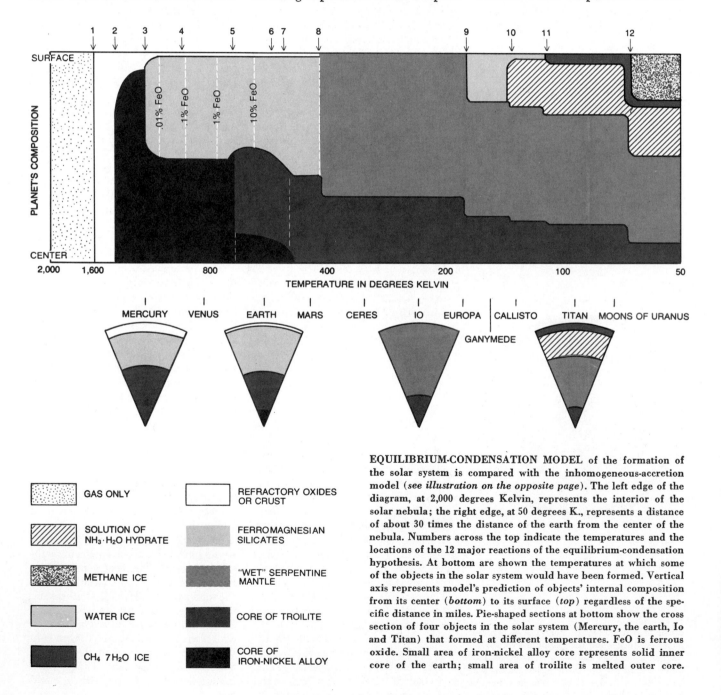

EQUILIBRIUM-CONDENSATION MODEL of the formation of the solar system is compared with the inhomogeneous-accretion model (see illustration on the opposite page). The left edge of the diagram, at 2,000 degrees Kelvin, represents the interior of the solar nebula; the right edge, at 50 degrees K., represents a distance of about 30 times the distance of the earth from the center of the nebula. Numbers across the top indicate the temperatures and the locations of the 12 major reactions of the equilibrium-condensation hypothesis. At bottom are shown the temperatures at which some of the objects in the solar system would have been formed. Vertical axis represents model's prediction of objects' internal composition from its center (bottom) to its surface (top) regardless of the specific distance in miles. Pie-shaped sections at bottom show the cross section of four objects in the solar system (Mercury, the earth, Io and Titan) that formed at different temperatures. FeO is ferrous oxide. Small area of iron-nickel alloy core represents solid inner core of the earth; small area of troilite is melted outer core.

they were formed according to their distance from the sun. It has often been argued that, by analogy with a hypothesis about the formation of meteorites, the different densities of the terrestrial planets are due to a physical process that fractionated the material in such a way that metals were preferentially accreted in one case (Mercury) and silicates were preferentially accreted in the others. Such preferential accretion, as it is seen in meteorites, might have been the result of differences in the magnetic or electrical properties of the material, or in how well the accreted matter adhered to the surface of the protoplanet. Since the sequence of reactions of the equilibrium-condensation model leads auto-

matically to the high density of Mercury, we are led to deny that any fractionation process affected the planets. Planets are so massive that they must have accreted their material by gravitation, a totally nonselective process. Temperature alone determined the composition of the accreted material. On the other hand, the parent bodies of meteorites, which could have had a mass as small as 10^{-12} times the mass of a planet, would have had only a feeble gravitational attraction. They probably accreted their material almost entirely through selective interparticle forces such as magnetism. Thus the paradox vanishes.

In summary, the equilibrium-condensation model shows very simply how the

density and the volatile content of the solid material in the solar system are related to the temperature at which the material was formed. The model leads to a large variety of quantitative predictions that can be tested in the laboratory, by astronomical observations and by planetary probes. Within a single chemical sequence we can now interrelate such seemingly diverse matters as the bulk composition of the solar nebula, the internal structure of the planets and the content of volatile compounds in the planets.

Perhaps the most striking result of this approach, however, is the underlying vision of an intimate interrelationship among all the objects in the solar

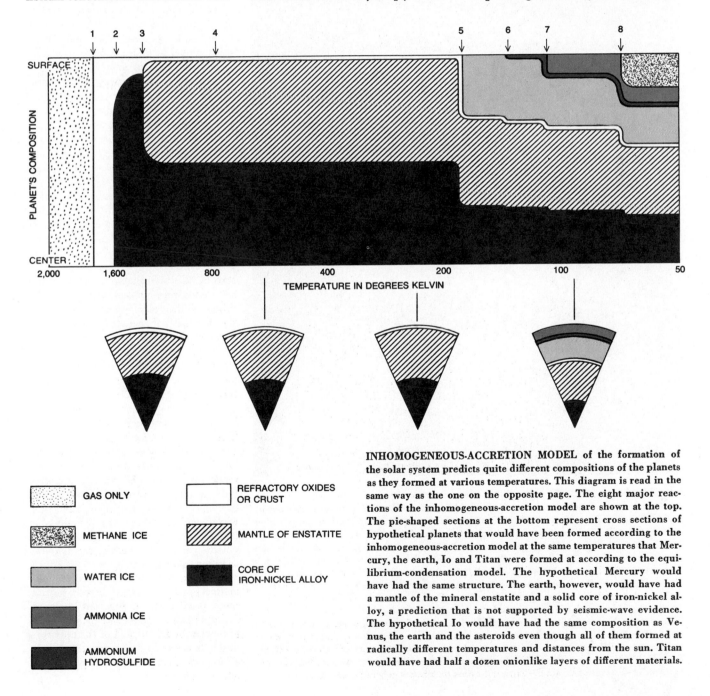

GAS ONLY

METHANE ICE

WATER ICE

AMMONIA ICE

AMMONIUM HYDROSULFIDE

REFRACTORY OXIDES OR CRUST

MANTLE OF ENSTATITE

CORE OF IRON-NICKEL ALLOY

INHOMOGENEOUS-ACCRETION MODEL of the formation of the solar system predicts quite different compositions of the planets as they formed at various temperatures. This diagram is read in the same way as the one on the opposite page. The eight major reactions of the inhomogeneous-accretion model are shown at the top. The pie-shaped sections at the bottom represent cross sections of hypothetical planets that would have been formed according to the inhomogeneous-accretion model at the same temperatures that Mercury, the earth, Io and Titan were formed at according to the equilibrium-condensation model. The hypothetical Mercury would have had the same structure. The earth, however, would have had a mantle of the mineral enstatite and a solid core of iron-nickel alloy, a prediction that is not supported by seismic-wave evidence. The hypothetical Io would have had the same composition as Venus, the earth and the asteroids even though all of them formed at radically different temperatures and distances from the sun. Titan would have had half a dozen onionlike layers of different materials.

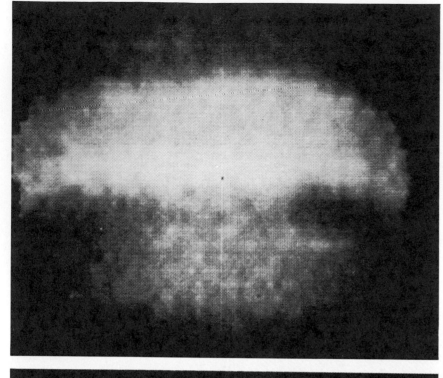

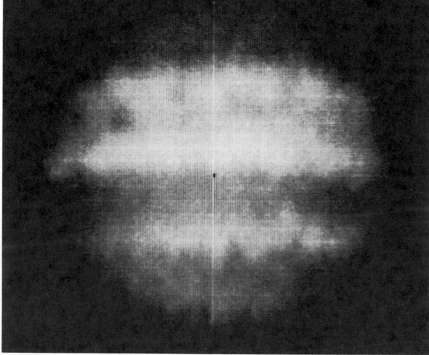

HEAT MAPS OF JUPITER at two different wavelengths were made from *Pioneer 10* with the infrared radiometer experiment of the California Institute of Technology, directed by Guido Munch. The radiometer is a reflecting telescope with an aperture of three inches mounted rigidly on the spacecraft. The spin of *Pioneer 10* and its motion in its trajectory past Jupiter allowed the radiometer to scan the planet and build up its images. The top picture was made at a wavelength of 20 microns and measured the heat being emitted at a distance of 40 kilometers below the top of Jupiter's atmosphere; the bottom picture was made at a wavelength of 40 microns and measured the heat being emitted at a distance of 50 kilometers below the top of the atmosphere. Both were made on December 3, 1973, during the spacecraft's closest approach to the planet. The warmest (lightest) areas are eight degrees K. warmer than the coolest (darkest) areas. The warmest areas correspond to the dark belts of Jupiter; coolest areas correspond to the bright belts. Planet is warmest near equator. It is estimated that the planet has an average temperature close to 128 degrees K. It radiates more than twice the amount of thermal energy it receives from the sun, confirming observations from the earth indicating that planet generates some heat of its own.

system. We see that there is strong evidence that the planets formed from a medium of uniform elemental composition. We see that the same qualitative processes of condensation and accretion are at work in the same general way throughout the entire system. The presence or absence of sulfur gases in the atmosphere of Venus, the density of the core of Mars, the electrical conductivity of the mantle of Callisto, the atmospheric pressure at the surface of Titan, the radius and mass of the satellites of Uranus—all tangibly influence our comprehension of the origin, composition and structure of our own planet. In this era in the history of geology we have seen how important it is to study the origins of all the chemical elements in order to understand the origin of any one of them, and we have been obliged to study global tectonics in order to understand the origin of any one mountain. Now we must be prepared to learn from comparative studies of the planets exactly what processes affect planets in general and how planets evolve. From this kind of investigation we can learn for the first time how the earth originated, evolved and attained its present structure.

Another long-range benefit of such comparative studies will be to provide insight into how frequently planets might be formed around other stars, and what the composition of those planets might be. Since the composition of most stars is indistinguishable from that of the sun, and since present theories of the origin of the solar system suggest that the formation of planets is a normal byproduct of the formation of stars, it would not be surprising to find that planets similar to those in the solar system are common throughout our galaxy and other spiral galaxies. It is interesting to speculate whether or not planets belonging to some compositional classes that are not seen in the solar system actually exist in other planetary systems. One such example would be a planet made up of iron and refractory oxides that was far denser than Mercury; another would be a very earthlike planet that was devoid of water; a third would be a true planet composed of material rich in volatile compounds such as the material ascribed to the asteroids.

The next few years will provide numerous opportunities for directly testing these predictions. A vigorous program for exploring the planets is the only way to realize the full benefit of the opportunities. It is the logical next step in our efforts to understand the planet on which we live.

BIBLIOGRAPHIES

I ENERGY

1. The Energy Resources of the Earth

MAN AND ENERGY. A. R. Ubbelohde. Hutchinson's Scientific and Technical Publications, 1954.

ENERGY FOR MAN: WINDMILLS TO NUCLEAR POWER. Hans Thirring. Indiana University Press, 1958.

ENERGY RESOURCES. M. King Hubbert. National Academy of Sciences—National Research Council, Publication 1000-D, 1962.

RESOURCES AND MAN: A STUDY AND RECOMMENDATIONS. Committee on Resources and Man. W. H. Freeman and Company, 1969.

ENVIRONMENT: RESOURCES, POLLUTION AND SOCIETY. Edited by William W. Murdoch. Sinauer Associates, 1971.

2. The Flow of Energy in an Industrial Society

ENERGY IN THE UNITED STATES: SOURCES, USES, AND POLICY ISSUES. Hans H. Landsberg and Sam H. Schurr. Random House, 1968.

AN ENERGY MODEL FOR THE UNITED STATES, FEATURING ENERGY BALANCES FOR THE YEARS 1947 TO 1965 AND PROJECTIONS AND FORECASTS TO THE YEARS 1980 AND 2000. Warren E. Morrison and Charles L. Readling. U.S. Department of the Interior, Bureau of Mines, No. 8384, 1968.

THE ECONOMY, ENERGY, AND THE ENVIRONMENT: A BACKGROUND STUDY PREPARED FOR THE USE OF THE JOINT ECONOMIC COMMITTEE, CONGRESS OF THE UNITED STATES. Environmental Policy Division, Legislative Reference Service, Library of Congress. U.S. Government Printing Office, 1970.

ENERGY CONSUMPTION AND GROSS NATIONAL PRODUCT IN THE UNITED STATES: AN EXAMINATION OF A RECENT CHANGE IN THE RELATIONSHIP. National Economic Research Associates, Inc., 1971.

3. Geothermal Power

PROCEEDINGS OF THE UNITED NATIONS CONFERENCE ON NEW SOURCES OF ENERGY. VOL. II, GEOTHERMAL ENERGY: I. United Nations, 1964.

PROCEEDINGS OF THE UNITED NATIONS CONFERENCE ON NEW SOURCES OF ENERGY. VOL. III, GEOTHERMAL ENERGY: II. United Nations, 1964.

THE ECONOMIC POTENTIAL OF GEOTHERMAL RESOURCES IN CALIFORNIA. Geothermal Resources Board, State of California, January, 1971.

GEOTHERMAL RESOURCE INVESTIGATIONS, IMPERIAL VALLEY OF CALIFORNIA—STATUS REPORT. United States Department of the Interior, Bureau of Reclamation, April, 1971.

UNITED NATIONS SYMPOSIUM ON THE DEVELOPMENT AND UTILIZATION OF GEOTHERMAL RESOURCES (A SUMMARY REPORT). United Nations ST/TAO/Ser. C/126, 1971.

THE EXPLOITATION OF GEOTHERMAL RESOURCES IN THE PRESENT AND FUTURE. Gunnar Bodvarsson. United Nations, in press.

II   ENVIRONMENTAL CYCLES

4. The Atmosphere and the Ocean

AN INTRODUCTION TO PHYSICAL OCEANOGRAPHY. William S. von Arx. Addison-Wesley Publishing Company, Inc., 1962.

THE GULF STREAM: A PHYSICAL AND DYNAMICAL DESCRIPTION. Henry Stommel. University of California Press and Cambridge University Press, 1965.

THE INFLUENCE OF FRICTION ON INERTIAL MODELS OF OCEANIC CIRCULATION. R. W. Stewart in *Studies on Oceanography*, edited by Kozo Yoshida. University of Washington Press, 1965.

ENCYCLOPEDIA OF OCEANOGRAPHY. Edited by Rhodes W. Fairbridge. Reinhold Company, 1966.

DESCRIPTIVE PHYSICAL OCEANOGRAPHY. G. L. Pickard. Pergamon Press, 1968.

5. The Control of the Water Cycle

ON THE GLOBAL BALANCE OF WATER VAPOR AND THE HYDROLOGY OF DESERTS. V. P. Starr and J. P. Peixoto in *Tellus*, Vol. 10, No. 2, pages 189–194; May, 1958.

POLE-TO-POLE MOISTURE CONDITIONS FOR THE IGY. Victor P. Starr, José P. Peixoto and Robert G. McKean in *Pure and Applied Geophysics*, Vol. 75, pages 300–331; 1969/IV.

FEASIBILITY OF A HELIOHYDROELECTRIC POWER PLANT ON THE EASTERN SHORE OF SAUDI ARABIA. M. A. Kettani and L. M. Gonsalves in *Proceedings of the 6th Annual Conference of the Marine Technology Society*, Vol. 2; 1970.

THE WATER CYCLE. H. L. Penman in *Scientific American*, Vol. 223, No. 3, pages 98–108; September, 1970.

6. The Oxygen Cycle

THE ATMOSPHERES OF THE PLANETS. Harold C. Urey in *Handbuch der Physik, Vol. LII, Astrophysics III: The Solar System*, edited by S. Flügge. Springer-Verlag, 1959.

HISTORY OF MAJOR ATMOSPHERIC COMPONENTS. L. V. Berkner and L. C. Marshall in *Proceedings of the National Academy of Sciences*, Vol. 53, No. 6, pages 1215–1226; June, 1965.

ATMOSPHERIC AND HYDROSPHERIC EVOLUTION ON THE PRIMITIVE EARTH. Preston E. Cloud, Jr., in *Science*, Vol. 160, No. 3829, pages 729–736; May 17, 1968.

DISSOCIATION OF WATER VAPOR AND EVOLUTION OF OXYGEN IN THE TERRESTRIAL ATMOSPHERE. R. T. Brinkmann in *Journal of Geophysical Research*, Vol. 74, No. 23, pages 5355–5368; October 20, 1969.

THE EVOLUTION OF PHOTOSYNTHESIS. John M. Olson in *Science*, Vol. 168, No. 3930, pages 438–446; April 24, 1970.

7. The Carbon Cycle

GEOGRAPHIC VARIATIONS IN PRODUCTIVITY. J. H. Ryther in *The Sea: Ideas and Observations on Progress in the Study of the Seas. Vol. II: The Composition of Sea-Water—Comparative and Descriptive Oceanography*, edited by M. N. Hill. Interscience Publishers, 1963.

THE INFLUENCE OF ORGANISMS ON THE COMPOSITION OF SEA-WATER. A. C. Redfield, B. H. Ketchum and F. A. Richards in *The Sea: Ideas and Observations on Progress in the Study of the Seas. Vol. II: The Composition of Sea-Water—Comparative and Descriptive Oceanography*, edited by M. N. Hill. Interscience Publishers, 1963.

THE ROLE OF VEGETATION IN THE CARBON DIOXIDE CONTENT OF THE ATMOSPHERE. Helmut Lieth in *Journal of Geophysical Research*, Vol. 68, No. 13, pages 3887–3898; July 1, 1963.

GROSS-ATMOSPHERIC CIRCULATION AS DEDUCED FROM RADIOACTIVE TRACERS. Bert Bolin in *Research in Geophysics, Vol. II: Solid Earth and Interface Phenomena*, edited by Hugh Odishaw. The M.I.T. Press, 1964.

IS CARBON DIOXIDE FROM FOSSIL FUEL CHANGING MAN'S ENVIRONMENT? Charles D. Keeling in *Proceedings of the American Philosophical Society*, Vol. 114, No. 1, pages 10–17; February 16, 1970.

PHOTOSYNTHESIS. E. Rabinowitch and Govindjee. John Wiley & Sons, Inc., 1969.

8. The Nitrogen Cycle

AUTOTROPHIC MICRO-ORGANISMS: FOURTH SYMPOSIUM OF THE SOCIETY FOR GENERAL MICROBIOLOGY HELD AT THE INSTITUTION OF ELECTRICAL ENGINEERS, LONDON, APRIL, 1954. Cambridge University Press, 1954.

DENITRIFICATION. C. C. Delwiche in *A Symposium on Inorganic Nitrogen Metabolism: Function of Metallo-Flavoproteins*, edited by William D. McElroy and Bentley Glass. The Johns Hopkins Press, 1956.

NITROGEN FIXATION IN PLANTS. W. D. P. Stewart. Athlone Press, 1966.

SYMBIOSIS: ITS PHYSIOLOGICAL AND BIOCHEMICAL SIGNIFICANCE. Edited by S. Mark Henry. Academic Press, 1966.

FIXATION OF NITROGEN BY HIGHER PLANTS OTHER THAN LEGUMES. G. Bond in *Annual Review of Plant Physiology: Vol. XXVIII*, edited by Leonard Machlis. Winslow R. Briggs and Roderic B. Park. Annual Reviews, Inc., 1967.

9. The Global Circulation of Atmospheric Pollutants

INTERNATIONAL SYMPOSIUM ON TRACE GASES AND NATURAL AND ARTIFICIAL RADIOACTIVITY IN THE ATMOSPHERE. *Journal of Geophysical Research*, Vol. 68, No. 13, pages 3745–4016; July 1, 1963.

PROCEEDINGS OF THE CACR SYMPOSIUM: ATMOSPHERIC CHEMISTRY, CIRCULATION AND AEROSOLS, AUGUST 15–25, 1965, VISBY, SWEDEN. *Tellus*, Vol. 18, No. 2–3, pages 149–684; 1966.

MAN'S IMPACT ON THE GLOBAL ENVIRONMENT: ASSESSMENT AND RECOMMENDATIONS FOR ACTION. Report of the Study of Critical Environment Problems (SCEP). The M.I.T. Press, 1970.

10. Mercury in the Environment

THE GENERAL PHARMACOLOGY OF THE HEAVY METALS. H. Passow, A. Rothstein and T. W. Clarkson in *Pharmacological Reviews*, Vol. 13, No. 2, pages 185–224; June, 1961.

ABSORPTION AND EXCRETION OF MERCURY IN MAN: I–XIV. Leonard J. Goldwater *et al.* Papers published in *Archives of Environmental Health*, 1962–1968.

CHEMICAL FALLOUT: CURRENT RESEARCH ON PERSISTENT PESTICIDES. Edited by Morton W. Miller and George C. Berg. Charles C Thomas, Publisher, 1969.

MERCURY IN THE ENVIRONMENT. Geological Survey Professional Paper 713. United States Government Printing Office, 1970.

III   PLATE TECTONICS, SEA-FLOOR SPREADING, AND CONTINENTAL DRIFT

11. Plate Tectonics

SPECULATIONS ON THE CONSEQUENCES AND CAUSES OF PLATE MOTION. D. P. McKenzie in *Geophysical Journal of the Royal Astronomical Society*, Vol. 18, No. 1, pages 1–32; September, 1969.

MOUNTAIN BELTS AND THE NEW GLOBAL TECTONICS. John F. Dewey and John M. Bird in *Journal of Geophysical Research*, Vol. 75, No. 14, pages 2625–2647; May 10, 1970.

12. The Evolution of the Andes

SEISMOLOGY AND THE NEW GLOBAL TECTONICS. Bryan Isacks, Jack Oliver and Lynn R. Sykes in *Journal of Geophysical Research*, Vol. 73, No. 18, pages 5855–5899; September 15, 1968.

MOUNTAIN BELTS AND THE NEW GLOBAL TECTONICS. John F. Dewey and John M. Bird in *Journal of Geophysical Research*, Vol. 75, No. 14, pages 2625–2647; May 10, 1970.

RELATIONS OF ANDESITES, GRANITES, AND DERIVATIVE SANDSTONES TO ARC-TRENCH TECTONICS. William R. Dickinson in *Reviews of Geophysics and Space Physics*, Vol. 8, No. 4, pages 813–860; November, 1970.

PLATE TECTONIC MODEL FOR THE EVOLUTION OF THE CENTRAL ANDES. David E. James in *Geological Society of America Bulletin*, Vol. 82, pages 3325–3346; December, 1971.

GEOSYNCLINES, MOUNTAINS AND CONTINENT-BUILDING. Robert S. Dietz in *Scientific American*, Vol. 226, No. 3, pages 30–38; March, 1972.

13. The Evolution of the Indian Ocean

THE EARTH AS A DYNAMO. Walter M. Elsasser in *Scientific American*, Vol. 198, No. 5, pages 44–48; May, 1958.

THE EAST PACIFIC RISE. Henry W. Menard in *Scientific American*, Vol. 205, No. 6, pages 52–61; December, 1961.

THE PRINCIPLES OF PHYSICAL GEOLOGY. Arthur Holmes. The Ronald Press Company, 1965.

REVERSALS OF THE EARTH'S MAGNETIC FIELD. Allan Cox, G. Brent Dalrymple and Richard R. Doell, in *Scientific American*, Vol. 216, No. 2, pages 44–54; February, 1967.

SEA-FLOOR SPREADING. J. R. Heirtzler in *Scientific American*, Vol. 219, No. 6, pages 60–70; December, 1968.

THE ORIGIN OF THE OCEANIC RIDGES. Egon Orowan in *Scientific American*, Vol. 221, No. 5, pages 102–119; November, 1969.

14. Continental Drift and the Fossil Record

THE ORIGIN OF CONTINENTS AND OCEANS. Alfred L. Wegener. Methuen, 1966.

THE BEARING OF CERTAIN PALAEOZOOGEOGRAPHIC DATA OF CONTINENTAL DRIFT. Anthony Hallam in *Palaeogeography Palaeoclimatology Palaeoecology*, Vol. 3, pages 201–241; 1967.

CONTINENTAL DRIFT AND THE EVOLUTION OF THE BIOTA ON SOUTHERN CONTINENTS. Allen Keast in *The Quarterly Review of Biology*, Vol. 46, No. 4, pages 335–378; December, 1971.

ATLAS OF PALAEOBIOGEOGRAPHY. Edited by Anthony Hallam. American Elsevier Publishing Co., in press.

15. Plate Tectonics and Mineral Resources

EXPLORATION METHODS FOR THE CONTINENTAL SHELF: GEOLOGY, GEOPHYSICS, GEOCHEMISTRY. P. A. Rona. National Oceanic and Atmospheric Administration Technological Report ERL 238-AOML 8, U.S. Government Printing Office; 1972.

SULFIDE ORE DEPOSITS IN RELATION TO PLATE TECTONICS. F. Sawkins in *Journal of Geology*, Vol. 80, pages 377–397; 1972.

HYDROTHERMAL MANGANESE IN THE MEDIAN VALLEY OF THE MID-ATLANTIC RIDGE. Martha R. Scott, Robert B. Scott, Andrew J. Nalwalk, P. A. Rona and Louis W. Butler in *EOS: American Geophysical Union Transactions*, Vol. 54, No. 4, page 244; April, 1973.

IV   GEOLOGY

16. The Evolution of Reefs

REVOLUTIONS IN THE HISTORY OF LIFE. Norman D. Newell in *Uniformity and Simplicity: A Symposium on the Principle of the Uniformity of Nature*. The Geological Society of America, Special Paper 89, edited by Claude C. Albritton, Jr.; 1967.

AN OUTLINE HISTORY OF TROPICAL ORGANIC REEFS. Norman D. Newell in *American Museum Novitates*, No. 2465; September 21, 1971.

REEF ORGANISMS THROUGH TIME. Proceedings of the North American Paleontological Convention, edited by Ellis Yochelson. Allen Press, Inc., 1971.

17. When the Mediterranean Dried Up

THE TECTONICS AND GEOLOGY OF THE MEDITERRANEAN SEA. William B. F. Ryan, Daniel J. Stanley, J. B. Hersey, Davis A. Fahlquist and Thomas D. Allan in *The Sea*, Vol. IV, Parts II and III, edited by A. E. Maxwell. John Wiley & Sons, Inc., 1971.

INITIAL CRUISE REPORTS OF THE DEEP SEA DRILLING PROJECT. W. B. F. Ryan, K. J. Hsü, *et al.* U.S. Government Printing Office, Washington, 1972.

18. River Meanders

FLUVIAL PROCESSES IN GEOMORPHOLOGY. Luna B. Leopold, M. Gordon Wolman and John P. Miller. W. H. Freeman and Company, 1964.

RIVER MEANDERS. Luna B. Leopold and M. Gordon Wolman in *Bulletin of the Geological Society of America*, Vol. 71, No. 6, pages 769–794; June, 1960.

RIVERS. Luna B. Leopold in *American Scientist*, Vol. 50, No. 4, pages 511–537; December, 1962.

19. Lateritic Soils

HUMAN GEOGRAPHY: AN ECOLOGICAL STUDY OF HUMAN SOCIETY. C. Langdon White and George T. Renner. Appleton-Century-Crofts, 1948.

LATERITE. S. Sivarajasingham, L. T. Alexander, J. G. Cady and M. G. Cline in *Advances in Agronomy*, Vol. XIV. Edited by A. G. Norman. Academic Press, 1962.

TROPICAL AND SUBTROPICAL AGRICULTURE. J. J. Ochse, M. J. Soule, Jr., M. J. Dijkman and C. Wehlburg. The Macmillan Company, 1961.

20. Mars from *Mariner* 9

Science, Vol. 175, No. 4019; January, 21, 1972.

MARS: THE VIEW FROM MARINER 9. Carl Sagan in *Astronautics & Aeronautics*, Vol. 10, No. 9, pages 26–41; September, 1972.

Icarus, Vol. 17, No. 2; October, 1972.

V   GEOPHYSICS AND PLANETARY RESEARCH

21. The Earth as a Dynamo

HYDROMAGNETIC DYNAMO THEORY. Walter M. Elsasser in *Reviews of Modern Physics*, Vol. 28, No. 2, pages 135–163; April, 1956.

HYDROMAGNETISM. I: A REVIEW. Walter M. Elsasser in *American Journal of Physics*, Vol. 23, No. 9, pages 590–609; December, 1955.

HYDROMAGNETISM. II: A REVIEW. Walter M. Elsasser in *American Journal of Physics,* Vol. 24, No. 2, pages 85–110; February, 1956.

22. Tides and the Earth-Moon System

HISTORY OF THE LUNAR ORBIT. Peter Goldreich in *Reviews of Geophysics,* Vol. 4, No. 4, pages 411–439; November, 1966.

ONCE AGAIN—TIDAL FRICTION. Walter Munk in *The Quarterly Journal of the Royal Astronomical Society,* Vol. 9, No. 4, pages 352–375; December, 1968.

ANCIENT ASTRONOMICAL OBSERVATIONS AND THE ACCELERATIONS OF THE EARTH AND MOON. Robert R. Newton. Johns Hopkins Press, 1970.

DYNAMICAL ASPECTS OF LUNAR ORIGIN. William M. Kaula in *Reviews of Geophysics and Space Physics,* Vol. 9, No. 2, pages 217–238; May, 1971.

23. The Fine Structure of the Earth's Interior

INTRODUCTION TO THE THEORY OF SEISMOLOGY. Keith E. Bullen. Cambridge University Press, 1963.

THE DENSITY DISTRIBUTION NEAR THE BASE OF THE MANTLE AND NEAR THE EARTH'S CENTER. Bruce A. Bolt in *Physics of the Earth and Planetary Interiors,* Vol. 5, pages 1–11; 1972.

OBSERVATIONS OF PSEUDO-AFTERSHOCKS FROM UNDERGROUND EXPLOSIONS. Bruce A. Bolt and A. Qamar in *Physics of the Earth and Planetary Interiors,* Vol. 6, pages 100–200; 1972.

24. The Chemistry of the Solar System

AN INTRODUCTION TO PLANETARY PHYSICS: THE TERRESTRIAL PLANETS. W. M. Kaula. John Wiley & Sons, Inc., 1968.

MOONS AND PLANETS: AN INTRODUCTION TO PLANETARY SCIENCE. William K. Hartmann. Bogden & Quigley, Inc., Publishers, 1972.

SPACE SCIENCE REVIEWS, Vol. 14; 1973.

INDEX